Marine Ecosystems and Climate Variation

Marine Ecosystems and Climate Variation

The North Atlantic
A Comparative Perspective

EDITED BY

Nils Chr. Stenseth,
Centre for Ecological and Evolutionary Synthesis (CEES),
Department of Biology, University of Oslo, Norway
and *Institute of Marine Research, Bergen, Norway*

Geir Ottersen
Institute of Marine Research, Bergen, Norway

IN COLLABORATION WITH

James W. Hurrell
National Center for Atmospheric Research, Boulder, USA

Andrea Belgrano
University of New Mexico, Department of Biology, Albuquerque, NM, USA

OXFORD
UNIVERSITY PRESS

OXFORD

UNIVERSITY PRESS

Great Clarendon Street, Oxford OX2 6DP

Oxford University Press is a department of the University of Oxford.
It furthers the University's objective of excellence in research, scholarship,
and education by publishing worldwide in

Oxford New York

Auckland Cape Town Dar es Salaam Hong Kong Karachi
Kuala Lumpur Madrid Melbourne Mexico City Nairobi
New Delhi Shanghai Taipei Toronto

With offices in

Argentina Austria Brazil Chile Czech Republic France Greece
Guatemala Hungary Italy Japan Poland Portugal Singapore
South Korea Switzerland Thailand Turkey Ukraine Vietnam

Oxford is a registered trade mark of Oxford University Press
in the UK and in certain other countries

Published in the United States
by Oxford University Press Inc., New York

A catalogue record for this title is available from the British Library

Library of Congress Cataloging in Publication Data
(Data available)

Typeset by Newgen Imaging Systems (P) Ltd., Chennai, India
Printed in Great Britain
on acid-free paper by Antony Rowe Ltd., Chippenham

ISBN 0-19-850748-8   978-0-19-850748-2

ISBN 0-19-850749-6   978-0-19-850749-9 (pbk.)

10 9 8 7 6 5 4 3 2 1

Preface

Coupled large-scale atmospheric and oceanic circulation patterns are well known to exhibit variability at timescales ranging from days to millennia. Variations in climate have a profound influence on a variety of biological processes. Consequently, patterns of abundance and dynamics are affected at the population as well as the community and ecosystem levels. The ecological responses may be direct in space and time or mediated through complex pathways, affecting other parts of the ecological systems perhaps years after the initial cause.

Climate variability can take the form of dramatic changes in the atmosphere–ocean system such as the El Niño-Southern Oscillation (ENSO) phenomenon, origining in the tropical Pacific but with effects worldwide. ENSO has occurred at semi-regular intervals throughout historical times. Related variations of the Aleutian Low pressure system in the North Pacific or the Asian–Australian complex are known to strongly affect the ecosystems of these regions.

The North Atlantic also undergoes important climatic variability. The North Atlantic Ocean is a particularly active sector and plays a key role in the global ocean circulation and the regulation of climate. One of the dominant features of climatic variability at interdecadal scale is the North Atlantic Oscillation (NAO), an alternation in the atmospheric pressure difference between the Arctic and subtropical Atlantic. The NAO strongly influences the speed and orientation of westerly winds from the Gulf of Mexico to northern Europe, partially regulating air and sea-surface temperatures (SSTs) throughout the North Atlantic region and driving long-term variations in patterns of circulation and convection.

This edited book focuses on the influence of climate variability on the marine biology of the North Atlantic. This is the first attempt to summarize the consequences of climatic fluctuations for the marine ecosystems in the region. We address the ecological impact of North Atlantic climate variability on population dynamics at the full range of trophic levels, from phytoplankton through zooplankton and fish to marine birds. We attempt to describe the range of scales at which the climate influences marine species. Climate effects on biodiversity and community structure are also studied.

Throughout the book a number of examples highlight different aspects of climate-marine ecology links. Fluctuations in windiness and wind direction, sea temperature and salinity, horizontal circulation patterns and vertical mixing are all demonstrated to play important roles. Trends in phytoplankton distribution in the North Atlantic over the last 50 years are probably a response to climate forcing. The first analysis of the ecological consequences of the NAO documented the impact on the zooplankton of the eastern North Atlantic and the North Sea. Fluctuations in several of the most important fish stocks in the North Atlantic region, including herring, cod, and salmon are linked to climatic variability. Typically being top predators, seabirds like for instance guillemots, puffins, gulls, and kittiwakes, are influenced by climate through prey availability, but also through physiological effects.

As is the case of other oceanic basins, these examples show how the ecosystem of the North Atlantic can respond to climate variability. Large-scale atmospheric and oceanographic fluctuations can explain a significant part of the variability in the marine ecosystem. We are just at the beginning of understanding these connections and the possible consequences of climate change on the North Atlantic ecosystem. A global understanding of the links between climatic oscillations and ecological processes can be regarded as a new challenge for oceanographers and for ecosystems ecologists.

Although several books have recently been published on related topics (for instance on ENSO and the NAO and their impacts), none provide the combined focused and broad perspective that

NAO

we do in this book. At the same time, although we focus on the marine environments of the North Atlantic, we provide—through commentary chapters—a broader outlook to other regions and ecological systems. For the North Atlantic we also include a chapter linking the marine with the terrestrial system.

We have also put much effort into making this an integrated book—not only a collection of papers (referred to as chapters). Rather than producing only an edited book, we have wanted to produce a book more like a multi-authored monograph.

For that purpose, we have included section introductions linking the various chapters within the five parts of the book together—as well as linking that particular part together with the rest of the sections in the book. Although some cross-referencing naturally is done within each of the individual chapters, most of this linking together of the various chapters—and parts—of the book is done in a series of section introductions.

Nils Chr. Stenseth and Geir Ottersen
January 2004

Contents

Contributors

Main editors

Nils Chr. Stenseth,   Centre for Ecological and Evolutionary Synthesis (CEES), Department of Biology, University of Oslo, P.O. Box 1050 Blindern, Norway; and Institute of Marine Research, Flødevigen Marine Research Station, N-4817 His, Norway

Geir Ottersen,   Institute of Marine Research, P.O. Box 1870 Nordnes, 5817 Bergen, Norway. *Current address*: Centre for Ecological and Evolutionary Synthesis (CEES), Department of Biology, University of Oslo, P.O. Box 1050 Blindern, Norway.

Collaborating editors

Andrea Belgrano,   University of New Mexico, Department of Biology, Albuquerque, New Mexico, USA; *Current address*: NCGR-National Center for Genome Resources, 2935 Rodeo Park Drive East, Santa Fe, NM 87505, USA.

James W. Hurrell,   The National Center for Atmospheric Research, Climate Analysis Section, P.O. Box 3000, Boulder, Colorado 80307, USA.

Contributing authors

Jürgen Alheit,   The Baltic Sea Research Institute, Seestraße 15, 18119 Warnemünde, Germany.

Tycho Ander-Nilsen,   Norwegian Institute for Nature Research, Tungasletta 2, N-7485 Trondheim, Norway.

Alfredo Aretxabaleta,   Department of Marine Sciences, CB# 3300, University of North Carolina, Chapel Hill, N.C. 27599–3300, USA.

Kevin M. Bailey,   Alaska Fisheries Science Center, 7600 Sand Point Way N.E., Building 4, Seattle, Washington 98115, USA.

Gregory Beaugrand,   Sir Alister Hardy Foundation for Ocean Science (SAHFOS) 1 Walker Terrace, West Hoe, Plymouth, PL1 3BN, UK.

Andrea Belgrano,   University of New Mexico, Department of Biology, Albuquerque, New Mexico, USA; *Current address*: NCGR-National Center for Genome Resources, 2935 Rodeo Park Drive East, Santa Fe, NM 87505, USA.

David G. Borkman,   Graduate School of Oceanography, University of Rhode Island, Kingston RI 02881, USA.

Robert R. Dickson,   The Centre for Environment, Fisheries and Aquaculture Sciences, Pakefield Road, Lowestoft, Suffolk NR33 OHT, UK.

Kenneth F. Drinkwater,   Institute of Marine Research, P.O. Box 1870 Nordnes, 5817 Bergen, Norway

Joël M. Durant,   Centre for Ecological and Evolutionary Synthesis (CEES), Department of Biology, Division of Zoology, University of Oslo, P.O. Box 1050, Blindern, Norway.

Karen Pehrson Edwards,   Department of Marine Sciences, CB# 3300, University of North Carolina, Chapel Hill, N.C. 27599–3300, USA.

Kevin D. Friedland,   UMass/NOAA CMER Program, Blaisdell House, University of Massachusetts, Amherst, MA 01003–0040, USA.

Jean-Marc Fromentin,   IFREMER, 1 rue Jean Vilar, BP 171, 34203 Sete Cedex 3, France.

Charles H. Greene, Ocean Resources and Ecosystems Program, Cornell University, Ithaca, NY 14853, USA.

Jacob Hagberg, National Board of Fisheries, Institute of Marine Research, Turistgatan 5, 453 21 Lysekil, Sweden.

Eberhardt Hagen, The Baltic Sea Research Institute, Seestraße 15, 18119 Warnemünde, Germany.

Michael P. Harris, Institute of Terrestrial Ecology, Banchory Research Station, Hill of Brathens, Glassel Banchory, Kincardineshire AB31 4BY, Scotland.

Roger Harris, Plymouth Marine Laboratory, Prospect Place, Plymouth, PL1 3DH, UK

Dag O. Hessen, Centre for Ecological and Evolutionary Synthesis (CEES), Department of Biology, P.O. Box 1050, Blindern, N-0316 Oslo, Norway.

Anne B. Hollowed, Alaska Fisheries Science Center, 7600 Sand Point Way N.E., Seattle, WA 98115, USA.

James W. Hurrell, The National Center for Atmospheric Research, Climate Analysis Section, P.O. Box 3000, Boulder, Colorado 80307, USA.

Fabian M. Jaksic, Center for Advanced Studies in Ecology and Biodiversity (CASEB), Pontificia Universidad Católica de Chile, Casilla 114-D, Santiago, Chile.

Ingrid Kröncke, Forschungsstation Senckenberg am Meer (Rudolf Richter-Haus), Schleusenstraße 18, 26382 Wilhelmshaven, Germany.

Patrick Lehodey, Oceanic Fisheries Programme, Secretariat of the Pacific Community, BP D5, 98848 Noumea Cedex, New Caledonia.

Kyrre Lekve, Centre for Ecological and Evolutionary Synthesis (CEES), Department of Biology, Division of Zoology, University of Oslo, P.O. Box 1050, Blindern, Norway.

Mauricio Lima, Center for Advanced Studies in Ecology and Biodiversity (CASEB), Pontificia Universidad Católica de Chile, Casilla 114-D, Santiago, Chile.

Geir Ottersen, Institute of Marine Research, P.O. Box 1870 Nordnes, 5817 Bergen, Norway.

Current address: Centre for Ecological and Evolutionary Synthesis (CEES), Department of Biology, University of Oslo, P.O. Box 1050 Blindern Norway.

Andrew J. Pershing, Ocean Resources and Ecosystems Program, Cornell University, Ithaca, NY 14853, USA.

Benjamin Planque, IFREMER, Rue de l'île d'Yeu, BP 21105, 44311 NANTES Cedex 03, France.

Eric Post, Penn State University, Department of Biology, 208 Mueller Lab, University Park, PA 16802–5301, USA.

Phillip C. Reid, Sir Alister Hardy Foundation for Ocean Science (SAHFOS), 1 Walker Terrace, West Hoe, Plymouth, PL1 3BN, UK.

Theodore Smayda, Graduate School of Oceanography, University of Rhode Island, Kingston, R.I.02881, USA.

Dietmar Straile, Limnological Institute, University of Konstanz, 78457 Konstanz, Germany.

Nils Chr. Stenseth, Centre for Ecological and Evolutionary Synthesis (CEES), Department of Biology, University of Oslo, P.O. Box 1050 Blindern Norway.

Paul Thompson, Department of Zoology, University of Aberdeen, Tillydrone Avenue, Aberdeen, AB24 2TZ, Scotland.

Björn Tunberg, Smithsonian Marine Station, 701 Seaway Drive, Fort Pieree, Florida 3Y949, USA.

Sarah Wanless, Institute of Terrestrial Ecology, Banchory Research Station, Hill of Brathens, Glassel, Banchory, Kincardineshire AB31 4BY, Scotland.

Francisco E. Werner, Department of Marine Sciences, CB# 3300, University of North Carolina, Chapel Hill, N.C. 27599–3300, USA.

Gunther Wieking, Forschungsstation Senckenberg am Meer (Rudolf Richter-Haus), Schleusenstraße 18, 26382 Wilhelmshaven, Germany.

Warren S. Wooster, School of Marine Affairs, University of Washington, Seattle, WA 98105–6715, USA.

Acronym list

ACI: Atmospheric circulation index
ADCP: Acoustic Doppler current profiler
AO: Arctic Oscillation
AIC: Aikaike information criterion
Aleutian low pressure system: Climatological centre of low air pressure near the Aleutian Islands off Alsaka.
ASIOs: Anamolous southward intrusions of the Oyashio Current (off Japan)
ATSW: Atlantic Temperate Slope Water
AW: Atlantic Water

BATS: Bermuda Atlantic Time-series site
BIOSYNOP: Biological Synoptic Ocean Prediction
BMR: Basal metabolic rate

CIL: Cold Intermediate Layer
CLIVAR: Climate variability and predictability research programme
CO_2: Carbon dioxide
COADS: Comprehensive ocean-atmosphere data set
CPR: Continuous plankton record
CF: *Costatum fusus*
CT: *Costatum tripos*
CSWS: Coupled Slope Water System (off NE US and Canada)
CZCS: Coastal Zone color scanner

DOC: Dissolved organic carbon
DOM: Dissolved organic material
DON: Dissolved organic nitrogen
DSOW: Denmark straight overflow water

EGC: East Greenland Current
Ekman divergence: Division of upper-layer waters towards opposite directions, related to upwelling
Ekman convergence: Gathering of upper-layer waters from opposite directions towards a common area, related to downwelling

ENSO: El Niño-Southern Oscillation
ESAI: East Shetland Atlantic Inflow

Fe: Iron
FMR: Field metabolic rate
FST: Faroe–Shetland Transport

GAM: Generalized Additive Model
GFDL: Geophysical Fluid Dynamics Laboratory
GODAE: Global Ocean Data Assimilation Experiment
GLOBEC: Global Ocean Ecosystem Dynamics research program
GOOS: Global Ocean Observing System
GSA: Great Salinity Anomaly
GSI: Gulf Stream Index

HABs: Harmful algal blooms

ICES: International council for exploration of the sea
IBM: Individual-based models
ICOS: Investigation of *Calanus finmarchicus* migrations between Oceanic and Shelf seas
IMBER: Integrated Marine Biochemistry and Ecosystem Research Program

JGOFS: Joint Global Ocean Flux Study research program

KMF: Kristineberg Marine Research Station

LSW: Labrador Sea Water
LSSW: Labrador Subarctic Slope Water
LZWWs: low zonal westerly winds

Match-mismatch: Hypothesis describing how differences in the temporal and spatial match between predator and prey may generate variability in predator survival rates.

MOC: Meridional overturning circulation
MOM: modular ocean model
MOC: Atlantic meridional overturning circulation
MT: Metric tonnes

NA: North Atlantic Ocean
NAO: North Atlantic Oscillation
NAOI: North Atlantic Oscillation Index
NAOWI: North Atlantic Oscillation winter index
NEADW: North East Atlantic Deep Water
NH: Northern Hemisphere
NH_4: Ammonium
NO_3: Nitrate
NOAA: National Oceanic and Atmospheric Administration
NWAC: Norwegian Atlantic Current
NSAIW: Norwegian Sea arctic intermediate water layer
NSDW: Norwegian Sea deep water
NPZD: Nutrients-phyto-zooplanktin-detritus

OGCM: Physical ocean general circulation model
OSSE: Observation systems simulation experiments
OWSI: Ocean weather station India

P:B: Production to biomass ratio
PCA: Principal component analysis
PDO: Pacific Decadal Oscillation
PE: Potential energy anomaly

P/F: Primary production to forage ratio
POC: Particulate organic carbon
POM: Particulate organic matter
PON: Particulate organic nitrogen
PO_4: Phosphate

Russell Cycle: Shifts in species composition in the English Channel related to climate variability
RSWT: Regional Slope Water Temperature (off NE US and Canada)

SBC: Schwartz's Bayesian Criterion
SiO_4: Silicate
SLP: Sea-level pressure
SST: Sea-surface temperature
SSTAs: Sea surface temperature anomalies
SAHFOS: Sir Alister Hardy Foundation for Ocean Science
SEC: South Equatorial Current
SOI: Southern Oscillation Index
SEPODYM: Spatial environmental population dynamics model

TIN: Total inorganic nitrogen
TNZ: Thermoneutral zone
TOGA: Tropical Ocean – Global Atmosphere research program

WOCE: The World Ocean Circulation Experiment

The North Atlantic Region

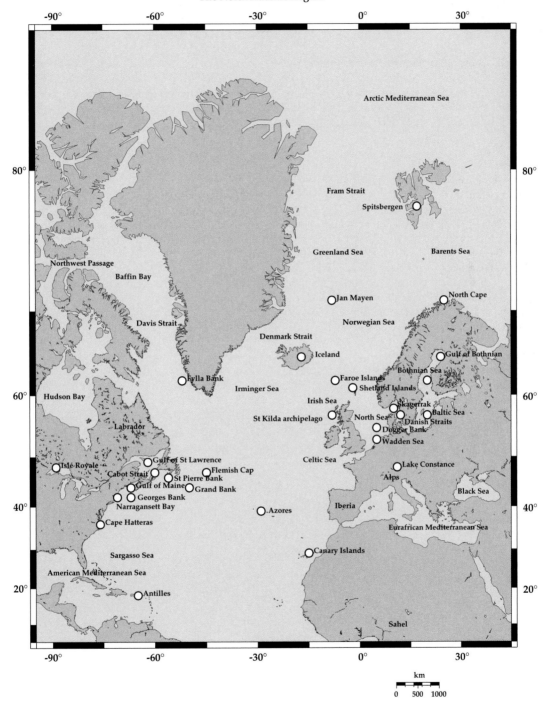

PART I

Background to climate variation and its effects on marine ecology

This part of the book summarizes the climate background together with the more conceptual framework for the climate–ecology link—a general introduction regarding responses of marine ecosystems to climate fluctuations, describing climate variability over the North Atlantic and introducing us to the modelling of marine ecosystems. In this part, we also provide a chapter summarising the modelling of climate fluctuations and their effects on large-scale ecological processes. One main objective of this part is to give a background for applications throughout the various chapters to come.

We, as book-editors, outline in Chapter 1 the main currents and hydrography of the North Atlantic including a brief account of the Atlantic as a whole and how it is linked to other oceans. A short summarizing account of some of the main circulation patterns in the world's oceans and the role of large-scale climate variability is also included. This is followed by an overview of oceanographic processes that affect marine life in the North Atlantic. Much of the chapter is, through examples, devoted to broadening the understanding of how diverse the responses of ecology are to atmospheric and ocean climate variability. As part of this introductory chapter, we also provide a summary account of how climate may affect ecological processes—linearly versus non-linearly, direct versus delayed, etc. (See also discussion in Chapter 10.) That chapter is closed by a synoptic account of how climate effects on ecological processes and patterns may be studied through methods within the field of time series analyses—methods discussed in several subsequent chapters (e.g. 6, 15, and 16).

The patterns and processes involved in climate variability—the topic of Chapter 2—are not easy to summarize: the issue is on the one side broad, on the other side rather complex making a full in-depth description difficult within limited space. Here, we try to strike a balance between breadth of coverage and depth of detail. To accomplish this some issues had to be left out to allow for coverage of others. Thus, Hurrell and Dickson focus Chapter 2 on a thorough presentation of the North Atlantic Oscillation (NAO), a north–south alternation in atmospheric mass between the subtropical atmospheric high-pressure centre over the Azores and the atmospheric sub-polar low-pressure centre over Iceland. Special attention is paid to the forcing of the North Atlantic Ocean by the NAO, both directly at the atmosphere–ocean interface, as well as gradually more indirect effects towards intermediate and abyssal ocean depths (several authors will, in later parts, cover responses of various parts of North Atlantic ecosystems to the NAO, in particular Chapter 5 dealing with effects on zooplankton and Chapter 13 with a freshwater point of view). Hurrell and Dickson also provide important background information relating to the mean state and variability of both the ocean and atmosphere.

A chapter on modelling the marine ecosystem and how it is affected by climate variability closes this Part. In Chapter 3, Werner and co-workers provide an outline of some of the models developed to increase the understanding of how marine ecosystems "work". The chapter deals both with lower trophic levels, with a focus on carbon fluxes, and higher levels, with a focus on pelagic fish (fluxes of carbon and other key elements is a main topic in Chapter 14, which takes a limnological point of view; population dynamics of fish will be covered in Chapter 6, Community ecology in Chapter 10 and in Chapter 11 a population dynamics model including environmental effects for large pelagics, Pacific tuna, is described). Overviews of existing physical–biological models and biological transport models are given with classification by geographical region, dimension, spatial and temporal scale, and other criteria. Moreover, paradigms explaining patterns, abundance, and variability in fish populations are described (in Chapter 12 complexity in fisheries dynamics and climate interactions is studied). Chapter 3 also covers the topic of data assimilation, the integration of models with data to improve the estimation of a system's state. The data assimilation system consists of three elements: a set of observations (i.e. a model linking data to the model), a dynamical model, and a data assimilation scheme. While data assimilation techniques have been used actively for some time in connection with purely physical numerical models, applications in North Atlantic marine ecosystem models are still rare. Looking ahead, as assimilation procedures progress, Werner and co-workers anticipate the greatest limitation to be scarcity of data.

Climatic fluctuations and marine systems: a general introduction to the ecological effects

Geir Ottersen, Nils Chr. Stenseth, and James W. Hurrell

Climate profoundly influences a variety of ecological processes and, consequently temporal and spatial patterns of population and species abundance. Responses to climate fluctuations are reflected in the productivity of marine ecosystems from phytoplankton to the dynamics of fish populations (Cushing 1982). These effects operate through variations in local weather and climate phenomena, such as temperature, wind, and residual currents as well as interactions among these. In the extra-tropics, local variations in weather are often coupled over large geographic areas through the transient behaviour of atmospheric planetary-scale waves. These large-scale variations drive temporally and spatially averaged exchanges of heat, momentum, and water vapour (Namias and Cayan 1981), which ultimately determines growth, recruitment, and migration patterns.

There is a tendency in all of ecology to believe that measurement of one or a few environmental variables (often those most easily measured) can serve to characterize all of environment. One way to, at least partially, take care of this is to include indices on large-scale climate patterns instead of focusing entirely on a few local weather descriptors such as temperature and wind. By applying the latter approach, an important dimension is missed; namely, the holistic nature of the climate system. Large-scale patterns of climate variability can be seen as a composed function involving a variety of climatic parameters over time and space. Indeed, they may be said to represent 'a package of weather' (see Stenseth *et al.* 2003).

To set the scene for what is to come, we will in this introductory chapter, first provide a brief account of the Atlantic as a whole and how it is linked to other oceans, before we focus on the main currents and hydrography of the North Atlantic. This is followed by a brief summarizing account of some of the main circulation patterns in the worlds oceans and the role of large-scale climate variability. We continue by providing an overview of oceanographic processes that are believed to be of particular importance to marine ecology. Finally, we present examples on the basis of which we provide an account of how diverse the responses of ecology are to atmospheric and ocean climate variability. Throughout we link our discussion to the chapters ahead.

1.1 The North Atlantic

The Atlantic is, after the Pacific, the world's second largest ocean. Several topographic features distinguish the Atlantic from other oceans such as the Pacific and Indian Oceans. First of all, the Atlantic Ocean extends both into the Arctic and Antarctic, giving it a total meridional extent (when including the Atlantic part of the southern ocean) of over 21,000 km from the Bering Strait through the Arctic mediterranean Sea to the Antarctic continent. In comparison, the largest zonal distance, between the Gulf of Mexico and the coast of northwest Africa, spans only a little more than 8300 km (Tomczak and Godfrey 1994).

3

Atlantic has [larger] n. of adjacent seas.

Arctic Med sea important for Deep Water

Second, the Atlantic has the largest number of adjacent seas, including mediterranean seas influencing the characteristics of its waters. This is particularly true for the Arctic Mediterranean Sea, which plays a crucial role in the formation of Deep Water not only for the Atlantic but for all the oceans in the world. The remaining adjacent seas can be divided on geographical arguments into four groups. The first group contains the waters connected to the Atlantic Ocean proper through the Labrador Sea and consists of the Davis Strait, Baffin Bay, the northwest Passage, and Hudson Bay. The second group is located between Europe, Africa, and Asia and contains the Eurafrican Mediterranean Sea (which includes the Black Sea). The third group is found near the junction of North and South America and contains the American Mediterranean Sea with its subdivisions the Caribbean Sea and the Gulf of Mexico. The shallow European seas make up the fourth and last group, which contains the Irish and North Seas and the Baltic Sea with its approaches.

Third and finally, the Atlantic Ocean is divided rather equally into a series of eastern and western basins by the Mid-Atlantic ridge, which in many parts rises to less than 1000 m depth, reaches the 2000-m depth contour nearly everywhere, and consequently has a strong impact on the circulation of the deeper layers (Tomczak and Godfrey 1994).

When all adjacent seas are included, the Atlantic covers an area of 107×10^6 km^2. Without the Arctic Mediterranean and the Atlantic part of the southern ocean, its size amounts to 74×10^6 km^2. Although its abyssal basins are deeper than 5000 m and most extend beyond 6000 m, the average depth of the Atlantic Ocean is, due to the fact that shelf seas, including its adjacent and mediterranean seas, account for over 13% of the surface area, 3300 m, less than the mean depth of both the Pacific and Indian Oceans (Tomczak and Godfrey 1994).

The upper water circulation of the Atlantic Ocean consists in its gross features of two great anticyclonic circulations or gyres, a counterclockwise one in the South Atlantic and a clockwise one in the North Atlantic (Fig. 1.1). The two gyres are driven separately, each by the trade winds in its own hemisphere, and they are separated over part of the equatorial zone by the eastward flowing Counter Current (Pickard and Emery 1990). The main features of the circulation of the North Atlantic were described already by Iselin (1936) and Sverdrup et al. (1946). The clockwise gyre may be considered to start with

the North Equatorial Current driven by the northeast trade winds. This current flows to the west and is joined from the south by that part of the South Equatorial Current, which has turned across the equator into the North Atlantic. Part of this combined current flows towards northwest as the Antilles Current, the other part into the Gulf of Mexico. From here it escapes between Florida and Cuba into the North Atlantic as the Florida Current. This current joins up with the Antilles Current off the coast of Florida and from about Cape Hatteras, where the joint current breaks away from the North American shore, it is called the Gulf Stream. The Gulf Stream flows northeast to the Grand Banks of Newfoundland at about 40°N, 50°W, after which it continues east and north as the North Atlantic Current. This divides and partly turns south past Spain and North Africa to complete the North Atlantic Gyre and feed into the North Equatorial Current (Pickard and Emery 1990).

The remainder of the North Atlantic Current flows northeast between Scotland and Iceland and enters the Nordic Seas, where the Faroe Islands separate the two main inflowing branches. The major part of the Atlantic Water (AW) continues as a Norwegian Atlantic Current (NWAC) north along the coast of Norway, which branches into the North Sea and also to the more central parts of the Nordic Seas. At the western boundary of the Barents Sea, the NWAC further bifurcates into the North Cape Current flowing eastwards into the Barents Sea, and the West Spitsbergen Current flowing northwards into the Fram Strait (Furevik 2001).

At the western side of the Nordic Seas, there is a cold and fresh flow of Polar Water originating in the Arctic Ocean. The major part of this water leaves the Nordic Seas as the East Greenland Current (EGC) through the Denmark Strait. The remaining part flows into the central areas of the Nordic Seas as the Jan Mayen Current and the East Icelandic Current. Here it gradually mixes with the AW, and modified by atmospheric forcing, it is an important factor in the ventilation of the intermediate and deep waters of the Nordic Seas, and to the thermohaline circulation of the North Atlantic (Furevik 2001). The EGC flows round the southern tip of Greenland into the Labrador Sea and continues up the west coast. The inflow to this area is balanced by the southward flow, along the west side of Baffin Bay, which continues south as the Labrador Current down the west side of the

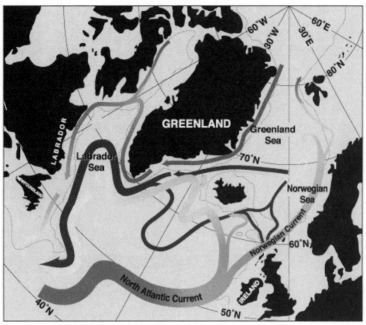

[Handwritten margin note: most mov of the oceans are caused by differences in density bw water masses ↓ THERMOHALINE CIRCULATION]

Figure 1.1 The main northward flowing warm-water routes and the cold deep southward return flows that form the North Atlantic thermohaline circulation (McCartney *et al.* 1996). Downloaded from the CLIVAR home page, www.clivar.org.

Labrador Sea back into the Atlantic proper (Pickard and Emery 1990).

1.2 Oceanic circulation and large-scale climate variability

Ocean water circulates in currents that transfer heat from tropical regions to polar regions, influence weather and climate, distribute nutrients, and scatter organisms (Garrison 2002). Coastal regions and the uppermost part of the ocean is the main focus of this book. Here, above the pycnocline, the zone in which density increases rapidly with depth, wind is the primary driving force creating the surface currents. While the effect of wind typically is horizontal movement of water (Fig. 1.1), winds can sometimes induce vertical movement in the surface water, upwelling or downwelling. Wind blowing parallel to shore or offshore can cause coastal upwelling when the removed surface water is replaced by water rising along the shore. Because the new surface water is often rich in nutrients, prolonged wind can result in increased biological productivity. Although we will not directly deal with the ocean interior in this book, it should be noted that most

movements of the oceans water masses are caused not by wind energy, but by differences in density between water masses. This slow *thermohaline* circulation is responsible for most of the vertical movement of ocean water and the circulation of the global ocean as a whole (Garrison 2002).

A schematic account of the global circulation pattern linking the major ocean basins of the planet is shown in Fig. 1.2. The large region of open ocean in the Equatorial Pacific allows significant warming of water as it drifts to the west and past the northern edge of Australia. This warm surface water continues to drift westward through the Indian Ocean and around the southern tip of Africa. From here it turns northward, crossing the Equator (taking a more easterly direction than shown here), and creating the Gulf Stream off eastern North America. The warm surface water continually cools as it moves northward past Great Britain and into the Norwegian Sea. By this time the water is so cold and dense that it sinks to lower depths in the ocean. This creates the start of the global scale return current at low depths that moves southward across the Equator, back around Africa, past the southern edge of Australia and back to the central Pacific Ocean. A smaller branch of the return current splits

[Handwritten margin note: GULF STREAM]

[Handwritten note at bottom left: PYCNOCLINE: zone in which density increases rapidly w/ depth; wind is the primary driving force creating the surface currents.]

[Handwritten note at bottom right: COASTAL UPWELLING = wind blowing parallel to the shore when removed surface water is replaced by water rising along the shore +++ bio productivity]

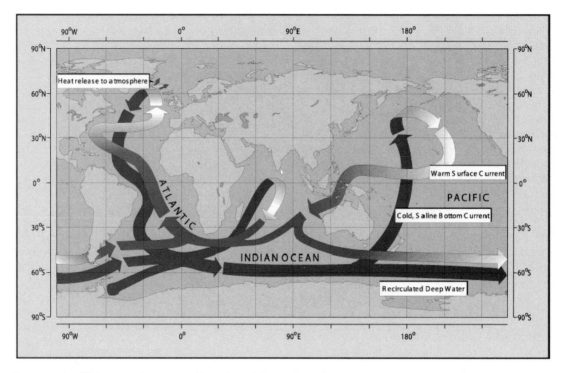

Figure 1.2 The global ocean circulation pattern 'The Conveyor belt'. Downloaded from the CLIVAR home page, www.clivar.org.

off after passing Africa and enters the Indian ocean where sufficient warming augments the warm surface current coming from the central Pacific (www.meteor.iastate.edu/gccourse/ocean/global.html).

The most well known large-scale climate mechanism operating on an inter-annual to decadal timescale is the El Niño-Southern Oscillation (ENSO; see, for example, Philander 1990 or Allan *et al.* 1996). This phenomena originates in the tropical Pacific but has impact worldwide. Ecological effects of ENSO are discussed in Chapters 11, 12, and 16. The first evidence of such patterns of significant simultaneous correlations between weather and climatic anomalies over widely separated regions, teleconnections[1], emerged from analyses of sea-level pressure and surface-air temperature (Walker and Bliss 1932). Near simultaneous fluctuations of

fish stocks in widely separated regions support the view that they are sometimes influenced by climate operating at a global scale (Schwartzlose *et al.* 1999).

Ecological regime changes may be linked to global climate variability, which in turn may be associated with an interdecadal cycle of dislocation of the convergence areas in the North Atlantic (Greenland and Labrador Seas) where thermohaline circulation starts (Kawasaki 1994). In the southeastern Atlantic, Benguela El Niño like episodes have been reported (Shannon *et al.* 1986; Gammelsrød *et al.* 1998). Benguela Niños advect warm, more saline water from the north onto the Namibian Shelf. This water may intrude about 600 km further south than in normal years. Off Japan anomalous southward intrusions of the Oyashio Current (ASIOs) have been described by Sekine (1991) and off Tasmania periods of low zonal westerly winds (LZWWs) also may cause a form of Tasmanian 'El Niño' during which temperature rises, nutrients become scarce in surface waters, new production decreases, and the biomass of larger zooplankton is drastically reduced (Harris *et al.* 1992). There is a

---

[1] The actual term 'teleconnections' was first used by Angstrom, A. 1935. Teleconnections of climate changes in present times. *Geographical Annals*, **17**, 242–58; and popularized by Bjerknes, J. 1969. Atmospheric teleconnections from the equatorial pacific. *Monthly weather review*, **97**, 163–72.

correspondence in time between the four Benguela Niños and the four LZWWs between 1949 and 1984. Information on ASIOs is only available since 1955, but the Japanese ASIOs occurred in each subsequent period of Benguela Niños off Namibia and Tasmanian LZWWs (Schwartzlose *et al.* 1999).

Influence on northern NE Atlantic climate by atmospheric teleconnections was suggested already by Izhevskii (1964) who argued for what he called a monophasic form of fluctuation throughout the Gulf Stream system (i.e. no time lag in water temperature fluctuations from the Florida current to the Barents Sea). More recent literature supports the view that atmospheric teleconnections influence marine climate and ecology across the North Atlantic (e.g. Taylor 1995; Rodwell *et al.* 1999; Ottersen *et al.* 2001; Stenseth *et al.* 2003). There is also strong support for much of the decadal variability in the North Atlantic sea surface temperature being explained as a local oceanic response to atmospheric variability (Deser and Blackmon 1993; Battisti *et al.* 1995; Houghton 1996).

1.3 Physical processes of importance to marine ecology

A number of physical processes may influence marine life at the individual, population, and community level. Processes that influence the reproductive habitat of many species of fish were, for example, grouped into three main classes by Bakun (1996):

(i) *Enrichment processes*: upwelling, mixing, cyclonic wind stress curl (Ekman divergence), cyclonic eddy formation;
(ii) *Concentration processes*: convergent frontal formation, anticyclonic wind stress curl (Ekman convergence), lack of dispersion by turbulent mixing processes, 'encounter-rate' increases/decreases as a result of variability in micro-scale turbulence;
(iii) *Retention processes*: lack of offshore transport in (1) Ekman field (near-surface and superficial layers), (2) geostrophic current (intermediate layers), and (3) offshore dispersion of eddy-like features (filaments) on the meso-scale; availability of enclosed gyral circulations, stability of current patterns to which life cycles are adapted.

Simplified, *Enrichment processes* make more nutrients available to biological productivity, *Concentration processes* enhance food availability for a predator through increasing the concentration of food particles and *Retention processes* contribute towards keeping individual members of a population in the appropriate place during the various parts of the life cycle. Examples of many of these mechanisms will be presented in subsequent chapters.

In Chapter 11 Patrick Lehodey points to the importance of upwelling for biological productivity in the Pacific, particularly along the equator and off the west coast of central South America. Since 90% of the worlds fisheries traditionally have been in 2–3% of the ocean area, mostly in coastal upwelling areas (Pond and Pickard 1983), we will dwell on this. Upwelling is, simply stated, a result of surface water skimmed away by the wind, being replaced by water masses from below, typically from depths not greater than 200–300 m. For a more thorough description of the processes briefly described here (Ekman transport and divergence) we refer to textbooks in physical oceanography, for example, Pond and Pickard (1983).

When upwelled water of high nutrient content is transported into the photic (light-rich) zone, the production of phytoplankton may increase dramatically, enhancing the production potential upwards throughout the food web. The main regions where large-scale coastal upwelling takes place are in eastern boundary currents off the west coasts of West and South Africa in the South Atlantic and North and South America in the Pacific. However, since subsurface waters do not always have a high content of nutrients, upwelling does not necessarily imply an increase in biological production. In the North Atlantic, northwest Africa and Iberia are major upwelling systems. Chapter 6 exemplifies how interannual variability in upwelling may influence fish recruitment in this region.

Upwelling may also take place in areas remote from physical boundaries, namely, where surface waters diverge along the equator. The prevailing easterly trade winds drag the surface water westward along the equator. The Earth's rotation deflects the westward-flowing current towards the right in the Northern Hemisphere and towards the left in the Southern Hemisphere, driving the surface water away from the equator and bringing up water from below. Chapter 11 illustrates how this system breaks down during El Niño events causing a marked decrease in the productivity of the eastern tropical Pacific.

A specific form of *Concentration process* is that described in the Encounter Rate hypothesis of

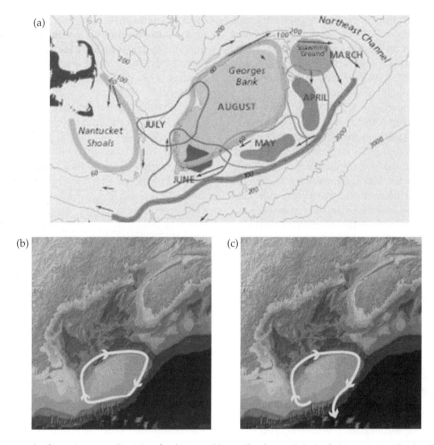

Figure 1.3 An example of retention versus dispersion of cod eggs and larvae. The characteristic circulation pattern on Georges Bank, northeast United States. (a) Currents, transport eggs, and larvae of cod from the spawning area on the northeastern part of the Bank to the southwest. Main spawning grounds and typical areas of main egg and larval concentration per month are indicated. Variation in the circulation pattern can result in (b) lesser or (c) greater advective losses from the Bank. Downloaded from Dr. Bruce C. Monger at Cornell University (http://www.eas.Cornell.edu).

ENCOUNTER RATE Hp: turbulence
increase encounter rates bw fishlarvae &
prey

Rothschild and Osborn (1988) dealing with small-scale turbulence and its effect on predator–prey encounter rates during the feeding process. The field study by Sundby and Fossum (1990) confirmed that moderate turbulence was found to enhance contact rates between cod larvae and their prey zooplankton nauplii, and hence increase the effective prey concentration available to larvae. Although, there still is some disagreement with regards to the parametrization, the main findings have later been supported by several modelling studies, recently by Werner *et al.* (2001). This topic is further discussed in Chapters 3 and 6.

Retention processes may be illustrated by an example from the waters off northeast United States.

The characteristic circulation pattern on Georges Bank transports eggs and larvae of cod from the spawning area on the northeastern part of the Bank to the southwest. Variation in the circulation pattern can result in greater or lesser advective losses from the Bank (Serchuk *et al.* 1994; see Fig. 1.3). A model study by Werner *et al.* (1993) showed that during the first 60 days circulation had a greater influence on the distribution of eggs and larvae than behaviour. The results of Lough *et al.* (1994) indicate a close connection between egg and larval distribution pattern and year-class strength. They demonstrate that years with good recruitment typically are associated with low losses from the bank due to favourable wind conditions, and vice versa.

Transport-related effects on recruitment are discussed further in Chapter 6. Furthermore, retention processes influencing stability of current patterns to which life cycles are adapted may be seen in the light of the member/vagrant hypothesis of Sinclair and Iles (1989) discussed in Chapter 3.

1.4 Ecological responses to climate variability

As we already have touched upon and which will be further demonstrated and discussed in later chapters, there is a great variety in the possible pathways by which climate variability may affect ecological processes. Here, we highlight some of the main topics of general concern. To be able to deal with this, it is, as pointed out by Ottersen *et al.* (2001), often a fruitful exercise to summarize the variety emerged from a number of studies into a restricted number of categories.

1.4.1 Direct or indirect response

Climate variability affects animals both directly through physiology, including metabolic and reproductive processes, as well as indirectly through affecting their biological environment (predators, prey, within population interactions, disease). The physical environment also affects feeding rates and competition through favouring one or another species (Fig. 1.4).

Chapter 6 exemplifies this for fish and shellfish, while in Chapter 7 the effect of climate variability on seabirds is described to follow two main lines: directly by physiological effect and indirectly through influence on food. Direct physiological effects include metabolic processes that happen during the life-cycle (e.g. reproduction and moult). Since their feed is composed of organisms with populations that fluctuate in response to climatic

changes, the main indirect influence of climate is for seabirds through regulation of food availability.

Ecological effects of the NAO were classified according to three major types: direct effects, indirect effects, and integrated effects by Ottersen *et al.* (2001). A similar categorization may be made for other climate phenomena.

1. The direct effects of the NAO are mechanisms that involve a direct ecological response to one of the environmental phenomena synchronized with the NAO. The effect of the NAO on metabolic rates via temperature is, for instance, of this type.
2. The indirect effects of the NAO are non-trivial mechanisms that either involve several physical or biological intermediary steps between the NAO and the ecological trait and/or have no direct impact on the biology of the population.
3. The integrated effects of the NAO involve simple ecological responses that can occur during and after the year of an NAO extreme. This is the case when a population has to be repeatedly affected by a particular environmental situation before the ecological change can be perceived (biological inertia) or when the environmental phenomenon affecting the population is itself modulated over a number of years (physical inertia, for example, reduction in the volume of North Sea Deep Water; Heath *et al.* 1999).

A problem with general classification schemes, as that above, is that there always are some cases that do not fit into any of the predefined categories, while sometimes several classes seem to fit. Thus, in Chapter 5 a new class of effects is defined and exemplified, *translations*. Translations involve movements of organisms from one place to another such as the advection of *Calanus finmarchicus* from the continental slope on to the shelf. These alterations are based entirely on the physical changes produced by climate variability.

The classification of response to change in climate as either direct or indirect makes sense also at the community level (Chapter 10). Changes in temperature that influence, for example, growth and recruitment, may lead to a direct response in fish species in the community, with the abundances of individual species changing in a predictable manner according to each species' optimum for growth and reproduction. On the other hand, climate shifts may lead to, for example, variability in oceanic circulation patterns and temperature-induced changes in prey abundance. Fish communities may then act

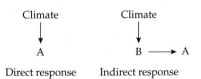

Climate   Climate

A   B ⟶ A

Direct response   Indirect response

Figure 1.4 Direct or indirect response. (A and B being biological units of interest.)

indirectly to a change in climate by way of responding to changes in local abiotic or biotic conditions.

1.4.2 Temporally lagged responses

Related to the issue of direct and indirect climate effects is the question of time lags in ecological responses to climatic variation (Fig. 1.5). In Chapter 15, a discussion is provided of how the existence of both immediate and lagged responses to climate introduces conceptual and analytical challenges to the study of ecological consequences of climatic variability.

An example of a temporally lagged response to a large-scale climate signal, not included in Chapter 15, is provided by Ottersen *et al.* (2001). They show how some of the effects of the NAO may be carried by a biological population over a number of years following a particular NAO situation. The increase in survival through the vulnerable early stages of Barents Sea cod during warm, high NAO years historically results in stronger year classes also at later, catchable, stages (see 'the cohort-effect' discussed by Stenseth *et al.* 2002). When the year class matures, the number of spawners as well as their individual size may be increased, enhancing the potential for high recruitment to the next generation. On the other hand, the year-class strength of cod in the North and Irish Seas is inversely related to a positive NAO phase and high temperature. This is possibly a result of limitation in energy resources necessary to achieve higher metabolic rates during warm years (Planque and Fox 1998). In both cases, the effects of the NAO are perceived in the fisheries with a lag of several years. (See Chapter 6 for more on population dynamics of cod and other fish and Chapter 10 for fish community ecology).

1.4.3 Continuous fluctuations versus episodic events: linear and non-linear responses

The environment may undergo slow continuous change, or more rapid, episodic events may transfer

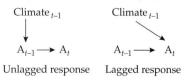

Figure 1.5 Temporally unlagged or lagged response. (*t−1* and *t* indicate the immediate past and the present, respectively.)

conditions between different states and regimes. The most well-known example of the latter is the switching between El Niño, La Niña, and 'normal' conditions in the tropical Pacific, an other is the Pacific Decadal Oscillation between a 'warm' and a 'cold' phase (more on this in Chapter 12).

These two categories may be exemplified by the mechanisms that seem to initiate and sustain regimes of sardine and anchovy worldwide. There may be a continous modification of habitat, for example, a trend to warming that permits an expansion in spawning range and enhanced egg production (Lluch-Belda *et al.* 1992). Alternatively, there may be episodic environmental events that trigger changes in populations and ecosystems in well separated areas. Formation of powerful year-classes could cause a population to expand quickly and, for short-lived species, a few poor year-classes could result in a rapid population decrease. Empirical evidence exists for both these forms of environmental influence at a global scale, and it is quite possible that several factors may operate at any given time (Schwartzlose *et al.* 1999).

By far most studies of links between climate and marine ecology—by assumption—only consider *linear* relations, while challenges related to *nonlinearity* (i.e. that episodic events may be more important than mean conditions), are neglected. While the relation between climate and ecology indeed may be linear, non-linear responses may be introduced in several ways (Fig. 1.6). The relationship between large-scale and local climatic variables may change over time (i.e. non-stationarity; Jones *et al.* 2001), and hence limit the time period for which effects on ecosystems can be predicted from known relationship with climate patterns. The relatively clear link between the NAO (as estimated by traditional two-station indices) and regional climate in the Barents Sea that has been observed since around 1970 was, for example, not present during the preceding 50 years (Dickson *et al.* 2000; Ottersen *et al.* 2003).

A study with starting point in terrestrial ecology, but with consequences also for a marine setting by Mysterud *et al.* (2003) examines this issue, with special emphasis to the NAO. They point to two different ways in which the effect of the NAO on ecological systems may be expected to be non-linear. First, the NAO may not be linearly related to local climatic variables, and plankton and fish are indeed expected to respond to the climate they experience at a local scale. Second, the plant's or

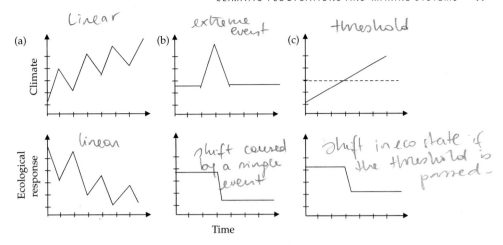

Handwritten annotations on figure: Linear; extreme event; threshold; linear; shift caused by a single event; shift in eco state if the threshold is passed.

Figure 1.6 Differential relations between climate signal and ecological response: (a) Linear climate signal and linear ecological response. (b) Single climate event causes shift in ecological state or regime. (c) Linear climate signal that causes shift in ecological state or regime when climate threshold value is passed.

Handwritten note: there is evidence that CC may lead to a higher frequency of more extreme events as a consequence of increased variance & not just change in average.

animal's response to changes in local climate may not always be linear. It should be emphasized that the issue of non-linear responses may at least in part be an issue of scale, as transforming the predictor and/or the response variable may linearize the relationships. Non-monotonous relationships, as found in Mysterud *et al.* (2001), cannot be linearized, however.

There is evidence that climate change may lead to a higher frequency of more extreme events, as a consequence of increased variance and not just change in average. It is generally regarded that the spatio-temporal distribution of extreme events are often ecologically more significant than seasonal mean values. Increased environmental variance has in general a negative effect on population growth, and non-linear relationships between climate and population processes imply that a change in climatic variability will affect the mean value of a process, even if the mean value of the climatic parameter remains the same (Mysterud *et al.* 2003).

It may further be useful to distinguish between linear and non-linear responses of fish *communities* to a climate signal (see Chapter 10). The linearity in the former case is understood as linear among species, that is, most species are affected in a similar fashion, without changing interaction effects among species. However, if changes in climate render the environment inhabitable for key species in the community, large (non-linear) community effects may be observed.

1.4.4 The match–mismatch hypothesis

Handwritten note: Cushing in the 1980's

Climatic fluctuations may affect the relative timing of food requirement and food availability. Survival of a predator depends on its ability to encounter and eat a sufficient quantity of suitable prey in order to avoid starvation and to grow. Differences in the temporal and spatial match between predator and prey may thus generate variability in predator survival rates, including interannual variability. This 'match–mismatch' hypothesis was first presented by D. Cushing some 30 years ago and later updated (Fig. 1.7; Cushing 1990; Cushing 1996).

Match–mismatch type mechanisms have been reported for different regions and at different trophic levels, often in its original context, relating cod larvae to their zooplankton nauplii prey (e.g. Ellertsen *et al.* 1989 for Barents sea cod), but also for other species, for example herring in the St Lawrence estuary (Fortier and Gagne 1990). However, it should be noted that it has been suggested that the survival-food availability coupling may be less precise than proposed in the 'match– mismatch' hypothesis (Leggett and Deblois 1994). Still, recent work by Brander *et al.* (2001) tested the hypothesis by examining the relation between modelled production of chlorophyll and copepod eggs, driven by meteorological forcing, and cod recruitment. They concluded that the interannual variability in *Calanus* egg production did have a significant effect on cod recruitment in the Irish Sea and at Iceland.

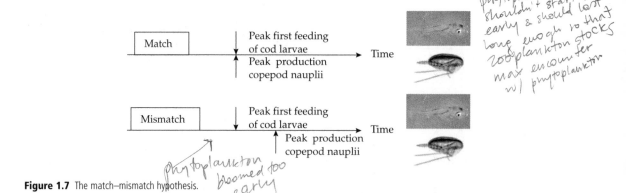

[handwritten annotation:] phytoplankton bloom shouldn't start too early & should last long enough so that zooplankton stocks max encounter w/ phytoplankton

Figure 1.7 The match–mismatch hypothesis.

[handwritten annotation:] Phytoplankton bloomed too early

In polar marine ecosystems variability in biological stocks and productivities is pronounced on all timescales because of the marked variability in the environmental forcing (meteorological and hydrographical) factors on all timescales. Polar pelagic ecosystems are, in fact, notoriously unstable (Sakshaug 1997).

One seasonal feature is, however, quite predictable in regions like the Barents Sea. All parts experience a spring phytoplankton bloom sometime during April–July. This bloom is based on the 'new' nutrients admixed to the upper layers during the winter. The timing of the spring bloom depends on the supply of light to the phytoplankton in spring, which in turn critically depends on the mixing characteristics of the upper water masses. Deep mixing causes phytoplankton to spend a major fraction of their life in very low light, causing severe light limitation and thus delays in the onset of a bloom (Sverdrup 1953).

Different processes may generate the stabilization of the upper layers necessary to allow phytoplankton prolonged access to the photic zone. Typically, the earliest blooms (in mid-April) arise in Atlantic waters close to the oceanic Polar Front—this happens in 'cold years' when the ice cover reaches that far south. In such situations, a meltwater-generated (salinity dependent) stabilization of the upper layers occurs early because the sea ice starts melting early as a consequence of the warming effects of the Atlantic waters. North of the oceanic Polar Front, stability trails the retreating ice as a consequence of the supply of fresh melt-water. This establishes a 30–50 km wide phytoplankton bloom zone that trails the ice edge. Such an effect is known from all seasonally ice-covered polar seas. South of the Polar Front, that is in the permanently ice-free areas that are characterized by Atlantic surface waters, stability typically depends instead on the formation of

a temperature-dependent thermocline. This thermocline is created fairly late, thus the spring bloom is not triggered until mid-May or early June (Sakshaug 1997).

Differences in the timing of spring blooms affect the match between phytoplankton and zooplankton maxima. To maximize grazing and minimize sedimentation ('match'), phytoplankton blooms should not begin too early and they should be of long duration so that the probability of zooplankton stocks encountering phytoplankton blooms is maximized (Sakshaug and Skjoldal 1989; Hassel *et al.* 1991). The late and protracted phytoplankton blooms in the permanently ice-free Atlantic waters may represent good examples of a 'match' whereas the early blooms that arise where sea ice overlies Atlantic water may be textbook examples of a 'mismatch' (Sakshaug 1997).

Match or mismatch is a fundamental issue also in several other systems, for instance for sea birds as discussed more comprehensively in Chapter 7. The ambient conditions in the sea (such as water temperature and currents) and large-scale climatic and hydrographic processes that affect these variables, generate variation in the production, distribution, and abundance of organisms on which seabirds feed and hence on the birds themselves. A climate-induced delay in food-production could create a mismatch between the prey production and the predator requirement that might have heavy impact on reproduction success.

In Chapter 3, Werner *et al.* examine several hypotheses providing explanations to how physical processes affect fish recruitment. The match–mismatch hypothesis is here seen in the light of the Hydrographic Containment hypothesis, which is a combination of match–mismatch and the migration triangle hypothesis of Harden–Jones (1968).

1.4.5 Modelling the ecological effects of climate fluctuations: a time series approach

Any ecological impact will depend on a variety of factors affecting the population biology of the particular species in question (Fig. 1.8). It is common when analysing the response of populations, communities, or ecosystems to environmental variables to implicitly or explicitly assume that other factors remain unchanged. However, response to a climate signal will depend on other ecological factors like density-dependence and interactions. It is therefore too simplistic to predict population abundances and demographic rates only based on, for example, the NAO—particularly if the responses are non-linear (Mysterud *et al.* 2003).

Figure 1.8 Interaction with other factors.

Recently, it has become rather common to search for signals of climate fluctuations in time-series data (e.g. see Stenseth *et al.* 2002, 2003). Figure 1.9 summarizes various ways climate effects may appear in time-series models—and how this may be interpreted ecologically. Panel A summarizes the (somewhat theoretical) case where only one population lives in an area without any closed feedback interactions with any other part (or species) of that system (this might, for instance, be exemplified by a species not being food-limited and only being exposed to some nomadic generalized predator (e.g. see Stenseth *et al.* 1999). Panel A.i depicts the case where there, within such a system, is no climate effect—that is, what is typically encountered in the non-climate oriented part of the ecological literature. Panels A.ii and A.iii depict two cases where climate affects the population—one (panel A.ii) where the climate effect is non-additive (on log-scale) implying, for instance, that climate affects the food-supply in a non-dynamic manner, whereas the other (panel A.iii) shows the case where climate fluctuations affect the way individuals interact within the population. Among these, the former

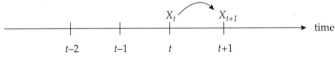

Figure 1.9 Expected signals of various types of climate effects on ecological systems. See text for discussion. $x_t = \ln(X_t)$.

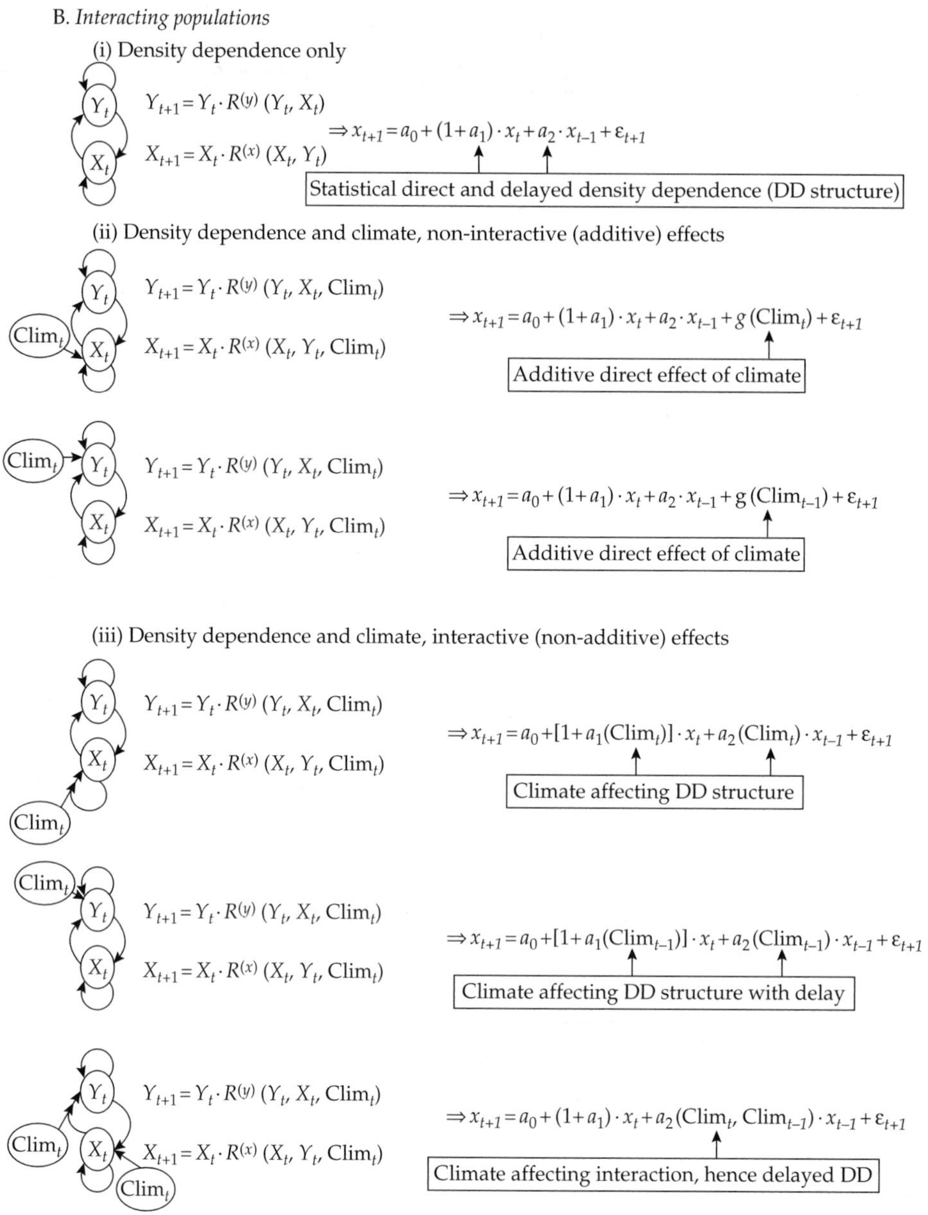

B. *Interacting populations*

(i) Density dependence only

$Y_{t+1} = Y_t \cdot R^{(y)}(Y_t, X_t)$

$X_{t+1} = X_t \cdot R^{(x)}(X_t, Y_t)$

$\Rightarrow x_{t+1} = a_0 + (1 + a_1) \cdot x_t + a_2 \cdot x_{t-1} + \varepsilon_{t+1}$

Statistical direct and delayed density dependence (DD structure)

(ii) Density dependence and climate, non-interactive (additive) effects

$Y_{t+1} = Y_t \cdot R^{(y)}(Y_t, X_t, \mathrm{Clim}_t)$

$X_{t+1} = X_t \cdot R^{(x)}(X_t, Y_t, \mathrm{Clim}_t)$

$\Rightarrow x_{t+1} = a_0 + (1 + a_1) \cdot x_t + a_2 \cdot x_{t-1} + g(\mathrm{Clim}_t) + \varepsilon_{t+1}$

Additive direct effect of climate

$Y_{t+1} = Y_t \cdot R^{(y)}(Y_t, X_t, \mathrm{Clim}_t)$

$X_{t+1} = X_t \cdot R^{(x)}(X_t, Y_t, \mathrm{Clim}_t)$

$\Rightarrow x_{t+1} = a_0 + (1 + a_1) \cdot x_t + a_2 \cdot x_{t-1} + g(\mathrm{Clim}_{t-1}) + \varepsilon_{t+1}$

Additive direct effect of climate

(iii) Density dependence and climate, interactive (non-additive) effects

$Y_{t+1} = Y_t \cdot R^{(y)}(Y_t, X_t, \mathrm{Clim}_t)$

$X_{t+1} = X_t \cdot R^{(x)}(X_t, Y_t, \mathrm{Clim}_t)$

$\Rightarrow x_{t+1} = a_0 + [1 + a_1(\mathrm{Clim}_t)] \cdot x_t + a_2(\mathrm{Clim}_t) \cdot x_{t-1} + \varepsilon_{t+1}$

Climate affecting DD structure

$Y_{t+1} = Y_t \cdot R^{(y)}(Y_t, X_t, \mathrm{Clim}_t)$

$X_{t+1} = X_t \cdot R^{(x)}(X_t, Y_t, \mathrm{Clim}_t)$

$\Rightarrow x_{t+1} = a_0 + [1 + a_1(\mathrm{Clim}_{t-1})] \cdot x_t + a_2(\mathrm{Clim}_{t-1}) \cdot x_{t-1} + \varepsilon_{t+1}$

Climate affecting DD structure with delay

$Y_{t+1} = Y_t \cdot R^{(y)}(Y_t, X_t, \mathrm{Clim}_t)$

$X_{t+1} = X_t \cdot R^{(x)}(X_t, Y_t, \mathrm{Clim}_t)$

$\Rightarrow x_{t+1} = a_0 + (1 + a_1) \cdot x_t + a_2(\mathrm{Clim}_t, \mathrm{Clim}_{t-1}) \cdot x_{t-1} + \varepsilon_{t+1}$

Climate affecting interaction, hence delayed DD

Figure 1.9 (*Continued*)

(panel A.ii) is fairly commonly treated in the ecological literature, whereas the other is only now being approached (e.g. see Stenseth *et al.* 2002).

Panel B of Fig. 1.9 illustrates time-series models for interacting species, but for which we only are studying one of the species (due to, for example, only data being available on this species and not the other). Again the distinction between additive and non-additive effect of climate is illustrated—the latter representing quite a bit of methodological challenges to statisticians. Indeed, it is our conviction that bringing specialists within the field of time-series analysis into the field of climate-ecology impacts will be very rewarding.

CHAPTER 2

Climate variability over the North Atlantic

James W. Hurrell and Robert R. Dickson

2.1 Introduction

The climate of the Atlantic sector and surrounding continents exhibits considerable variability over a wide range of timescales. It is manifest as coherent fluctuations in ocean and land temperature, rainfall, and surface pressure with a myriad of impacts on society and the environment. Of central importance is the North Atlantic Oscillation (NAO), which dictates much of the climate variability from the eastern seaboard of North America to Siberia and from the Arctic to the subtropical Atlantic, especially during boreal winter. The NAO refers to a redistribution of atmospheric mass between the Arctic and the subtropical Atlantic, and swings from one phase to another to produce large changes in the mean wind speed and direction over the Atlantic, the heat and moisture transport between the Atlantic and the neighbouring continents, and the intensity and number of storms, their paths, and their associated weather. Such variations have a significant impact on the wind- and buoyancy-driven ocean circulation, as well the site and intensity of water mass transformation, so that the strength and character of the Atlantic meridional overturning circulation (MOC) is substantially influenced.

In this chapter, we provide a broad review of the NAO and its forcing of the North Atlantic Ocean. Of particular interest is the long, irregular amplification of the oscillation towards one extreme phase during winter over recent decades. This climatic event, which is unprecedented in the modern instrumental record of NAO behaviour, has produced a wide range of effects on North Atlantic ecosystems. Some attention will also be given to the climatic impacts of periods of atypical NAO behaviour, such as the spatial displacement of the main centres of action in some winters, or to periods when other patterns of large-scale Atlantic climate variability are more pronounced. An in-depth treatment of the full range of Atlantic climate variability is beyond the scope of this chapter; however, the interested reader is encouraged to pursue the many references to scientific works included herein. For a comprehensive and multidisciplinary overview of material (theory, observations, and models) related specifically to the NAO, the reader should also consult *The North Atlantic Oscillation: Climatic Significance and Environmental Impact* (Hurrell *et al.* 2003). For an exposition of how the NAO is used in the ecological literature, see Stenseth *et al.* (2002, 2003).

2.2 The North Atlantic Atmosphere

2.2.1 Mean state

The long-term (1899–1999) distribution of sea-level pressure (SLP) over the Northern Hemisphere (NH) is illustrated in Fig. 2.1. Large changes from boreal winter (December–February) to boreal summer (June–August) are evident. Perhaps most noticeable are those over the Asian continent related to the development of the Siberian anticyclone during winter and the monsoon cyclone over southeast Asia during summer. Over the northern oceans, subtropical anticyclones dominate during summer, with the Azores high-pressure system covering nearly all of the North Atlantic. These anticyclones weaken and move towards the

15

(a)

DJF climatology

1899–1999   hPa

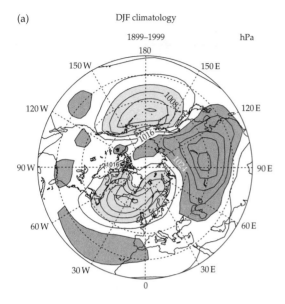

(b)

JJA climatology

1899–1999   hPa

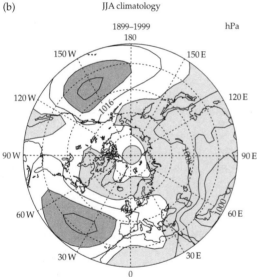

Figure 2.1 Mean (1899–1999) distribution of SLP (hPa) over the NH during boreal winter (December–February = DJF) and summer (June–August = JJA). The contour increment is 4 hPa, and values less (greater) than 1012 hPa (1024 hPa) are indicated by the light (dark) shading.

equator by winter, when the high-latitude Aleutian and Icelandic low-pressure centres predominate.

Because air flows counterclockwise around low pressure and clockwise around high pressure in the NH, westerly flow across the middle latitudes of the Atlantic occurs throughout the year. The vigour of

the flow is related to the pressure gradient, so the surface winds are strongest during winter when they average between 5 and 10 m s^{-1} from the eastern United States across the Atlantic onto northern Europe. The middle latitude westerlies extend throughout the troposphere and reach their maximum (up to 40 m s^{-1}) at a height of about 12 km. This 'jet stream' roughly depicts the path of storms (atmospheric disturbances operating on timescales of about a week or less) travelling between North America and Europe. Over the subtropical Atlantic the prevailing northeasterly winds are relatively steady but strongest during boreal summer.

2.2.2 Variability

What is the NAO?
Monthly mean surface pressures vary markedly about the long-term mean SLP distribution (Fig. 2.1). This variability occurs in well-defined spatial patterns (Wallace and Gutzler 1981; Barnston and Livezey 1987) particularly during boreal winter over the NH. Such variations are commonly referred to as 'teleconnections' in the meteorological literature, since they result in simultaneous variations in weather and climate over widely separated points on earth. One of the most prominent patterns is the NAO. Meteorologists have noted its pronounced influence on the climate of the Atlantic basin for more than two centuries (van Loon and Rogers 1978).

The NAO refers to a north–south oscillation in atmospheric mass between the Icelandic low- and the Azores high-pressure centres (Walker and Bliss 1932). It is most clearly identified when time-averaged data (monthly or seasonal) are examined, since time averaging reduces the 'noise' of small-scale and transient meteorological phenomena not related to large-scale climate variability. The spatial signature and temporal variability of the NAO are usually defined through the regional SLP field, for which some of the longest instrumental records exist. It is also readily apparent in meteorological data to the lower stratosphere, however, where the 'seesaw' in mass between the polar cap and the middle latitudes is much more zonally symmetric. This more annular mode of variability has been termed the Arctic Oscillation (AO) by Thompson and Wallace (1998). That the NAO and AO reflect essentially the same mode of surface variability is emphasized by the similarity of their time series,

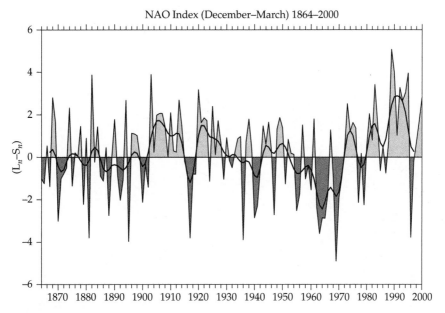

NAO Index (December–March) 1864–2000

Figure 2.2 Winter (December–March) index of the NAO based on the difference of normalized SLP between Lisbon, Portugal, and Stykkisholmur/Reykjavik, Iceland from 1864 through 2000. The indicated year corresponds to January (e.g. 1950 is December 1949–March 1950). The average winter SLP data at each station were normalized by division of each seasonal pressure by the long-term mean (1864–1983) standard deviation. The heavy solid line represents the index smoothed to remove fluctuations with periods less than 4 years.

with differences depending mostly on the details of the analysis procedure (Deser 2000; Wallace 2000).

A time series (or index) of more than 100 years of NAO variability is shown in Fig. 2.2[1], and a schematic of the spatial signature and climate impacts of the oscillation is shown in Fig. 2.3. Although the NAO is evident throughout the year, these plots illustrate conditions during boreal winter when the atmosphere is dynamically the most active. During the months of December through March, for instance, the NAO accounts for more than one-third of the total variance in SLP over the North Atlantic, substantially more than any other pattern of variability (Barnston and Livezey 1987; Rogers 1990).

Differences of more than 15 hPa in SLP occur across the North Atlantic between the two phases of the NAO in winter (Hurrell 1995a). In the so-called positive phase, higher than normal surface pressures south of 55°N combine with a broad

region of anomalously low pressure throughout the Arctic and subarctic (Fig. 2.3(a)). Consequently, this phase of the oscillation is associated with stronger-than-average westerly winds across the middle latitudes of the Atlantic onto Europe, with anomalous southerly flow over the eastern United States and anomalous northerly flow across western Greenland, the Canadian Arctic, and the Mediterranean. The easterly trade winds over the subtropical North Atlantic are also enhanced during the positive phase of the oscillation. During the negative phase, both the Icelandic low- and Azores high-pressure centres are weaker-than-normal, so both the middle latitude westerlies and the subtropical trade winds are also weak (Fig. 2.3(b)).

There is little evidence for the NAO to vary on any preferred timescale (Fig. 2.2). Large changes can occur from one winter to the next, and there is also a considerable amount of variability within a given winter season (Nakamura 1996). This is consistent with the notion that much of the atmospheric circulation variability in the form of the NAO arises from processes internal to the atmosphere, in which various scales of motion interact with one another to produce random (and thus

[1] More sophisticated and objective statistical techniques, such as eigenvector analysis, yield time series and spatial patterns of average winter SLP variability very similar to those shown in Figs 2.2 and 2.3.

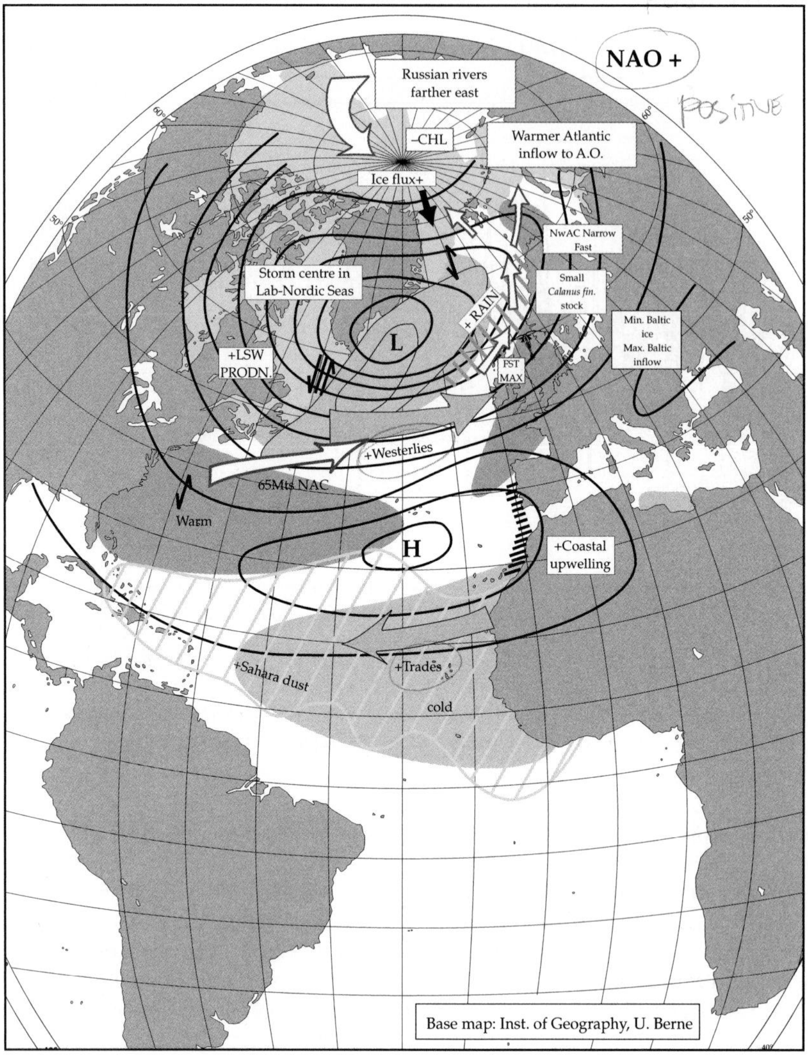

Figure 2.3 A schematic of the Atlantic–Arctic sector under (a) NAO-positive and (b) NAO-negative conditions. The following will explain the abbreviated text and symbols and provide references to the sources of information. (See Plate 1(a) and (b)).

Note:

1. Bold (weak) up-down arrows (↑ ↓ and ↕) indicate intense open (capped) ocean convection and strong (weak) heat flux. Boxes labelled '+18°W Prodn', '+LSW Prodn' and '+GSDW Prodn' indicate the location and NAO state associated with the maximal production of the three main Atlantic Mode Waters—18° Water, Labrador Sea Water and Greenland Sea Deep Water respectively. This relationship and the coordination of Mode Water formation by the NAO are described in Dickson *et al.* (1996); Dickson (1997); Talley (1996); Joyce *et al.* (2000).
2. 'Storm Centre off US Coast' and 'Storm Centre in Lab-Nordic Seas' boxes are from Rogers (1990); see also Bacon and Carter (1993); Kushnir *et al.* (1997); Alexandersson *et al.* (1998), and Carter (1999).
3. SST distributions are the cold = blue, warm = pink areas from Cayan (1992c), see also Becker and Pauly (1996), Cayan (1992a,b), Dickson and Turrell (1999).
4. 'LC + 1 SV' refers to the fact that the transport of the Labrador Current is thought to increase by 1 Sverdrup in NAO-negative extrema (Myers *et al.* 1989; see also Marsh *et al.* (1999) and Drinkwater *et al.* (1999).
5. '50 Mts⁻¹ NAC' and '65 Mts⁻¹ NAC' refer to the calculated increase in transport of the North Atlantic Current by 30% as NAO− changes to NAO+. The entire gyre circulation is strengthened, so that it would be equally appropriate to refer to a strengthening of the subtropical and subpolar gyre recirculations as the NAO amplifies. There is recent German evidence from their repeat hydrography during WOCE that the subpolar gyre expands eastwards as the NAO has amplified. See Curry and McCartney (2001), and also Bersch *et al.* (1999) and McCartney *et al.* (1997).
6. 'Ice flux −' and Ice flux +' in the Fram Strait refer to the observation that changes in the annual volume flux of ice from the Arctic via the western Fram Strait are associated with the NAO (Vinje and Finnekasa 1986; Hilmer *et al.* 1998; Vinje *et al.* 1998; Kwok and Rothrock 1999; Dickson *et al.* 2000). Dickson *et al.* (2000) and Hilmer *et al.* (1998) suggest this association may not have applied in earlier decades.

(b)

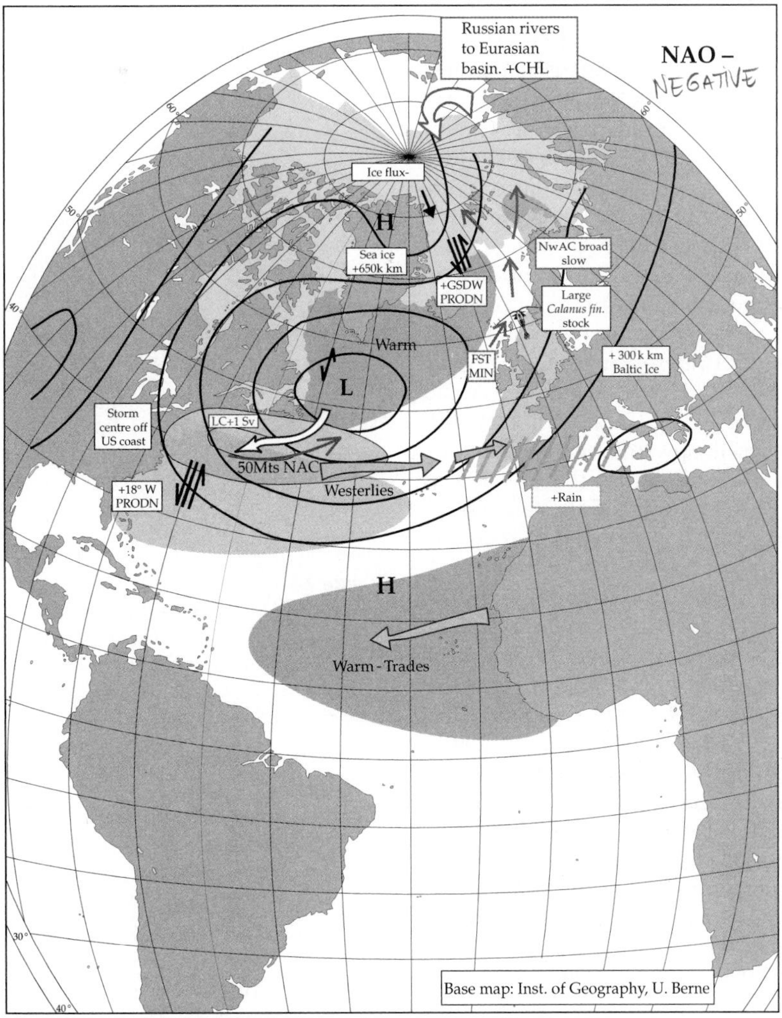

Russian rivers to Eurasian basin. +CHL
NAO –
NEGATIVE
Ice flux-
H
Sea ice +650k km
+GSDW PRODN
NwAC broad slow
Large Calanus fin. stock
Warm
FST MIN
+ 300k km Baltic Ice
L
Storm centre off US coast
LC+1 Sv
50Mts NAC
Westerlies
+Rain
+18° W PRODN
H
Warm - Trades
Base map: Inst. of Geography, U. Berne

weak centres that generate weak westerlies & weak easterlies

Figure 2.3 *(Continued)*

7. 'Sea ice + 650,000 km^2' in the 'NAO−' schematic represents the retraction of the median sea-ice border at the end of winter (April) between the most extreme 7-year runs of NAO-negative winters (1963–69) and NAO-positive winters (1989–95). The distributional change is illustrated in Dickson *et al.* (2000) while the correlation between ice concentration and the NAO index is most clearly indicated by Deser *et al.* (2000).

8. '+Rain' and its changing distribution (Dickson *et al.* 2000; see also Cayan and Reverdin 1994; Hurrell 1995a; Hurrell and van Loon 1997).

9. 'Large Calanus fin. Stock' and 'Small Calanus fin. Stock' (see Fromentin and Planque 1996).

10. 'NwAC Broad slow' and 'NwAC Narrow Fast' are from Blindheim *et al.* (2000) who use the 35 isohaline to define the width of the Norwegian Atlantic Current west of Norway.

11. 'FST MAX' and 'FST MIN' refer to the Faroe—Shetland Transport as calculated in an inverse treatment of two historic Scottish hydrographic sections (Dye 1999).

12. Baltic ice and inflow, from B. Mackenzie, personal communication.

13. '+ Saharan Dust' and the dust distribution under NAO+ conditions (Moulin *et al.* 1997).

14. Changes in middle latitude westerlies and subtropical trade winds, primarily from Cayan (1992c).

15. We have chosen to use SLP-anomaly distributions rather than sea level pressure *per se* since they better explain the sense of the anomalous airflow (based on Rogers 1990; see also Walsh *et al.* 1996; Serreze *et al.* 1997; Dickson *et al.* 2000).

16. 'Warmer Atlantic Inflow to the AO' refers to the increase in the temperature, extent and perhaps volume transport of Atlantic water inflow to the Arctic Ocean (Quadfasel *et al.* 1991; Grotefendt *et al.* 1998; Morison *et al.* 1998a; Dickson *et al.* 2000; see also Carmack *et al.* 1995; Carmack *et al.* 1997; Swift *et al.* 1997; Morison *et al.* 2001).

17. 'Russian Rivers Further East', '− CHL' on NAO + and 'Russian Rivers to Eurasian Basin; + CHL' from Steele and Boyd (1998) and Wieslaw Maslowski (personal communication).

unpredictable) variations (Wallace and Lau 1985; Lau and Nath 1991; Ting and Lau 1993; Hurrell 1995b). There are, however, periods when anomalous NAO-like circulation patterns persist over many consecutive winters. In the Icelandic region, for instance, SLP tended to be anomalously low during winter from the turn of the century until about 1930 (positive NAO index), while the 1960s were characterized by unusually high surface pressure and severe winters from Greenland across northern Europe (negative NAO index). A sharp reversal has occurred over the past 30 years, with strongly positive NAO index values since 1980 and SLP anomalies across the North Atlantic and Arctic that resemble those in Fig. 2.3(a). In fact, the magnitude of the recent upward trend is unprecedented in the observational record (Hurrell 1995a; Thompson *et al.* 2000) and, based on reconstructions using paleoclimate and model data, perhaps over the past several centuries as well (Osborn *et al.* 1999; Stockton and Glueck 1999). Whether such low frequency (interdecadal) NAO variability arises from interactions of the North Atlantic atmosphere with other, more slowly varying components of the climate system such as the ocean (Rodwell *et al.* 1999; Mehta *et al.* 2000; Hoerling *et al.* 2001), whether the recent upward trend reflects a human influence on climate (Corti *et al.* 1999; Fyfe *et al.* 1999; Osborn *et al.* 1999; Shindell *et al.* 1999; Ulbrich and Christof 1999; Gillett *et al.* 2000, 2001; Monahan *et al.* 2000), or whether the longer timescale variations in the relatively short instrumental record simply reflect finite sampling of a purely random process (Wunsch 1999), are topics of considerable current interest.

2.2.3 The NAO and surface temperature variability

The NAO exerts a dominant influence on wintertime temperatures across much of the NH. Surface air temperature and sea-surface temperature (SST) across wide regions of the North Atlantic Ocean, North America, the Arctic, Eurasia, and the Mediterranean are significantly correlated with NAO variability[2] (see also Section 2.3.2). Such changes in surface temperature (and related

---

[2] SSTs are used to monitor surface air temperature over the oceans because intermittent sampling is a major problem and SSTs have much greater persistence.

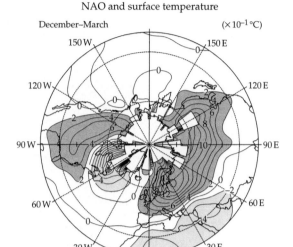

NAO and surface temperature

December–March   ($\times 10^{-1}$ °C)

Figure 2.4 Changes in land surface and SST ($\times 10^{-1}$ °C) corresponding to a unit deviation of the NAO index for the winter months (December–March) from 1935–99. The contour increment is 0.2 °C. Temperature changes >0.2 °C are indicated by dark shading, and those <−0.2 °C are indicated by light shading. Regions of insufficient data are not contoured.

changes in rainfall and storminess) can have significant impacts on a wide range of human activities as well as on marine and terrestrial ecosystems. When the NAO index is positive, enhanced westerly flow across the North Atlantic during winter moves relatively warm (and moist) maritime air over much of Europe and far downstream across Asia, while stronger northerlies over Greenland and northeastern Canada carry cold air southward and decrease land temperatures and SST over the northwest Atlantic (Figs 2.4 and 2.6(b)). Temperature variations over North Africa and the Middle East (cooling), as well as North America (warming), associated with the stronger clockwise flow around the subtropical Atlantic high-pressure centre are also notable.

The pattern of temperature change associated with the NAO is important. Because the heat storage capacity of the ocean is much greater than that of land, changes in continental surface temperatures are much larger than those over the oceans, so they tend to dominate average NH (and global) temperature variability. Given the especially large and coherent NAO signal across the Eurasian continent from the Atlantic to the Pacific (Fig. 2.4), it is not surprising that NAO variability explains about one-third of the NH interannual surface temperature variance during winter, and that the trend towards the positive NAO phase in recent decades has contributed significantly to observed global warming (Hurrell 1996; Thompson *et al.* 2000).

2.2.4 The NAO, storms and precipitation variability

Changes in the mean circulation patterns over the North Atlantic are accompanied by changes in the intensity and number of storms, their paths, and their associated weather. During winter, a well-defined storm track connects the North Pacific and North Atlantic basins, with maximum storm activity over the oceans. The details of changes in storminess differ depending on the analysis method and whether one focuses on surface or upper-air features. Generally, however, positive NAO index winters are associated with a northeastward shift in the Atlantic storm activity (Fig. 2.3(a)), with enhanced activity from southern Greenland across Iceland into northern Europe and a modest decrease in activity to the south (Rogers 1990, 1997; Hurrell and van Loon 1997; Serreze *et al.* 1997). The latter is most noticeable off the east coast of the United States and from the Azores across the Iberian Peninsula and the Mediterranean. Positive NAO winters are also typified by more intense and frequent storms in the vicinity of Iceland and the Norwegian Sea (Serreze *et al.* 1997; Deser *et al.* 2000).

Changes in the mean flow and storminess associated with swings in the NAO are also reflected in pronounced changes in the transport and convergence of atmospheric moisture and, thus, the distribution of precipitation. Anomalously low precipitation rates occur over much of Greenland and the Canadian Arctic during high NAO index winters (Fig. 2.3), as well as over much of central and southern Europe, the Mediterranean, and parts of the Middle East. In contrast, more precipitation than normal falls from Iceland through Scandinavia (Hurrell 1995a; Dai *et al.* 1997; Dickson *et al.* 2000).

This spatial pattern, together with the upward trend in the NAO index since the late 1960s (Fig. 2.2), is consistent with recent observed changes in precipitation over much of the Atlantic basin. One of the few regions of the world where glaciers have not exhibited a retreat over the past several decades is in

Scandinavia (Hagen 1995; Sigurdsson and Jonsson 1995) where more than average amounts of precipitation have been typical of many winters since the early 1980s. In contrast, over the Alps, snow depth and duration in recent winters have been among the lowest recorded this century, and the retreat of Alpine glaciers has been widespread (Frank 1997). Severe drought has persisted throughout parts of Spain and Portugal as well. Finally, as far eastward as Turkey, river runoff is significantly correlated with NAO variability (Cullen and deMenocal 2000).

2.2.5 Atypical NAO winters and other atmospheric variations

While the NAO explains a substantial portion of the variability over the North Atlantic, the chaotic nature of the atmospheric circulation means that, at most times, there are significant departures from the schematic in Fig. 2.3. Even during periods of strongly positive or negative NAO index winters, the atmospheric circulation typically exhibits significant local departures from the idealized NAO pattern. For instance, Hilmer and Jung (2000) have shown that the centres of maximum interannual variability in SLP associated with the NAO have been located further to the east since the late 1970s, when the NAO winter index has mostly been positive. Such longitudinal displacements affect the NAO-related interannual variability of sea ice, surface temperature, surface heat fluxes, precipitation, and storms (see also Section 2.3.6). Another example is the winter of 1996, which was characterized by a very negative NAO index (Fig. 2.3). Conditions during the 1996 winter over much of Europe were severe; however, the anomalous anticyclonic circulation was located well to the east of the canonical NAO pattern (Fig. 2.3(b)), with positive SLP anomalies of more than 9 hPa centred over Scandinavia. Such persistent high-pressure anomalies are a typical feature of the North Atlantic climate and are referred to as 'blocks'. The longitudes of 150°W and 15°W are particularly favoured for the development of blocking highs, which occur when the westerly flow across the middle latitudes of the NH is weak and typified by an exaggerated wave pattern. North Atlantic blocking is most typical during boreal spring and early summer, but it occurs throughout the year. Over recent decades, for instance, increased anticyclonicity and easterly flows from the warm, summer continent have brought anomalously warm and dry conditions over much of northern Europe and the United Kingdom (Wright *et al.* 1999).

2.3 The North Atlantic Ocean

2.3.1 Mean state and variability

By the end of the nineteenth century, a surprisingly modern distribution of hydrographic stations had already traced out the basic physical geography of the North Atlantic Ocean and Nordic Seas. The properties of the upper ocean reflect a two-gyre structure with warm waters circulating anticyclonically in a subtropical gyre to the south and east and cold waters circulating cyclonically in a subpolar gyre to the north and west (Fig. 2.5). The main North Atlantic

Current flows east along the boundary between the two gyres, where there is an abrupt change of water properties (the Oceanic Arctic Front).

Determining the *rate* of the Atlantic circulation has proved more elusive. In the absence of direct observations, evidence is largely restricted to episodes when some anomalous ocean-climate 'signal' has propagated through the system. The dates in Fig. 2.5 provide one well-known example, tracing the slow ($\approx$3 cm s^{-1}) spread of the great salinity anomaly (GSA) through the Atlantic gyre circulation in the late 1960s and 1970s (Dickson *et al.* 1988).

It is difficult, therefore, to provide separate descriptions of the mean state of the ocean and its variability since both have coloured the observational record. Here, instead, the differential response of the surface-, intermediate- and deep-layers of the North Atlantic Ocean to the recent

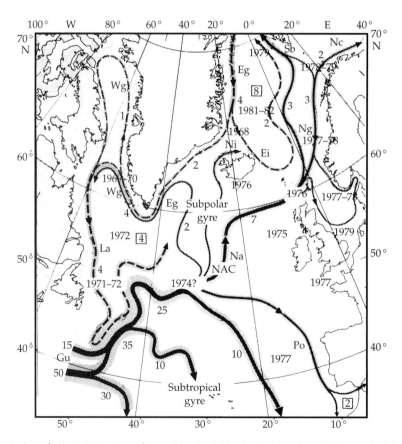

Figure 2.5 Transport scheme for the 0–1000 m layer of the northern North Atlantic with dates of arrival of the great salinity anomaly (GSA) superimposed (Dickson *et al.* 1988). Jakobsson (1992) concluded the GSA 'has probably generated more variability in fisheries during the last quarter of a century than any other hydrographic event in recent years'.

NAO trend is described (Fig. 2.2), as are the responses to two episodes of atypical NAO behaviour. Most attention is focused on the northwest Atlantic and Labrador Sea, which have special climatic sensitivity and importance.

2.3.2 Direct response to recent NAO forcing at the ocean surface

It has long been recognized that fluctuations in SST and the strength of the NAO are related (Bjerknes 1964). The pattern of SST variability associated with the winter NAO consists of a tri-polar structure marked, in the positive NAO phase, by a cold anomaly in the subpolar North Atlantic, warm anomalies extending from Cape Hatteras in the west Atlantic to Biscay in the east and poleward along the Norwegian coast, and a cold subtropical anomaly between the equator and 30°N (Figs 2.3(a) and 2.6(b)). This structure suggests that the SST anomalies are primarily driven by changes in the surface wind and air–sea heat exchanges associated with NAO variations (Cayan 1992c). Indeed, the relationship is strongest when the NAO index leads an index of the SST variability by several weeks (Deser and Timlin 1997).

The pattern of correlation between the winter NAO index and scalar wind speed reflects the redistribution of storm activity described earlier (Section 2.2.4). The positive correlation between the NAO index and wind strength over much of the Atlantic midlatitudes and Nordic Seas and extending into the North Sea reflects both the northeastward extension of the storm track under NAO-positive conditions and the development of storms to maturity along that track (Figs 2.3(a) and 2.6(a)). Extreme cold and dry air streams out from the Canadian Arctic across the Labrador Sea during positive NAO index winters, increasing the monthly sea-to-air flux of sensible and latent heat by approximately 200 W ms^{-1} compared to negative NAO index winters (Cayan 1992c). The result is the development of extremely cold SSTs in Atlantic midlatitudes from the Davis Strait and West Greenland Banks across the Labrador and Irminger Seas to Iceland and the Faroe Islands (Fig. 2.6(b)). By contrast, the warm, moist southerly airflow that is directed along the eastern boundary of the North Atlantic produces very warm SSTs in the North Sea and along the Norwegian coast to the Barents Sea, in spite of the strong wind speeds there (see also

figures 7a and 8a of Cayan 1992c). West of Africa, anomalously strong northeasterly trade winds during positive NAO winters increase the surface fluxes of heat to the atmosphere, thereby cooling the ocean surface (Figs 2.3(a) and 2.6(b)).

On longer timescales, the recent upward trend towards positive NAO index winters and the corresponding northeastward extension and increase of winter storm activity (e.g. Alexandersson 1998) have been associated with an increase in wave heights over the northeast Atlantic and a decrease south of 40°N (Bacon and Carter 1993; Kushnir et al. 1997; Carter 1999). These changes are also illustrated by the remarkable increase in mean scalar wind speeds at Utsire in the north-central North Sea during the whole of the year (Fig. 2.7). Such changes have a range of fundamental consequences; for the vernal blooming of phytoplankton (Sverdrup 1953), for the operation and safety of shipping, for offshore industries such as oil and gas exploration, and for coastal development.

Variations in the NAO are also coupled to changes in Arctic sea ice, where the strongest interannual variability occurs in the North Atlantic sector (Deser et al. 2000). Sea ice fluctuations display a seesaw in ice extent between the Labrador and Greenland Seas (Fig. 2.8). Strong interannual variability is evident in the sea ice changes over the North Atlantic, as are longer-term fluctuations including a trend over the past 30 years of diminishing (increasing) ice concentration during boreal winter east (west) of Greenland (Chapman and Walsh 1993; Maslanik et al. 1996; Cavalieri et al. 1997; McPhee et al. 1998; Parkinson et al. 1998). Associated with the sea ice fluctuations are large-scale changes in SLP that closely resemble the NAO (Fig. 2.8).

When the NAO is in its positive phase, the Labrador Sea ice boundary extends farther south while the Greenland Sea ice boundary is north of its climatological mean extent. Given the implied surface wind changes (Fig. 2.3(a)), this is qualitatively consistent with the notion that sea ice anomalies are directly forced by the atmosphere, either dynamically via wind-driven ice drift anomalies, or thermodynamically through surface air temperature anomalies (Prinsenberg et al. 1997). The relationship between the NAO index (Fig. 2.2) and an index of the North Atlantic ice variations is indeed strong, although it does not hold for all individual winters (Deser et al. 2000). This last point illustrates the importance of the regional atmospheric circulation in forcing the extent of sea ice.

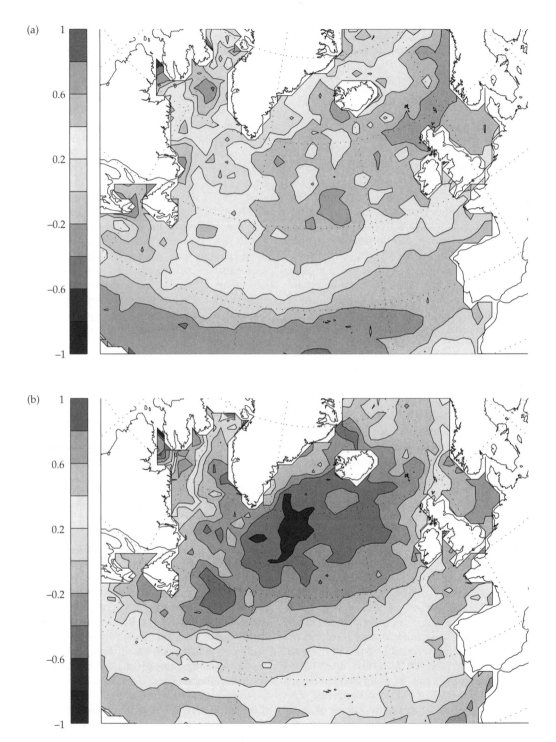

Figure 2.6 Spatial distribution of the Pearson correlation coefficient between the NAO index and (a) local scalar windspeed and (b) local SST for winter (January–March) over 1950–95 (courtesy of Ben Planque). (See Plate 2.)

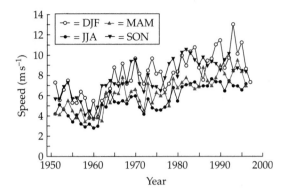

Figure 2.7 The observed increase in seasonal mean scalar windspeeds at Utsire (59 19′N 4 53′E) in the northern North Sea during the long-term amplification of the NAO between the mid-1960s and the mid-1990s.

Figure 2.8 Winter SLP (contours) and ice concentration (shading) anomalies associated with the leading pattern of ice variability. The SLP anomaly field is in units of hectopascals (40 year)$^{-1}$, contoured every two units. Light (dark) shading denotes ice concentration anomalies greater than 8% (40 year)$^{-1}$ [less than -8% (40 year)$^{-1}$]. From Deser *et al.* (2000).

2.3.3 Direct and indirect response to NAO forcing at intermediate ocean depths

Subsurface ocean observations more clearly depict long-term climate variability, because the effect of the annual cycle and month-to-month variability in the atmospheric circulation decays rapidly with depth. Although these measurements are much more limited than surface observations, they reveal a decadal 'system' to the temperature evolution of the upper North Atlantic Ocean, with large organized SST anomalies circulating around the gyres coincident with decadal changes in the NAO (Hansen and Bezdek 1996; McCartney *et al.* 1997; Curry *et al.* 1998). These recurrent surface features track the movement of deep-seated anomalies formed by winter convection and the formation of vertically extensive and homogeneous intermediate-depth water masses known as 'mode waters'.

Mode waters form where the water mass structure, ocean circulation, and winter surface heat exchange favour deep-reaching convection. Since the same shifting patterns of winter storm activity and surface heat exchange that affect the surface also affect the formation of mode waters, the depth of vertical exchange also varies greatly in both time and space: to depths of 1000–3500 m in the Greenland Sea, 1000–2300 m in the Labrador Sea, and a few hundred meters in the Sargasso Sea, where 18° Water forms. The regional importance of these localized sites is that the winter renewal of mode waters provides a mechanism for carrying the signal of climate change to intermediate and greater depths, and throughout the ocean basin by horizontal spreading. It is of more than local importance also that the scale of NAO forcing appears to have imposed a degree of coordination on convective activity at the three aforementioned sites (Dickson *et al.* 1996). Deep convection over the Labrador Sea, for instance, was at its weakest and shallowest in the postwar instrumental record during the late 1960s. Since then, Labrador Seawater has become progressively colder and fresher, with intense convective activity to unprecedented depths (>2300 m) in the early to mid-1990s (Fig. 2.9). In contrast, warmer and saltier deep waters in recent years are the result of suppressed convection in the Greenland and Sargasso Seas, whereas ventilation was a maximum at these sites during the 1960s.

The reasons for the synchronization with the NAO are beginning to emerge. During the negative NAO index years of the 1960s, an enhanced land–sea temperature gradient in winter contributed to the formation of more storms than normal off the eastern seaboard of the United States (Dickson and Namias 1976; Hayden 1981). Offshore, the cold, stormy conditions caused maximum formation and ventilation of the 18° Water

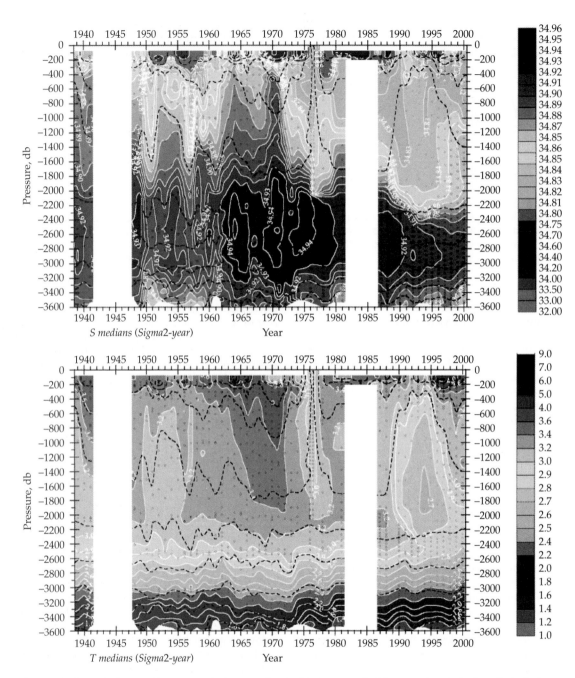

Figure 2.9 Changes in the potential temperature (θ; lower panel) and salinity (upper panel) of the watercolumn in the Central Labrador Sea over the complete period of the hydrographic record since 1938. The data set was selected to lie within the 3300-m isobath of the Labrador Sea, and the plots represent the median values of vertical property profiles, binned according to σ_2 density intervals. Kindly provided by Igor Yashayaev. (See Plate 3.)

pycnostad (Jenkins 1982; Talley and Raymer 1982; Talley 1996). With winter storm activity concentrated at the US eastern seaboard, storminess decreased to a postwar minimum further north so that Labrador Sea convection became increasingly suppressed and freshwater built up at the surface (Lazier 1980, 1988, 1995).

The tendency towards opposite conditions prevailed over the subsequent 20 years (Fig. 2.9; see also Dickson *et al.* 1996; Dickson *et al.* 2000), so that by the early 1990s, Labrador Sea Water (LSW) was fresher, colder, denser, and deeper than at any other time in the history of deep measurements in the region. From 1966 to 1994, spanning most of the period of the trend towards more positive NAO index winters, the overall freshening of the water column of the Labrador Sea was equivalent to mixing down an extra 7 m of freshwater from the sea surface, and its cooling was equivalent to an increased loss rate of 8 W m^{-2} (Lazier 1995). These are among the largest changes observed in oceanography.

Such altered water mass properties have value in tracing-out the rates and pathways by which LSW spreads across the basin. It is this cold, fresh, and dense new vintage of LSW that Sy *et al.* (1997) use, together with its chlorofluorocarbon signature, to derive modern estimates of LSW spreading rates that are an order of magnitude greater than published values within the Labrador-Irminger Basin, and perhaps three to four times greater than previous trans-ocean estimates (e.g. Read and Gould 1992). The influence of these changes on the Atlantic gyre circulation itself, however, is likely to be of even greater importance.

The main North Atlantic Current is driven by the gradient of potential energy anomaly (PE') across the mutual boundary between the subtropical and subpolar gyres (Curry and McCartney 2001). Since PE' reflects the vertical density structure and heat content of the upper ocean to well below the wind-driven layer, coordinated changes of opposite sign in the production and characteristics of the mode waters in each gyre will have the potential to drive deep-seated changes in the PE' gradient and, hence, in the strength of the Atlantic gyre circulation. If these changes in the density and heat content of mode waters are attributable to NAO forcing, moreover, the amplification of the NAO to extreme values over recent decades is likely to have been accompanied by a corresponding multi-decadal spin-up of the Atlantic gyre circulation. Specifically,

Curry and McCartney (2001) calculate a 30% increase to the mid-1990s in the 0–2000 dbar east-going baroclinic mass transport along the gyre boundary, with both gyres contributing equally. Moreover, if the link to NAO forcing proves valid, probably at no other time over the twentieth century did the North Atlantic gyre circulation exceed its strength during the early 1990s.

2.3.4 Indirect response to NAO forcing at abyssal depths

Below the LSW layer in the Labrador Sea at depths of 2300–3500 m, repeat hydrography has indicated a steady freshening over the past three to four decades (Fig. 2.9(a)). At these depths, beyond the reach of deep convection, such a change cannot be due to local forcing. There is growing evidence that, instead, it reflects a large-scale freshening of the upper Nordic Seas passed on via the dense northern overflows that cross the Greenland–Scotland Ridge through the Denmark Strait and the Faroe Bank Channel. The freshening of the Arctic and subarctic seas is, however, itself associated with the multi-decadal NAO variability that forms the focus of this chapter. As such, the changes at abyssal depths may be viewed as an indirect response to atmospheric forcing. The factors outlined below provide an impressive illustration of the varied mechanisms and locations by which a change in NAO forcing may translate into a large-scale ocean response.

1. The direct export of sea-ice from the Arctic Ocean is one such cause of recent subarctic freshening. A combination of current measurements, upward-looking sonar and satellite imagery reveals an increased annual efflux of ice through the western Fram Strait to a record volume-flux of 4687 km^3 year^{-1} in 1994–95 (Vinje *et al.* 1998; Kwok and Rothrock 1999). Although the relationship is not robust in the longer-term, each 1σ increase in the NAO index since 1976 has been associated with an approximately 200 km^3 increase in the annual efflux of ice (Dickson *et al.* 2000).
2. Throughout the marginal ice-zone of the Nordic Seas, a steady decrease in the local winter production of sea-ice at the end of winter has been associated with the trend in the NAO over the past 40 years (Deser *et al.* 2000).
3. Precipitation along the Norwegian Atlantic Current increases during NAO-positive conditions

by approximately 15 cm per winter compared with NAO-negative conditions, associated with the aforementioned extension of storm activity to the Nordic Seas (Dickson *et al.* 2000).

Other factors and mechanisms have undoubtedly contributed to the long and gradual but dramatic freshening of the European subarctic seas in recent decades, most of them associated in some way with the amplifying NAO. Blindheim *et al.* (2000) describe a range of factors internal to the Nordic Seas. These include an increased freshwater supply from the East Icelandic Current, a narrowing of the salty Norwegian Atlantic Current towards the Norwegian Coast, a dual change towards reduced deep water formation and increased Arctic Intermediate Water formation in the Greenland Sea (Dickson *et al.* 1996; Verduin and Quadfasel 1999) that has had the effect of deepening the interface between the Arctic Intermediate Water and Deep Water in the Norwegian Sea. Other more-remote influences may even include an observed freshening of the Pacific inflow to the Arctic Ocean through the Bering Strait (Carmack 2000, personal communication), and a contribution from an apparent recent thinning of sea-ice in the Arctic Ocean (Rothrock *et al.* 1999). In turn, the warming and perhaps strengthening of Atlantic water inflow to the Arctic Ocean due to the amplifying NAO (Quadfasel *et al.* 1991; Grotefendt *et al.* 1998; Dickson *et al.* 2000), may itself have contributed to ice-thinning in the Eurasian Basin of the Arctic Ocean (Steele and Boyd 1998); by the late 1980s and early 1990s, both inflow streams were between 1 and 2 °C warmer than normal with a consequent warming, spreading, and shoaling of the Atlantic-derived sub-layer across the Eurasian Basin of the Arctic Ocean (Carmack *et al.* 1995; Aagaard *et al.* 1996; Carmack *et al.* 1997; Swift *et al.* 1997; Morison *et al.* 1998a,b; Morison *et al.* 2001).

While it may not be possible to partition the recent freshening of the Nordic Seas into its individual contributory components, it is clear from three of the longest observational records available that the climate change has occurred over a sufficiently deep layer (approximately 1–1.5 km) to affect the hydrographic character of the two dense overflows, which cross the Greenland–Scotland Ridge and ventilate the abyssal North Atlantic, thus helping to drive the MOC. Hydrographic sections monitoring the outflow of North East Atlantic Deep Water (NEADW) through the Faroe–Shetland Channel

confirm that NEADW salinities have decreased linearly by ≈0.01 per decade since the mid-1970s (Turrell *et al.* 1999; Dickson *et al.* 2001), precisely the rate, period, and steadiness of freshening recorded downstream in the NEADW layer of the abyssal Labrador Sea. A similar rate and period of freshening is also found in the deepest layers of the Labrador Sea, which is occupied by water that overflows the Greenland–Scotland Ridge via the Denmark Strait (DSOW). Both dense overflows, therefore, appear to have tapped and delivered to the headwaters of the THC the freshening signal of the upper Nordic Seas (Fig. 2.10). In addition, near-surface variability in temperature generated by the NAO in the eastern Fram Strait appears to re-circulate to the south, affecting the temperature of the deepening dense overflow of DSOW off of southeast Greenland and from there determining the density of abyssal depths of the Labrador Sea (Dickson *et al.* 1999, 2001).

The subarctic seas are thus observed to act in two capacities: first, as a source of change for the climatically-sensitive Arctic Ocean, and second as a mechanism for transferring such changes to the MOC. Directly, or indirectly, the signature of the trend in NAO toward its positive index state in recent decades can be identified in Atlantic Ocean variability over the full range of ocean depths.

2.3.5 Space- and time dependence in NAO forcing and ocean response

The climate record is noisy, reflecting a continuum of variability on all space and timescales. As Wunsch (1992) points out 'The ocean is a turbulent fluid in intimate contact with another turbulent fluid, the atmosphere. Although I am unaware of any formal theorems on the subject, experience with turbulent systems suggests that it is very unlikely that any components of such a complex nonlinear system can actually remain fully steady; . . . the frequency/wavenumber spectrum of the ocean circulation is almost surely everywhere filled.' As described in Section 2.2.5, the chaotic nature of the atmospheric circulation means that significant departures from the anomalies depicted schematically in Fig. 2.3 occur more often than not. Since the NAO is itself variable in both space and time, so too is the ocean (and ecosystem) response.

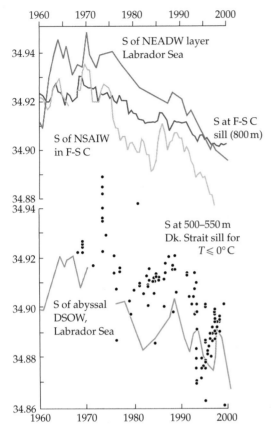

Figure 2.10 Evidence that the recent freshening of the upper Nordic Seas is passing across the Greenland–Scotland Ridge via both intermediate-depth overflows to affect the deep and abyssal layers of the North Atlantic. The upper group of curves indicate a freshening by ≈0.01 per decade at sill depth in the Faroe–Shetland Channel and in the Norwegian Sea Arctic Intermediate Water layer (NSAIW) overflowing the Faroe–Shetland Channel, together with a corresponding decrease downstream in the NEADW of the Labrador Sea. The lower curves show that a very similar rate of freshening has been observed at sill depth in the Denmark Strait (dots) and in the DSOW layer that occupies abyssal depths in the Labrador Sea. From Dickson et al. (2001).

Changes in the NAO amplitude over time
While the recent amplification of the NAO has brought a progressive warming to parts of the subpolar gyre and Barents Sea (e.g. Dickson et al. 2000, their figure 10), the NAO does not appear to have been the dominant cause of warming in this region during the middle decades of the twentieth century. Between the 1920s and 1960s, the correlation between the NAO index and the temperature of the Barents Sea dropped to a long-term minimum

(Dickson et al. 2000, their figure 14) and instead, abnormally warm and saline conditions appear to have passed through subpolar gyre from other causes, bringing an amelioration of the marine climate to northern waters from the West Greenland Banks to Iceland, Faroes, and the Barents Sea (Dickson and Brander 1993). It was not until the cold, fresh conditions of the GSA passed through the northern gyre in the late 1960s and 1970s (Dickson et al. 1988; Reverdin et al. 1997), that this warming came to an abrupt end.

The mechanisms and processes responsible for the so-called 'Warming in the North' are still under investigation (e.g. Delworth and Knutson 2000). Its signature, as well as its impact on the marine ecology, however, is unmistakable:

- The salinity of North Atlantic Water passing through the Faroe–Shetland Channel reached a century-long high (Dooley et al. 1984).
- Salinities were so high off Cape Farewell that they were thrown out as erroneous (Harvey 1962).
- A precipitous warming of more than 2 °C in the 5-year mean pervaded the West Greenland Banks (Dickson and Brander 1993).
- The northward dislocations of bio-geographical boundaries for a wide range of species from plankton to commercial fish, terrestrial mammals and birds were at their most extreme in the twentieth century.

The astonishing nature of these radical events is vivid in the contemporary scientific literature, as summarized in a comprehensive bibliography by Lee (1949) and reviewed in an ICES Special Scientific Meeting on 'Climatic Changes in the Arctic in Relation to Plants and Animals' in 1948. The rise and spread of the West Greenland cod fishery was one spectacular result, apparently reflecting a change in the effectiveness of egg and larval drift in the Irminger/West Greenland Current system (Buch and Hansen 1988; Dickson and Brander 1993; Schopka 1993, 1994). While similar 'cod periods' at West Greenland also occurred in the 1820s and 1840s (Hansen 1949), and perhaps much earlier (Fabricius 1780), the change in the middle decades of the twentieth century represented a return of cod to the region after a 50–70 year absence (Buch and Hansen 1988; Dickson et al. 1994), only to end again with the arrival of the GSA.

Radical changes in the Atlantic ecosystem were not confined to W. Greenland and the Davis Strait in these middle decades of the Century. The

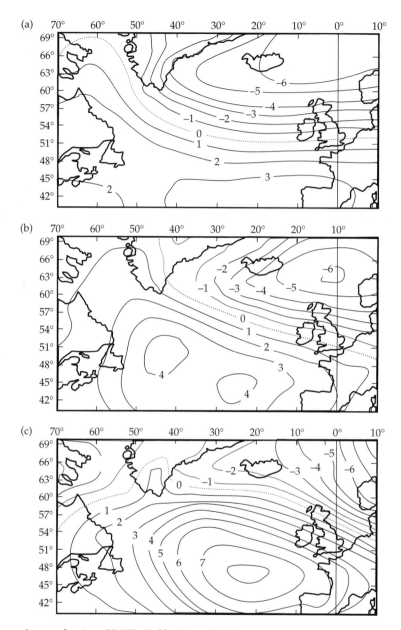

Figure 2.11 SLP anomaly pattern for winters (a) 1993–95, (b) 1999, and (c) 2000.

'Russell Cycle' (see, e.g. Chapter 10) brought a more-southerly community to the ecosystem of the Western Channel at about the same time and perhaps for related reasons (e.g. Cushing 1982). In the Nordic Seas, the spawning stock biomass of the Norwegian spring-spawning herring rose to a maximum in mid-Century as the wave of warmth passed through the northern Gyre (Toresen and Ostvedt 2000), and the range of its feeding-spawning-overwintering movements expanded to the west and north following the retreat of the Ocean Polar Front (since reversed; see Vilhjalmsson 1997). In the Barents Sea the wave of warmth brought a sustained increase in the yield, weight, liver-weight, roe weight, and recruitment of *skrei* from their pre-existing minima associated with the extreme cold in the early years of the Century (Helland-Hansen and Nansen 1909; Anon 1996).

2.3.6 Temporal shifts in the NAO pattern

As described earlier, NAO-positive winters are usually associated with chill, dry northwesterly winds across the Labrador Sea, and hence intense and deep-reaching convective renewal of LSW and a widespread distribution of chilled surface temperatures across the northwest Atlantic (Fig. 2.6(b)). For instance, correlations between the winter NAO index and both temperature and salinity on top of Fylla Bank, West Greenland, are strongly negative (table 10.1 of Buch 1995). However, this relationship has failed to hold following the temporary deep minimum of the NAO index in the winter of 1996 (Fig. 2.2). The NAO-positive conditions of the most recent winters have been accompanied by a shutdown of Labrador Sea convection and the warmest conditions on top of Fylla Bank evident in the 50-year record (Buch 2000). In just two or three winters, the long sustained cooling and freshening of LSW (Fig. 2.9), described in Section 2.3.4. as 'among the largest changes observed in oceanography', has been largely reversed. The suggested cause is shown in Fig. 2.11. Comparing the Atlantic SLP anomaly pattern for the extreme NAO-positive winters of 1993–95 with that for 1999–2000 reveals a slight east or northeast displacement in the more recent period. This subtle shift has little effect along the eastern boundary to the Barents Sea where widespread warming has continued. In the northwest Atlantic, however, the slight eastward displacement of the 'normal' NAO pattern has made an important difference to the marine climate of the West Greenland Banks and to the convective centre of the Labrador Sea. Instead of a chilling northwesterly flow of air, light anomalous southerly winds have prevailed across the Labrador Sea to West Greenland in 1999–2000.

The salient point is that the interannual 'noise' of the extratropical atmosphere is large, and even subtle shifts of recurrent climate patterns such as the NAO may have deep-reaching and long-lasting effects. Moreover, to the extent that anthropogenic climate change might influence modes of natural variability, perhaps making it more likely that one phase of the NAO is preferred over the other (e.g. Corti *et al.* 1999; Gillett *et al.* 2000, 2001), or that subtle shifts might occur in its spatial pattern (e.g. Ulbrich and Christoph 1999), it would be unwise to rely entirely on past experience in defining the 'typical' Atlantic Ocean (and ecosystem) response to NAO forcing. Schematics such as Fig. 2.3 should therefore be used with caution!

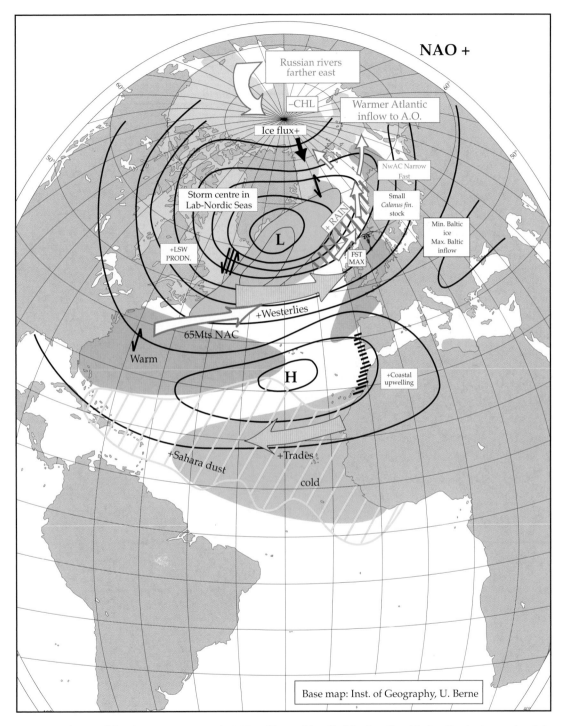

Plate 1a A schematic of the Atlantic–Arctic sector under NAO-positive conditions. The following will explain the abbreviated text and symbols and provide references to the sources of information.

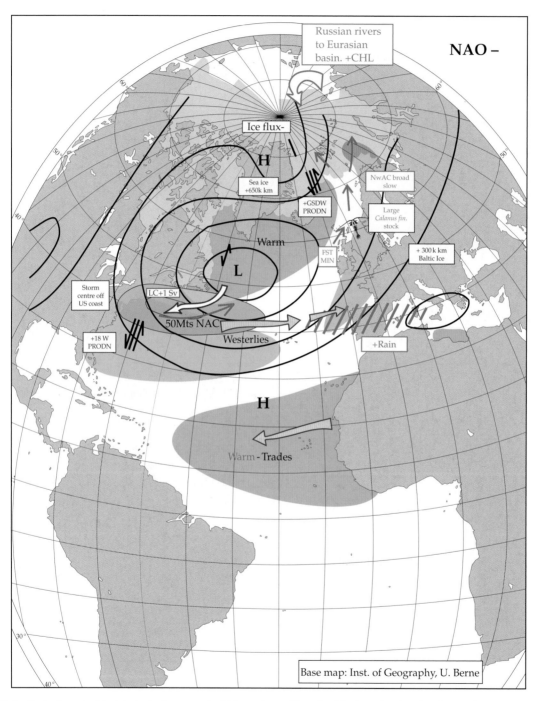

Plate 1b A schematic of the Atlantic–Arctic sector under NAO-negative conditions. The following will explain the abbreviated text and symbols and provide references to the sources of information.

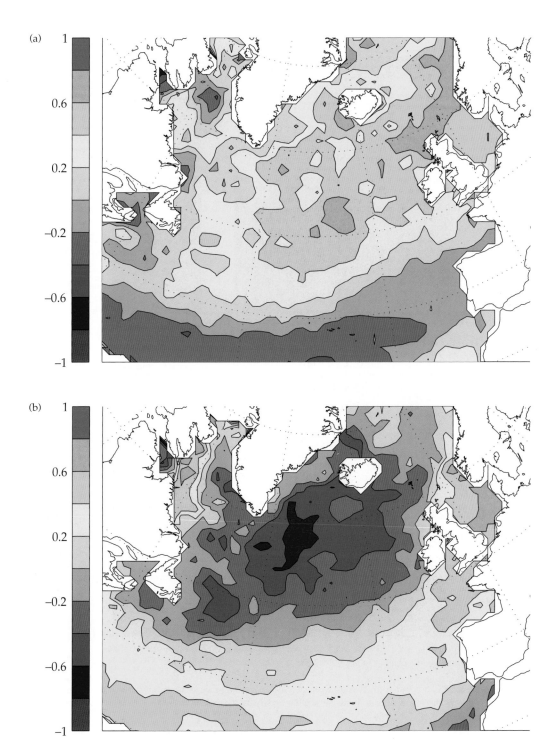

Plate 2 Spatial distribution of the Pearson correlation coefficient between the NAO index and (a) local scalar windspeed and (b) local SST for winter (January–March) over 1950–95 (courtesy of Ben Planque).

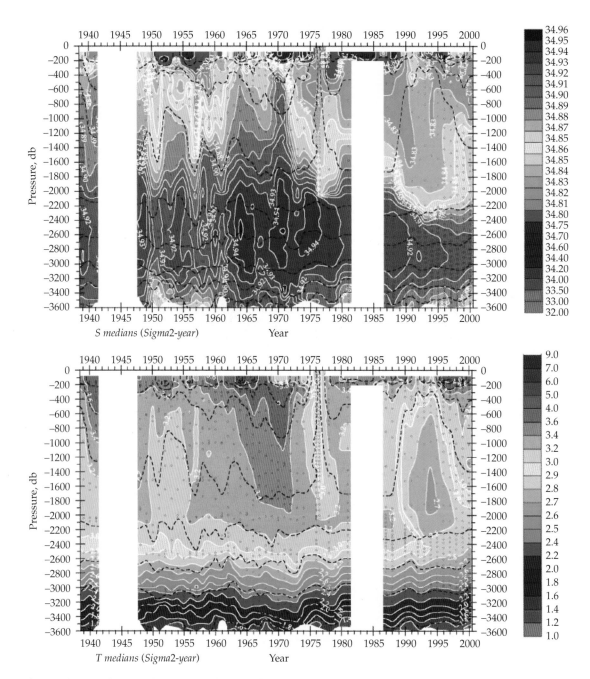

Plate 3 Changes in the potential temperature (θ; lower panel) and salinity (upper panel) of the watercolumn in the Central Labrador Sea over the complete period of the hydrographic record since 1938. The data set was selected to lie within the 3300-m isobath of the Labrador Sea, and the plots represent the median values of vertical property profiles, binned according to σ_2 density intervals. Kindly provided by Igor Yashayaev.

CHAPTER 3

Modelling marine ecosystems and their environmental forcing

Francisco E. Werner, Alfredo Aretxabaleta, and Karen Pehrson Edwards

3.1 Introduction

The development of our understanding of how marine communities and ecosystems are structured, and how they function, is of fundamental importance for several reasons. As we strive to understand the fate and production of biogenic materials, for example, as we attempt to construct quantitative budgets of CO_2 and determine its role in global change, we must determine the details of how the marine compartment—through its ecosystems—recycles, sequesters, or exports carbon. At the same time, and in light of recent collapses of some of our key fisheries, understanding and predicting the response of marine ecosystems to global climate change is essential as we attempt to develop sustainable management practices of our living marine resources.

Marine communities and ecosystems exhibit significant variability over a number of spatial and temporal scales. The observed variability is due to a complex interaction between natural biological processes, community interactions, fluctuations in environmental forcings, and anthropogenic stresses. While we have long identified the complexity of this problem (e.g. Hjort (1914) for the case of marine fisheries), we are just beginning to compile adequate data and to develop computational models that are able to couple physical and biological processes, that allow us to quantitatively probe and unravel some of the reasons for the observed fluctuations.

In the past decade, national efforts together with international programmes such as Joint Global Ocean Flux Study (JGOFS) and Global Ocean Ecosystem Dynamics (GLOBEC) have enabled important and exciting advances in our ability to observe and model marine ecosystems. Instruments and observational platforms can potentially provide us with continuous and uninterrupted time series, and in some instances, with significant synopticity. At the same time, numerical models have begun to realistically integrate environmental forcing functions and biological responses. In this chapter, we present a brief review and description of some of the issues and status of modelling efforts attempting to link and explain changes in climate and in pelagic marine communities with attention to the North Atlantic systems, but drawing from knowledge gained in studies in all oceans.

Figure 3.1 shows the ecosystem components we will consider. Lower trophic levels include the smaller size classes through the mesoplankton; higher trophic levels will refer to the mesoplankton and fish. Important missing components from this sketch are explicit linkages to benthic communities and to upper trophic levels such as marine mammals, sea birds, and humans. The chapter's structure

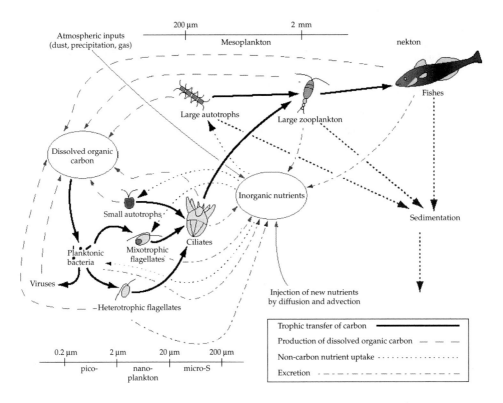

Figure 3.1 Components of the pelagic marine ecosystem considered in this chapter ranging from the microbial loop at the smallest scales to fish at the highest tropic levels. OEUVRE Per Jonsson, with revisions by committee.

is as follows. First, we briefly discuss aspects of the relationship between variability in the environment and variability in marine populations. We then present approaches that are being taken in understanding this variability and highlight areas where progress has been made. We close with recommendations and suggestions for new approaches necessary for further advances.

3.2 Relationship between environmental and marine population fluctuations: an issue of scales

Natural variations in marine ecosystems are a combination of community interactions (biotic processes) and variations in environment (abiotic processes). While presently these fluctuations may be additionally confounded by anthropogenic stresses, examples of fluctuations in fish populations predating increased human activities exist from historical accounts (Cushing 1982; Alheit and Hagen 1997), as well as from more recently

established quantitative measures like analysis of sediment records (e.g. Soutar and Isaacs 1974; O'Connell and Tunicliffe 2001). Our understanding of the relationships between variations in marine habitat and its populations can be synthesized in the form of space–time diagrams such as Fig. 3.2.

These diagrams (see Haury *et al.* 1978; Hofmann and Powell 1998) show overlaps in scales between physical processes and populations. At the smallest scales the ability of organisms to encounter their prey is affected by turbulent processes that occur at the scale of the individuals. Intermediate scales, such as the seasonal presence or absence of fronts or the strength of upwelling structures, may affect the regional distribution and the type of populations at selected locations over scales of hundreds of kilometres. Internannual fluctuations such as El Niño (see Chapters 11, 12, and 16 for more on ecological effects of El Niño), or the North Atlantic Oscillation (NAO; which was thoroughly introduced in Chapter 2) may affect the basin-scale variation of populations, on the order of thousands of kilometres, and introduce possible synchronicities

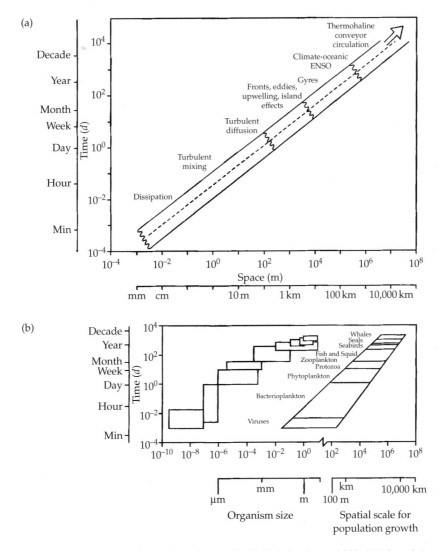

Figure 3.2 Diagrams of the dominant space and timescales in the ocean for (a) physical motions and (b) biological populations. From Hoffman and Powell (1998).

or asynchronicities across oceanic basins. At the largest scales, global variations play a role, such as global warming or the thermohaline ocean conveyor belt circulation.

One of the issues to keep in mind when interpreting diagrams such as Fig. 3.2 is that although the discussion and presentation of the processes and scales are discrete, the scales are not independent (Hofmann and Powell 1998). Changes at one scale are generally coupled—at least—to those at neighbouring scales. Processes occurring at different scales may lead, through their interactions, to

high or low food concentrations and thus variations in growth rates for particular species. Storms affect mixing, which in turn affect stratification, nutrient availability, and primary production. Ensuing trophic interactions will cascade upward and affect secondary producers (zooplankton), larval fish, and eventually fish stocks.

Despite that all scales may, at some point, be important in the quantitative description of the observed variability of marine ecosystems, the approach to studying these fluctuations has been, by necessity, focused on subsets of these scales—either as subsets of trophic scales, environmental scales, or both. However, as pointed out by Werner and Quinlan (2002), this approach may fail in the long-run in explaining and predicting the state of target populations. For example, a study that quantitatively captures the variability at intermediate scales may still be an inadequate predictor of the overall state of target populations if it misses a larger-scale change in circulation that affects the timing or location of a source of prey (Heath *et al.* 1999, 2001) or an overall temperature change and its impact on growth rates (Brander 1995; Michalsen *et al.* 1998; Brander 2000).

3.3 Approaches in modelling marine ecosystems

Modelling is necessary to integrate across the many disparate scales discussed above and is a tool that can be used to relate abundance, distributions, fluctuations, and production of living organisms to variations in the abiotic environment, including climate, food condition, and predation (Carlotti *et al.* 2000). Models may be simple or complex representations of the marine food web and they can have various objectives, including: the estimation of the energy and/or matter through an ecological entity, be it an organism or a community; the estimation of survival and persistence of populations in response to the factors that regulate their variability; and the study of behavioural traits and ecology. There is no universally correct approach. Details of the modelling approach are defined by the objectives of each modelling study, including what are appropriate simplifications. For example, can planktonic state-variables be lumped and modelled as $dQ/dt = \alpha Q$ (where Q is plankton biomass with α (generally) assumed constant), or does size- and age-structure need to be explicitly considered? In other words, the currency of the model may vary from lumped units of carbon to address mass balances, to increasingly detailed representations of size and class to address questions of community structure, where the currency may be numbers of individuals or density, if population abundance is the target.

Based on these different approaches, the global programmes JGOFS and GLOBEC (and the currently in planning Integrated Marine Biogeochemistry and Ecosystem Research, IMBER), have been established to examine different questions regarding marine ecosystems and their response to climate variability and change. The objective of JGOFS (www.igbp.kva.se/jgofs.html) is 'to improve our knowledge of the processes controlling carbon fluxes between the atmosphere, surface ocean, ocean interior and its continental margins, and the sensitivity of these fluxes to climate changes'. The objective of GLOBEC (www.globec.org) is 'to advance our understanding of the structure and functioning of the global ocean ecosystem, its major subsystems, and its response to physical forcing so that a capability can be developed to forecast the responses of the marine ecosystem to global change'. The primary goal of IMBER (see www.igbp.kva.se/obe)—a programme still under development—is 'to understand the sensitivity of the ocean to global change within the context of the broader Earth System, focusing on biogeochemical cycles, marine food webs and their interactions'.

Since the goals of the studies are different in several key respects, they provide different frameworks within which to study marine ecosystems. We highlight some of these in the following.

3.4 Modelling the lower trophic levels of marine ecosystems: understanding the fate of biogenic materials

One of the goals of process studies on ocean carbon (and other biologically active chemical substances) is to improve the mechanistic understanding of biological, chemical, and physical processes that control oceanic and atmospheric carbon dioxide (CO_2) levels (Koeve and Ducklow 2001; Doney *et al.* 2002). To achieve this goal, the role of the lower trophic levels of marine ecosystems must be identified to be able to determine: how anthropogenic CO_2 is taken up from the atmosphere into oceanic surface waters and then exported for storage in the deep oceans; how global climate change and CO_2 storage will cause changes in the physics and chemistry of the oceans, including future changes in CO_2 uptake; and how marine communities and ecosystems will respond to climate change, for example, to changes in physical and chemical properties of the ocean caused by increased CO_2 uptake.

Understanding the role of biological processes in regulating the historical atmosphere–ocean balance of CO_2 also provides indications of changes that might be expected in the future. There are several possible reasons for the differences between glacial and interglacial concentrations. Biological productivity may have been enhanced during glacial times, so that organisms near the sea surface removed more CO_2 from the atmosphere and, by sinking, carried it to the deep ocean. Or, it could be that the oceans might have been more alkaline, thereby drawing down atmospheric CO_2 as required by the physical chemistry of the carbonate system. These possibilities suggest that the magnitudes of oceanic ecosystem processes such as production and export may change in response to climate change (and possibly ocean circulation), or that climate-induced changes in ocean chemistry, such as increased rates of iron input, can affect ocean ecology and biogeochemical cycling. Quantification and understanding of these effects requires explicit consideration of certain components of the marine ecosystems. For example, changes in the rates of nitrogen fixation and denitrification, the rate of calcification, and the relative dominance of organisms such as diatoms that are efficient exporters of organic matter may all have significant influences on the production, remineralization, and partitioning of carbon and inorganic nutrients among oceanic compartments (Sarmiento *et al.* 1993).

The modelling studies undertaken to study these processes and develop the ability for prediction (see Doney *et al.* 2002) are based on solving a coupled system of equations based on the continuity/budget equation:

$$\frac{\partial c}{\partial t} + \mathbf{v}\cdot\nabla c = \nabla\cdot(\mathbf{D}\,\nabla c) + R$$

where c is the tracer concentration, $\mathbf{v}$ is the ocean current/advective velocity vector, $\mathbf{D}$ is a diffusivity tensor that accounts for the combined effect of molecular and turbulent mixing processes, and R is the combined net effect of all biogeochemical sources and sinks. This system of equations can be used model to ecosystem components such as phytoplankton and zooplankton as well as dissolved or suspended matter. The general structure of such 'mass balance' or 'budget' models was introduced in references such as Fasham *et al.* (1990). These models include various forms of nitrogen (nitrate, ammonia, labile dissolved organic nitrogen, and detritus) and relatively simplified, unstructured biotic components (phytoplankton, zooplankton, and bacteria). These formulations have been embedded in models of varying degrees of physical complexity, ranging from simplified mixed layer models, to three-dimensional ocean general circulation models (see review by Doney *et al.* (2002)).

Using the above formulations as a starting point, many studies have been undertaken with increased complexity during the past decade (see Koeve and Ducklow 2001; Doney *et al.* 2002; and references therein) to examine the processes by which CO_2 fixed in photosynthesis is transferred to the interior of the ocean where it may be permanently or temporarily stored (sequestered), that is, the biological pump (Fig. 3.3). Their impact on better understanding of marine ecosystems stems from the need to reduce the uncertainty in estimates of the magnitude of the biological pump for the whole ocean. The oceanic component of the carbon cycle is determined by the exchange of CO_2 with the atmosphere, the circulation of the ocean and the exchange of carbon between the surface and the deep ocean. Oceanic carbon is involved in a number of processes: chemical reactions in the water column and in the sediments, and biological activity of marine organisms. Biological processes increase the complexity of the oceanic carbon cycle, for example, the seasonal and regional variability in the biological processes makes it difficult to estimate the ambient distribution of carbon in the ocean.

During the development of these studies, models with multiple ecosystem compartments were constructed to represent current paradigms of the controls of production and export. Initially, the formulation by Fasham *et al.* (1990) included seven state variables (phytoplankton, zooplankton, detritus, bacteria, dissolved organic nitrogen, nitrate, and ammonium) and was based on the idea that production (both new production based on nitrate and regenerated production based on ammonium) was dominated by large phytoplankton species (diatoms) and export dominated by the sinking of faecal pellets produced by large zooplankton. Over the past decade, studies and models have evolved (e.g. see Fasham and Evans 2000) to represent marine ecosystem processes and descriptions whose importance has only recently been identified, including:

(1) the importance of small-celled phytoplankton species in the global productivity budget, including the role of microzooplankton and the bacterial loop (e.g. Anderson and Ducklow 2001);
(2) the importance to export of carbon by discrete algal blooms of large species such as diatoms and their aggregations (Jackson 2001; Jackson and Burd 2002);

(3) the importance of iron limitation in phytoplankton growth (Denman and Peña 1999; Moore *et al.* 2002a);

(4) the importance of nitrogen fixation and mesoscale eddies (e.g. McGillicuddy and Robinson 1997; Garçon *et al.* 2001) to the delivery of nitrogen to the ocean's mixed layer; and

(5) the importance of diel vertical migration of zooplankton (e.g. Flynn and Fasham 2002; Noguchi-Aita *et al.* 2003).

The inclusion of biodiversity and biological complex interactions in the ecosystem models is fundamental to understand the marine processes. This constitutes a daunting task considering our current understanding of the ecosystem. The use of

nitrogen-based ecosystem models for predicting the fate of carbon is being reconsidered. Observed stoichiometric ratios suggest that uptake in the surface ocean may be variable (Sambrotto *et al.* 1993; Karl 1999) and may be related to variability in ecosystem and community structure (species diversity) currently not included in models. As such new models must be able to provide for alternative controls of production and stoichiometric ratios, for example, domination in a given region by siliceous diatoms, calcifiers, or nitrogen fixers, or other organisms altogether, may determine the composition of organic matter exported from the surface. Models must also be able to predict when and where each of these controls will be dominant; and the consequences of the differences in export

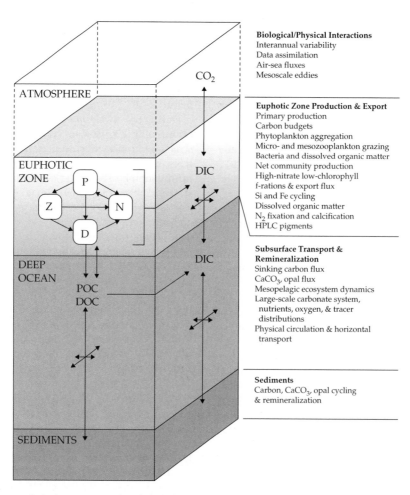

Figure 3.3 Carbon transfer in the upper ocean through the biological pump (from Doney *et al.* 2002).

processes for recycling rates in the water column and possible eventual burial. New approaches include the implementation of intermediate complexity models (e.g. Moore *et al.* 2002b).

3.5 Modelling the higher trophic levels of marine ecosystems—understanding fluctuations in (fish) communities

Estimating carbon fluxes through marine ecosystems provides an important framework within which to examine productivity at different trophic levels. However, quantitatively estimating flows of carbon in complex (and changing) food webs is difficult. In some cases, while overall budgets of either energy or matter remain relatively constant, the actual community structures do not, as suggested by community shifts (see also Chapter 10). Understanding alterations in species composition is the object of intense research in marine ecosystem modelling under the GLOBEC umbrella.

Although fluctuations in fisheries have been noted in historical records (see Cushing 1982), systematic studies of these fluctuations have only begun in the past century. These early studies did not only identify biotic variables, but rather identified links between physical and biological conditions as a key component in determinating the observed fluctuations (e.g. Helland Hansen and Nansen 1909).

Since the early 1900s, the field of fisheries has focused on the general problem of predicting the number of young at birth that will survive to some size or age, termed recruitment (Bradford 1992; Cushing 1996). Recruitment can be defined as numbers surviving to a certain age, reaching maturity, or becoming available to the fishery. Understanding the causes of variation in growth and survival of larval fish is important because much of the eventual interannual variation observed in recruitment in marine fish can be attributed to variation in growth and mortality during early life stages (Shepard *et al.* 1984; Rose and Summers 1992). Ecosystem models in this case are used to quantify of the causes of variation in larval growth and mortality.

One of the main differences in modelling approaches used to address 'mass balance or budget' versus 'fisheries' questions is the recognition that zooplankton and fish populations have structures that are integral to their dynamics (see Carlotti *et al.* 2000). State-variables cannot simply be modelled as $dQ/dt = \alpha Q$, with α a constant. In fact, α is a function of species, stage, temperature, and food, among others, and cannot be measured at the community or even the species level. Furthermore, 1 g of juvenile fish or zooplankton 'behaves' quite differently from 1 g of adults, and thus biomass alone is not adequate to represent populations of higher trophic level species. Increasingly structured representations of the populations is required (see Fig. 3.4). Just as in the lower trophic (budget) models, these higher trophic level model formulations are also embedded in realistic physical models characterizing flow features, hydrography, etc. (e.g. Backhaus *et al.* 1994;

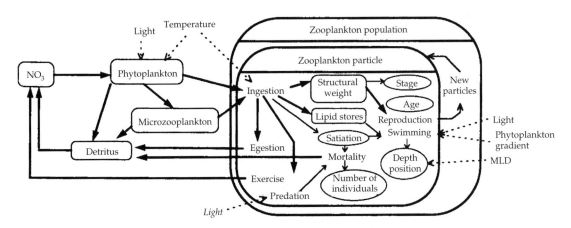

Figure 3.4 Schematic of a Lagrangian model formulation for the study of zooplanktonic ecosystem models. Stage, age, behaviour, and their links to biotic and abiotic variables are explicit. Taken from Carlotti and Wolf (1998).

Table 3.1 Characteristics of selected physical–biological models that have been developed for marine ecosystems

Reference	Region	Circulation model	Ecosystem model	Dimension	Space–Time Scale	Objective
Davis (1984)	Georges Bank	Prescribed advection, diffusion and temperature	Z	r, t	100 km, months	Copepod distribution
Flierl and Wroblewski (1985)	NE US	Prescribed advection	LF	x, t	100's km, months	Larval fish distribution
Franks et al. (1986)	NE US	Empirically derived stream function	N, P, Z	r, z, t	10's km, months	Plankton dynamics
Hofmann (1988)	South Atlantic Bight	Current meter observations	NO_3, NH_4, D, SP, LP, Z (5 stages)	x, y, z, t	10's km, weeks	N and C flux
Walsh et al. (1988)	Mid Atlantic Bight	Barotropic with Ekman imposed	NO_3, P	x, y, z, t	100's km, seasonal	Spring bloom fate
Walsh et al. (1989)	Gulf of Mexico	2-layer baroclinic, Ekman imposed	NO_3, P	x, y, z, t	100's km, seasonal	N and C flux
Ishizaka (1990a)	South Atlantic Bight	Current meter observations	N, P, Z, D	x, y, t	10's km, weeks	N and C flux
Radach and Moll (1993)	North Sea	1D mixed layer	PO_4, P, D	z, t	Point, annual	Response to weather patterns
Gupta et al. (1994)	Saco River, Maine	Primitive eq. reduced to 1D	Z	x, t	10's km, months	Copepod distribution
Pribble et al. (1994)	South Atlantic Bight	Primitive eq.	NO_3, P	x, y, z, t	100's km, weeks	Assess new production
de Young et al. (1994)	Conception Bay, Newfoundland	3D diagnostic	LF	x, y, z, t	10's km, weeks	Capelin larval dispersal
Baretta et al. (1995)	European Regional Seas (ERSEM)	Specified horizontal transport and dispersion	C, N, PO_4, SiO_4	x, z, t	100's km, seasonal	Nutrient and carbon cycling
McGillicuddy et al. (1995a)	North Atlantic	Quasi-geostrophic with surface boundary layer	NO_3, NH_4, P, HT	z, t	Point, days/weeks	Plankton bloom dynamics
McGillicuddy et al. (1995b)	North Atlantic	Quasi-geostrophic with surface boundary layer	NO_3, NH_4, P, HT	x, y, z, t	100's km, days/weeks	Mesoscale dynamical processes
Doney et al. (1996)	BATS	Oceanic planetary boundary layer model, turbulent mixing	NO_3, NH_4, P, D	z, t	Point, seasonal	N and C flux
Moll (1998)	North Sea	Baroclinic circulation model	PO_4, P, BD	x, y, z, t	100's km, seasonal	Seasonal variability
Hood et al. (2001)	BATS	Modified Price et al. mixed layer model	DIN, DON, P, Tr, HT, D	z, t	Point, interannual variability	Interannual variability in N_2 fixation

Abbreviations: NO_3—nitrate, NH_4—ammonium, PO_4—phosphate, SiO_4—silicate, B—bacterioplankton, P—phytoplankton, LP—phytoplankton >5 μm, SP—phytoplankton <5 μm, Z—zooplankton, CZ—carnivorous zooplankton, LF—larval fish, F—fish, D—detritus, DOC—dissolved organic carbon, POC—particulate organic carbon, AT—autotrophs, HT—heterotrophs, TIN—total inorganic nitrogen, BN—benthos, POM—particulate organic matter, DOM—dissolved organic matter, O_2—dissolved oxygen, DON—dissolved organic nitrogen, PON—particulate organic nitrogen, Tr—*Trichodesmium* spp., Fe—iron, BATS—Bermuda Atlantic Time-series Site, OGCM—Ocean General Circulation Model. In the entry "Dimension", r refers to radius, x and y to the two horizontal dimensions, z to the vertical dimension and t to time.

Table 3.2 Characteristics of biological transport models

Reference	Geographic region	Circulation model	Particle behaviour	Dimension	Space–Time scale	Species
Ishizaka and Hofmann (1988)	South Atlantic Bight	Current meter	Passive	x, y, t	10's km, weeks	General transport patterns
Ishizaka (1990a)	South Atlantic Bight	Current meter	Passive	x, y, t	10's km, weeks	General transport patterns
Bartsch (1993)	North Sea	Baroclinic	Vertical migration	x, y, z, t	100's km, months	Herring larvae
Werner et al. (1993)	Georges Bank primitive eq.	Diagnostic	Vertical migration	x, y, z, t	100's km, seasonal	Cod and haddock larvae
Tremblay et al. (1994)	Georges Bank primitive eq.	Diagnostic	Vertical migration	x, y, z, t	100's km, seasonal	Scallop larvae
Bernsten et al. (1994)	North Sea	Primitive eq.	Passive	x, y, z, t	10's km, interannual	Sandeel larvae
Bartsch and Knust (1994)	German Bight	Baroclinic circulation	Vertical migration	x, y, z, t	100's km, days	Sprat larvae
Werner et al. (1996)	Georges Bank	Prognostic finite element hydrodynamic	Vertical and horizontal swimming	x, y, z, t	100's km, weeks	Cod and haddock larvae
Bartsch and Coombs (1997)	Eastern North Atlantic	Primitive eq.	Vertical migration	x, y, z, t	1000's km, seasonal	Blue whiting larvae
Bryant et al. (1998)	Eastern North Atlantic	Hamburg Shelf Ocean Model	Vertical migration	x, y, z, t	100's km, seasonal	Calanus finmarchicus
Carlotti and Wolf (1998)	Norwegian Sea, OWSI	1D upper ocean circulation	Vertical migration	z, t	Meters, seasonal	Calanus copepods
Miller et al. (1998)	Georges Bank	Prognostic finite element hydrodynamic	Passive	x, y, z, t	100's km, seasonal	C. finmarchicus
Proctor et al. (1998)	Northwest European Shelf	2D circulation	Passive	x, y, t	100's km, days	Sandeel larvae
Blanton et al. (1999)	Southeastern US coastal inlets	High-resolution finite element hydrodynamic	Passive	x, y, z, t	10's km, days	Post larval white shrimp, blue crab megelopae
Hare et al. (1999)	Central US east coast	High-resolution finite element hydrodynamic	Vertical and horizontal swimming	x, y, z, t	100's km, seasonal	Menhaden and spot larvae
Rice et al. (1999)	North Carolina east coast	High-resolution finite element hydrodynamic	Vertical and horizontal swimming	x, y, z, t	100's km, seasonal	Menhaden larvae

Abbreviations: OWSI—Ocean Weather Station India, ADCP—acoustic Doppler current profiler. In the entry "Dimension", x and y refer to the two horizontal dimensions, z to the vertical and t to time.

Bryant et al. 1998; Miller et al. 1998). Moreover, the interaction of behaviour, for example, vertical migration that can result in transport to feeding areas, requires a more detailed description of the structure of the flow. As such, the physical models used in the study of these higher trophic levels are of greater resolution. These models frequently represent the target species, for example, zooplankton and fish larvae as particles (passive or with behaviours distinct from physical transport) in a Lagrangian framework, making these models part of a broader class known as individual-based models (IBMs; see Werner et al. 2001). An overview of some marine physical–biological models is given in Table 3.1, while some biological transport models are summarized in Table 3.2.

How to simulate movement that is 'active behaviour' and not simply passive is of fundamental

Table 3.3 Paradigms explaining population pattern, abundance, or variability

Hypothesis	Population pattern—factors that keep individual shocks distinct	Abundance—factors that maintain mean population level set	Variability—factors that affect fluctuations in abundance
Hydrographic containment (*migration triangle match–mismatch*)	The presence of tidal current streamlines between spawning and nursery grounds	Density-dependent growth and survival associated with food availability along larval drift route	Timing in onset of stratification and subsequent plankton bloom relative to spawning date
Stable ocean	(not explained)	(not explained)	Frequency and intensity of mixing such that prey aggregations at the pycnocline are disrupted
Encounter rate	(not explained)	(not explained)	The influence of small-scale turbulence on relative motion between predator and prey
Member/vagrant	Retentive hydrographic structures which result in limited dispersal of early life history stages	The size of the hydrographic structure associated with spawning location	Food-web and physical loss from appropriate habitat are both possible

Source: Werner and Quinlan (2002); adapted from Sinclair and Page (1995).

importance, yet remains unresolved. Active movement can be critical because such movement can greatly affect the physical environment and prey experienced by the individual organisms (Tyler and Rose 1994). We do not really know why larval fish move (Hare *et al*. 1999), or the triggers of diapause cycles of zooplankton (Heath 1999; Heath *et al*. 1999). Prescribed (and/or passive) behaviours may be unrealistic as the physical–biological models move more towards simulating the growth and survival of the icthyo- and zooplankton. Static approaches to movement will be likely replaced by model-derived behaviours that include components maximizing some biological characteristic, such as reproductive value (Giske *et al*. 1994; Fiksen and Giske 1995; Fiksen *et al*. 1995), survival to maturity (Railsback and Harvey 2002), or short-term tradeoffs between growth and mortality (Tyler and Rose 1997). The realism of predicted growth and mortality from models that must include active movement of organisms may very well rely on how well we can model their behaviour (Runge *et al*. 2003).

The issue remains whether the combination of such detailed biological statements embedded in realistic physical models can explain variability in recruitment of target species. As stated above, the interaction of physical and biological scales in the

oceans is quite complex. And, in different years, the interactions will likely be different. There are several key hypotheses on how physical processes affect recruitment (directly or indirectly) and explain population pattern, abundance and/or variability through transport, feeding environment, population integrity, and match–mismatch. While some of these ideas focus on physical processes affecting early life history stages, the matter of year-class strength being set during larval stages (e.g. Hjort 1914) or post-larval is unresolved (e.g. Leggett and DeBlois 1994). Table 3.3 summarizes four main paradigms relating physics to population pattern, abundance, or variability (see also Heath 1992; Cushing 1995; Sinclair and Page 1995). Two of the four (hydrographic containment and stable ocean hypotheses) rely on biological intermediates, the other two (encounter rate and member/vagrant hypotheses) have direct physical mechanisms in operation.

In both the budget-based studies, discussed previously, and the size or age-structured ecosystem models, it is clear that even the most complex models have significantly simplified the real ecosystem. Similarly, not all model formulations will be equally suited across target species. Rather than increasing the complexity of the models' formulation, a 'middle-out' modelling approach should be considered, wherein focus is placed first on the taxa or trophic level of primary interest, with decreasing resolution in detail in the links up to predators and down to prey (ICES 1993).

3.6 New directions: data assimilation for marine ecosystem modelling

Meteorologists have been using data assimilation for decades. This technique was developed with the goal of using observations to improve the forecasting skill of operational meteorological models by melding the observations into the model in order to obtain an improved initial condition (Malanotte-Rizzoli 1996). This methodology was later on introduced into the field of oceanography with more extensive goals. A general definition of the term data assimilation is provided by Hofmann and Friedrichs (2002), defining it as 'the systematic use of data to constrain a mathematical model'.

Data assimilation is a necessary component in the development of prediction systems because of the loss of predictability associated with non-linear coupled physical and biological dynamics in complex models (e.g. see Robinson and Lermusiaux 2002). One of the big limitations of interdisciplinary models for marine ecosystems is the choice of parameters. Disagreements between model results and observations can be the consequence of poor model representation of physics and growth dynamics, or deficient selection of the model parameters. The assignment of values to these parameters is done, in most cases, by using values from the literature or by trying to optimize the result of the model with estimates from laboratory or *in situ* experiments. The assimilation of data into ecosystem models provides a way of constraining these parameters, improving the total skill of the model. Using assimilation techniques, we can find the optimal adjustment of a specific model to an experimental data set. If the model still fits the observations poorly, we can assume the structure of the model presents some inadequacies (Fig. 3.5).

The data assimilation system consists of three elements: a set of observations, a dynamical model, and a data assimilation scheme. The choice of the scheme depends on the kind of model and data we are working with and is closely related to the type of error field present (GLOBEC 1999). The process of data assimilation produces information that can be used as a feedback to improve the deficiencies of the model (e.g. calibrating important sub-grid scale parameterizations). The data collected for assimilation is used for validation of the system, providing improved parameter estimates (forcing, initial and boundary conditions, and model parameters). See Table 3.4 for a compilation of approaches and Fig. 3.6 for a brief list of marine ecosystem model results since the 1990s derived from the thorough

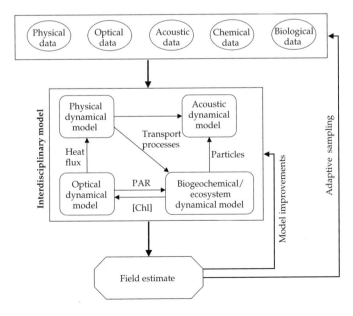

Figure 3.5 General data assimilation schematic (adapted from Robinson *et al.* 1998).

Table 3.4 Characteristics of data assimilative marine ecosystem models in the North Atlantic

Reference	Assimilation scheme	Data source
Ishizaka (1990a,b)	Data insertion	CZCS
Anderson and Sarmiento (1995)	Nudging	Phosphate distributions
Fasham and Evans (1995)	Powell's conjugate direction method	JGOFS NABE data sets
Armstrong et al. (1995)	Nudging	CZCS
Lawson et al. (1996)	Variational adjoint method	BATS data sets
Hurtt and Armstrong (1996)	Simulated annealing	BATS data sets
Semovski et al. (1996)	Optimal interpolation and modified adjoint method	CZCS
McGillicuddy et al. (1998)	Variational adjoint method	Copepod abundance and distribution
Spitz et al. (1998)	Variational adjoint method	BATS
Hurtt and Armstrong (1999)	Simulated annealing	BATS data sets
Evans (1999)	Powell's conjugate direction method	JGOFS NABE data sets
Fasham et al. (1999)	Powell's conjugate direction method	JGOFS NABE data sets
Vallino (2000)	Comparison of multiple methods	Mesocosm experiments
Anderson et al. (2000)	Optimal interpolation	BIOSYNOP cruise and GULFCAST data
McGillicuddy et al. (2001)	Nudging and adjoint methods	Simulated data (OSSE)
Lynch et al. (2001)	Adjoint method	Real time cruise data
Anderson and Robinson (2001)	Optimal interpolation	BIOSYNOP cruise and GULFCAST data
Popova et al. (2002)	Optimal interpolation	*Discovery 227* cruise data
Besiktepe et al. (2003)	Optimal interpolation	Massachusetts Bay

Abbreviations: CZCS: Coastal Zone Colour Scanner, BATS: Bermuda Atlantic Time-series Site, NABE: North Atlantic Bloom Experiment. Adapted from Hofmann and Friedrichs (2002).

reviews of Robinson and Lermusiax (2002) and Hofmann and Friedrichs (2002).

Another fundamental application of data assimilation is adaptive sampling. Requiring close cooperation between experimentalists and modellers from different fields (biological, chemical, and physical oceanography), the idea is to use observation systems simulation experiments (OSSE) to identify the optimum number of variables needed for the best result (GLOBEC 1999). These experiments can provide information about the kind of data required for assimilation and the spatial and temporal distribution of measurements that produce the best estimate. The use of adaptive sampling in real time, even at sea, provides significant help in the design of experiments (Lynch et al. 2001).

There remain uncertainties in our understanding of basin-scale marine ecosystem modelling mainly with respect to data requirements that allow us to fully comprehend ecosystem dynamics, particularly the more complex models. However, significant strides have been made to improve early

approaches and we have a large number of formulations both for models and data assimilation techniques. The process of comparison and calibration has just begun and the final result will be a reduced number of methods.

Presently, the scarcity of certain types of observations constitutes a limitation for data assimilation efforts, and may continue to be a limitation in the coming years (Fasham et al. 1999). Data from satellites is used in large-scale ecosystem models, but there is no vertical structure available for the same timescale and extension (Gunson et al. 1999). The development of real-time (in water) monitoring systems should greatly improve forecasts of marine ecosytems using data assimilative approaches (Dickey 2003). During the last decade a number of different efforts have been made towards the creation of global (e.g. Global Ocean Observing System (GOOS); http://ioc.unesco.org/goos/, Global Ocean Data Assimilation Experiment (GODAE); www.bom.gov.au/bmrc/ocean/godae) and coastal ocean prediction systems (e.g. Pinardi et al. 1999; Seim 2000). These systems

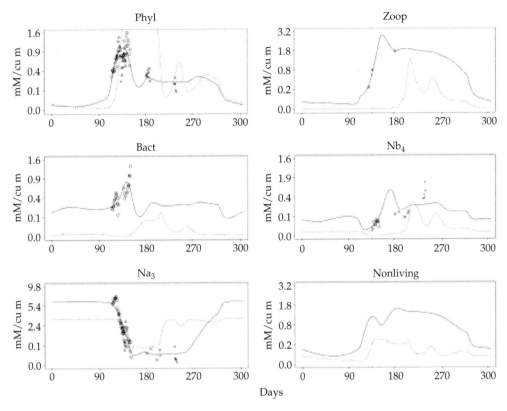

Figure 3.6 Concentrations measured during NABE (North Atlantic Bloom Experiment) in the upper mixed layer and the annual cycle predicted by the FDM model with two different sets of parameters. The dashed lines show the result using parameters from Fasham *et al.* (1990). The solid lines represent the best set of parameters for the same model using data assimilation. Data from Atlantis (°), Meteor (ρ), Discovery (×) and Tyro (+). The ordinate is plotted on a square-root scale. From Evans (1999).

will provide continuous and synoptic information of the physics and biology in areas and regions where the error is the largest or where processes of interest are occurring, in some instances, in real time.

While these results are very encouraging and promising, data assimilation should not be considered the solution to all problems in marine ecosystem modelling. Efforts to develop models that quantitatively capture the complexity of the biological and physical processes should remain as main goal of ecosystem modellers, with data assimilation as another tool to include complexity into the predictive system (GLOBEC 1999).

3.7 Conclusions and closing remarks

The modelling of marine ecosystems has reached a high level of maturity in the past decade.

We have made considerable advances in our ability to examine the coupling of physics and oceanic ecosystems in realistic settings, and we have improved our understanding of the role of environmental forcing on marine ecosystems. We have examined successfully the role of the lower trophic levels of the ecosystems in the oceanic sequestration of carbon, and we have determined the effect of circulation on the retention or dispersal of certain fish and zooplantonic populations. However, and for the most part, advances in these approaches have occurred separately rather than jointly.

It will be necessary in the coming years to link the 'mass balance' lower trophic, basin-scale modelling approaches, with the 'individual based' higher trophic level regional studies. This coupling will be necessary to achieve long-term qualitative (and eventually quantitative) predictive capability of the fluctuations of marine ecosystems. In the case of higher trophic levels, modelling studies have been regional and of short duration, typically of weeks to seasonal cycles over continental shelves and margins. While this may provide information on event-scales, at longer timescales, the variability of and exchanges with the neighbouring open ocean needs to be explicitly considered, as discussed by Heath *et al.* (2001) 'at least at a gross level, there must be some relationship between the amount of primary production in the sea and the amount of fish available for harvesting. However, the details of this relationship are very much less clear. This is partly because the turn over rates of carbon . . . are very much lower at the top of the web compared to the lowest trophic levels. Hence the production by fish reflects the long-term integral of the primary production'.

PART II

Plankton populations

Pioneering work by Colebrook showed the importance of relating changes in the abundance of planktonic species to fluctuations in the environmental conditions. Colebrook applied data collected by the continuous plankton recorder (CPR) survey in the northeast Atlantic and North Sea, still the most comprehensive and continuous data set of its kind. Also based on CPR data, more recent studies on zooplankton in the same regions by Fromentin and Planque demonstrated that closely related species responded differently to large-scale climate fluctuations, suggesting that the effect of climate variability on any given ecosystem might be very complicated with both direct and delayed effects throughout the trophic interactions.

This part of the book presents and discusses responses of populations of planktonic organisms, both phyto- and zooplankton, to climate variability. Although, the main focus is on large-scale climatic forcing, patterns at different spatial scales from basin down to estuary and fjord are explored.

In Chapter 4 Theodore Smayda *et al.* examine the response of phytoplankton populations to climate variability. They draw on results from diverse areas on both sides of the North Atlantic. Particular attention is paid to the role of the North Atlantic Oscillation (NAO) (the NAO was thoroughly described in Chapter 2). This is highlighted through a case study on Narragansett Bay, northeast United States, as well as overviews of phytoplankton responses to climate variation in both the North Sea, neighbouring Celtic Sea and Skagerrak, and central northeastern North Atlantic.

Finally, Smayda and coworkers point to connections between large-scale climate phenomena and harmful algal blooms. While most evidence of such links so far are with El Niño-Southern Oscillation

(ENSO) in the Pacific (see Chapters 11, 12, and 16 for ENSO effects), the authors underline the general importance of relating the occurrence of potentially toxic phytoplankton species in coastal waters to climate variability. Indeed, this is of particular interest in light of the recent tendency towards a global increase in harmful algal blooms.

Chapter 5 is also written by a team of authors covering both sides of the North Atlantic. While studying the responses to climate variability of different zooplankton populations and species Andrew Pershing and coauthors focus on calanoid copepods and *Calanus finmarchicus* in particular. This large copepod and its cogeners dominate the spring zooplankton biomass in areas across the North Atlantic. Early stages of *Calanus* constitute the main food resources for larvae and early juveniles of many fish species, regulating recruitment and thus heavily influencing fish population dynamics (see Chapter 6) and biodiversity (see Chapter 10). For this reason few, if any, North Atlantic zooplankton species have received more attention. The response of *C. finmarchicus* in the northeast Atlantic to the NAO is described, and several hypotheses outlined in Chapter 5. Particularly intriguing is the temporal development in the abundance of *C. finmarchicus* and its cogener *Calanus helgolandicus*. While there has been a pronounced decreasing trend in *C. finmarchicus,* the opposite is the case for *C. helgolandicus*. This has been explained by the general increase in water temperatures over the recent decades, favouring the more southern and warm-water *helgolandicus* species. The temporal and spatial abundance pattern of *C. finmarchicus* and *C. helgolandicus* were among the very first linked to the NAO in ecology.

While less is known about the response of zooplankton to climate forcing in the northwest

Atlantic, Pershing and coworkers provide a summary of recent results from the Gulf of Maine, northeastern United States. Mechanisms linking population dynamics of *C. finmarchicus* to the NAO in this region also are suggested. Such mechanisms may involve either local biological processes or circulation changes. This latter example leads towards one of the main findings of the chapter, a class of climate effects named *translations*, involving movements of organisms from one area to another through physical changes (Chapter 1).

We will return to plankton and climate in Chapter 8 where we will be covering the impact of climate on community structure and dynamics.

CHAPTER 4

Responses of marine phytoplankton populations to fluctuations in marine climate

Theodore J. Smayda, David G. Borkman, Gregory Beaugrand, and Andrea Belgrano

4.1 Phytoplankton and climate linkages

Pelagic environments and plankton processes are inherently variable, in large part because they are open to wind-, weather-, and climate-driven disturbances, which range in magnitude from micro-scale turbulence to storm events to global scale ocean–atmosphere interactions, such as the North Atlantic Oscillation (NAO) and El Niño-Southern Oscillation (ENSO) conditions. Operative at daily, seasonal, annual, and aperiodic frequencies, these disturbances modify phytoplankton growth-variables—temperature, irradiance, nutrient supply, mixed-layer depth, etc. The changes in phytoplankton production, abundance and dynamics, which can accompany habitat perturbations linked to local meteorology, have been well documented. Mini-blooms (see figure 4 in Conover 1954) and disrupted successional cycles (Smayda 1980), for example, are often weather-induced. Weather-driven habitat disturbances and associated phytoplankton responses are usually ephemeral, short-term events, easily measured, regionally localized, and generally occur on small spatial scales (Dickson 1995). Phytoplankton dynamics disrupted by weather are usually not permanently deflected into new trajectories from their 'baseline' behaviour. The mean state of the continuously perturbed system is set by the cumulative effects of the variability in abiotic and biotic processes, which occurs on many scales, rather than being set by the ephemeral

and localized weather-driven changes and responses, alone (McGowan 1990).

The phytoplankton habitat, niche structure, and community behaviour are subject to significant externally (i.e. weather and climate) forced disruptions and regulation. The internal (biological) buffering capacity (= autoregulation) of phytoplankton communities against weather-driven change appears to be relatively weak, unlike in terrestrial plant communities (Margalef 1978). Yet, chaotic behaviour is less than expected (Ascioti *et al.* 1993); rather, there is impressive, quasi-regular, and predictable annual recurrences in major blooms, seasonal cycles, species' successions, etc. (Smayda 1998). The internal mechanisms, which rectify the inherently high and continuous variability induced by external (weather) forcing, and transform these disruptions into the relatively coherent annual phytoplankton cycles and successions being observed, are enigmatic.

In contrast to the weather-driven, ephemeral changes in phytoplankton production, abundance and dynamics, the patterns, trends, and cycles altered in response to climate change are more durable and much more difficult to detect and to quantify. There is the problem of defining 'change'—it requires knowledge of the 'baseline' conditions of the plankton properties or processes suspected of being modified by climate change. Intrinsic trends and cycles, onto which climatic change effects may be superimposed, can confound interpretation

(Smayda 1998), and anthropogenic effects may be falsely attributed to climatic change. Two essential elements are required to evaluate the effects of climate change on phytoplankton. As discussed in Chapter 1, a time series of quantitative measurements of suitable duration must be available, from which the trends, frequencies, amplitudes, and direction of significant departures from long-term means of population behaviour can be calculated to derive measures of change (McGowan 1990). The measurements and variables used to gauge phytoplankton change, or to serve as proxies of climate change, should be quantitative. Otherwise, statistical tests to evaluate the relationships between climate and plankton behaviour will be compromised. Detectable statistical relationships suggestive of a climate-induced shift in plankton behaviour from some previous equilibrium must be treated as approximations, rather than definitively. The details of the intermediary mechanisms, including trophic interactions, needed to bring about the realized plankton behaviour in response to the climatic signal are not revealed; that is, cause and effect are only partially revealed. See Chapter 1 for a further discussion on time-series modelling in climate–ecology studies.

Just as there is the need to define 'plankton change', a suitable quantitative measure of 'climate change' based on a representative time series is also needed. The NAO is an alternation of atmospheric mass between the Subtropical and the Arctic–Atlantic (Hurrell 1995; see also Chapter 2). This climatic oscillation appears to have a major role governing certain hydro-climatic processes, which may in turn influence year-to-year and long-term changes in plankton composition and biomass. Note, as explained in Chapter 1, that there is a regionally variable component to the NAO–climate relationship such that in the north-western Atlantic from the Gulf of Maine southward (including the Narragansett Bay region) there is a positive correlation between NAO and sea temperature anomalies (Marshall et al. 2001; Ottersen et al. 2001; Wanner et al. 2001), whereas along the North European coastal zone there is a negative correlation between the NAO and the sea temperature anomalities.

Plankton communities, particularly zooplankton, in the northeast Atlantic appear responsive to climate change and variation, based on observed correlations with the NAO (Fromentin and Planque 1996; Planque and Reid 1998; Reid et al. 1998a,b; Stein et al. 1998; Reid and Beaugrand 2002.

The responses of phytoplankton, generally, and of plankton communities in the northwest Atlantic to climate change, however, are less well known, but anticipated, given that the NAO is a basin-scale phenomenon centred over the north Atlantic Ocean. This chapter focuses on the responses of phytoplankton as indicator species of climate variation and change in Narragansett Bay, located in the northwest Atlantic, and in the North Sea and neighbouring regions of the northeast Atlantic. Narragansett Bay is selected because of the availability of a 38-year (1959–96) quantitative time series of phytoplankton and habitat based on weekly measurement. Methodological details are provided in Borkman and Smayda (1998) and Li and Smayda (1998). The geographical location of Narragansett Bay and paucity of documented climate change effects on phytoplankton elsewhere also contributed to its selection. When not otherwise referenced, the results pertaining to Narragansett Bay are new. Relationships between phytoplankton and the NAO in the North Atlantic, the North Sea, Celtic Sea, and Skagerrak have been reported earlier (Reid et al. 1998a,b; Belgrano et al. 1999; Beaugrand et al. 2000); however, here we summarize some of the climate–ecology links and possible mechanisms. Finally, we will discuss the seemingly increasing problem of harmful algal blooms and their possible link to climate variability and change.

4.2 Climate and phytoplankton in the northwest Atlantic, the Narragansett Bay example

Narragansett Bay is a well-mixed, relatively shallow (mean depth = 8.8 m) estuary located (ca. 41°30'N, 71°20'W) southwest of Cape Cod on the northeast coast of the United States (Fig. 4.1), and contiguous with Rhode Island and Long Island Sounds. The annual phytoplankton and trophic cycles are dominated by diatom blooms (Karentz and Smayda 1998) and strong benthic–pelagic coupling. A singular feature of Narragansett Bay relevant to climate change issues is its location at the biogeographical boundary between Boreal (north of Cape Cod) and Temperate (south of Cape Cod) waters. It lies at the approximate southern boundary limits within the northwest Atlantic of the ecologically important, Boreal-Arctic diatom bloom species, *Detonula confervacea* and *Thalassiosira*

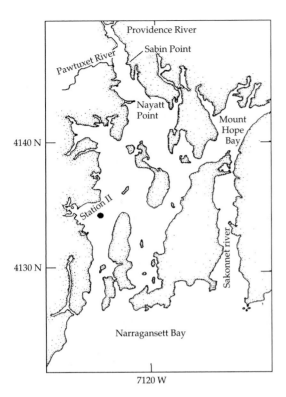

Figure 4.1 Narragansett Bay, Rhode Island, USA (ca. 41° 30′ N, 71° 20′ W) located southwest of Cape Cod on the northeast coast of the United States, showing location of long-term phytoplankton monitoring station ('Station II') in the lower West Passage of Narragansett Bay.

nordenskioeldii. These species, together with the diatoms, *Skeletonema costatum* and *Asterionellopsis glacialis*, are among the five most abundant phytoplankton species in Narragansett Bay (Karentz and Smayda 1998). Any effects of climate change on plankton responses are expected to be most readily detected at the transition zones between biogeographical boundaries.

The effects of climate change and variation (with NAO links) on plankton have been reported primarily for the northeast Atlantic, and principally zooplankton responses (Fromentin and Planque 1996; Planque and Reid 1998). There is a paucity of information on climate-driven responses of phytoplankton (Reid *et al.* 1998a). However, in Narragansett Bay, at least three distinct changes in phytoplankton dynamics are linked to the long-term increases observed in temperature and associated climate driven parameters accompanying the interannual fluctuations and upward trend in the NAO index

during the 38-year time-series. These are: a decrease in chlorophyll biomass, decreased occurrence, and abundance of a major winter–spring bloom diatom, *D. confervacea*, and a change in the season, duration, and magnitude of blooms of the diatom, *S. costatum*, the dominant phytoplankton species in Narragansett Bay.

4.2.1 Temperature, the NAO and chlorophyll

Annual sea surface temperature (SST) in lower Narragansett Bay during the 38-year (1959–96) time series ranged from −2 °C to 26 °C, with an annual mean of 11.4 °C (Borkman and Smayda 2003a). A warming trend began in the early 1960s leading to a long-term increase of ca. 2.3 °C from 1964 to 1996. The long-term warming trend was most pronounced in the winter quarter, when temperatures increased overall by 2.5–3 °C (see also Cook *et al.* 1998). The mean annual SST and the NAO index exhibit parallel cycles and trends, with peaks in the spectral energy at 9–11 and 8-years, respectively. A moderate, positive correlation occurred between winter quarter temperatures and the NAO index, but not for the other seasons (Fig. 4.2).

Annual mean chlorophyll levels in the Narragansett Bay time series decreased approximately fivefold from 1973 to 1992, then progressively increased until 1996, when levels reached the 1973 mean (Fig. 4.3(a); Li and Smayda 2001). The initial 20-year decline, excluding the prominent 1987 deviation, was characterized by modest interannual oscillations in mean chlorophyll, rather than a uniform decrease. The long-term (25-year) prolonged chlorophyll decrease, followed by the more recent increase, was inversely correlated with the

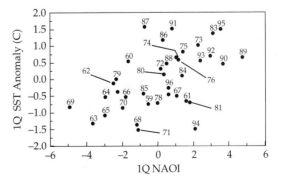

Figure 4.2 The winter NAO index versus mean first quarter (January–March) sea surface temperature anomaly in Narragansett Bay.

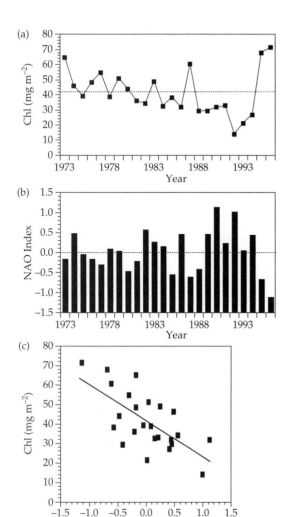

Figure 4.3 (a) Mean annual chlorophyll in Narragansett Bay during the period from 1973 to 1996 based on weekly measurement; (b) mean annual NAO index. (c) Mean annual NAO index versus mean annual chlorophyll concentration.

annual mean NAO index (Fig. 4.3(b,c)). A slight increase in NAO accompanied the decreasing chlorophyll trend, while very low NAO-index values accompanied the very high chlorophyll levels in 1995 and 1996.

As a proxy of climate, the NAO index incorporates a cluster of climatic variables whose associated long-term changes and variability are expected to be more directly responsible for the observed decline in chlorophyll biomass and other phytoplankton responses. The long-term decrease in

chlorophyll Narragansett Bay was also inversely correlated with irradiance and directly correlated with river flow. Sub-grouping of the variables into quarterly intervals yielded significant direct and inverse correlations between winter quarter NAO and temperature and chlorophyll, respectively (consistent with independent measurements by Keller *et al.* 1999).

Failure to establish correlations between annual NAO and temperature during both the 25-year chlorophyll time-series and the entire 38-year time-series, despite the correlations of each parameter with chlorophyll levels, reveals the inherent problems of identifying which of those parameters (and by what mechanisms), embedded within the climate change driver, actually is responsible for the observed plankton changes.

4.2.2 Long-term decline in *D. confervacea* linked to the warming trend

The Arctic-Boreal diatom, *D. confervacea*, was the second most abundant species in Narragansett Bay between 1959 and 1980, and remains among the top five species. It was a major winter–spring bloom species, with 80% of its detected occurrences found at temperatures <5 °C, and is not found at >12 °C (Borkman and Smayda 2003b). Narragansett Bay is near the southernmost boundary of *Detonula*'s distributional range in the northwest Atlantic. During the period from 1959 to 1996, the mean annual abundance of *D. confervacea* progressively declined by ca. sixfold (Fig. 4.4(a)). It's proportionate contribution to annual production decreased from an estimated 10% during its 1960s blooms to <1% during the later years of the time series (Borkman and Smayda 2003b).

The long-term increase in winter temperature explains much of the decline in abundance of *D. confervacea*. The cumulative deviations in temperature and in abundance of *D. confervacea* relative to their time series means presented in Fig. 4.4(b) show the evidence for thermal regulation of *D. confervacea*'s occurrences and gradual withdrawal from Narragansett Bay located near the southern edge of its distributional range. The trends in *D. confervacea*'s abundance show an opposite pattern to temperature: increasing abundance from 1959 to 1965, stationary around mean abundance from 1966 to 1976, and decreasing abundance thereafter. There was no significant correlation

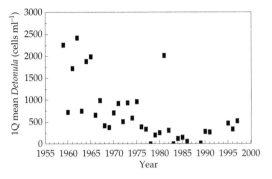

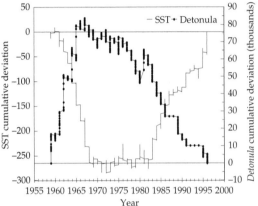

Figure 4.4 (a) Frequency of observed presence of *D. confervacea* versus temperature during its occurrences in Narragansett Bay, 1959–96. (b) Relationship between sea surface temperature anomaly and abundance of *D. confervacea* expressed as the cumulative deviations from weekly means during the 1959–96 time series. (From Borkman and Smayda 2003b.)

between NAO and abundance of *D. confervacea* (Borkman and Smayda 2003b). This does not compromise an interpretation that the observed change in *Detonula*'s behaviour is climate linked. Durbin and Durbin (1992) have demonstrated that increased winter temperatures reduce the development time of the major zooplankton grazer, the copepod *Acartia hudsonica*, then present. Thus, the influence of climate driven temperature changes on *Detonula*'s dynamics may not only be through reduced growth rate (see Smayda 1969), but also through increased grazing pressure. Both abundance (Durbin and Durbin 1992) and grazing rate (Deason 1980) of *Acartia* increase with temperature. There is further support for this proposed NAO—temperature—grazing linkage from the significant correlations found between the NAO

index and various features of zooplankton behaviour in Narragansett Bay. The increased abundance of the herbivorous *Acartia* species in the quarter prior to inception of *Detonula*'s winter–spring bloom reflects the effect of elevated temperatures accompanying positive NAO indices. Inception of winter–spring blooms in Narragansett Bay is dependent upon reduced grazing pressure (Pratt 1965; Martin 1970).

4.2.3 Decadal shifts in annual bloom patterns of *S. costatum*

Phytoplankton dynamics in Narragansett Bay primarily reflect the bloom behaviour of the diatom, *S. costatum*, which has frequently contributed >75% of the total annual phytoplankton cell density (see Smayda 1973). Dramatic long-term shifts have occurred in the magnitude of its annual abundance and seasonal timing of its blooms. Its mean level of abundance between 1959 and 1974 was nearly twice (ca. 2300 cells ml^{-1}) that during 1980–96 (Borkman 2002). *S. costatum* exhibited a bimodal bloom pattern during the 1960s, dominated by its winter–spring bloom in 80% of the decadal years. During the subsequent three decades, from 1970 to 1996, the frequency of winter–spring dominated years (i.e. annual maximum) decreased to ca. 25%, while the frequency of summer-dominated blooms increased to ca. 50% (i.e. the winter–spring bloom was diminished). Autumn blooms, which were previously infrequent, occurred in 25% of the years beginning in the 1980s.

The seasonal shift in occurrence of *S. costatum*'s major annual bloom from winter–spring to summer–fall, and accompanied by a reduction in mean annual abundance, represents a major trophic change. These long-term changes in bloom dynamics can be partly related to a change in climate. The NAO mean was significantly lower in years when *Skeletonema*'s major annual bloom was a winter–spring event, and much higher in those years when its summer–autumn bloom was dominant. Furthermore, in years when the winter–spring bloom dominated, the winter season was statistically significantly brighter, colder, and windier than in years when the summer–autumn bloom was dominant.

The climatic linkage of *S. costatum*'s altered dynamics in Narragansett Bay is further suggested by the contrasting bloom characteristics during the five highest, positive NAO index winters

Table 4.1 Some significant differences in mean bloom responses of *S. costatum* in Narragansett Bay in years (*n* = 5 each) of extreme (−) and (+) NAO index years Abundance units = cells per millilitre; w–s = winter–spring; 1Q = first quarter; 2Q = second quarter

Bloom response	(−) NAO years	(+) NAO years	p value
Annual maximum week	13	30	0.046
W–s bloom duration (weeks)	9	4	0.025
Maximum w–s abundance	24,967	1713	0.009
1Q mean abundance	5536	399	0.016
2Q mean abundance	2578	166	0.009
% weeks > 500 cells ml^{-1}	0.43	0.23	0.028

Source: Borkman (2002).

(i.e. warmest) and five lowest, negative NAO index winters (coldest). The week of the annual maximum during the five coldest winters occurred in week 13 (late March), on average, but during the five warmest winters was delayed until week 30 (late July)—a 4-month displacement from a winter–spring event to a summer event (Table 4.1). Duration of the winter–spring bloom was shortened, from 9 to 4 weeks, and maximal winter–spring abundance decreased from about 25,000 to 1700 cells ml^{-1}. The shift in *Skeletonema*'s seasonal maximum to a summer occurrence during warm-winter years was accompanied by reduced abundance year-round: the first and second quarter mean abundances decreased by about 15-fold, and the percentage of weekly occurrences exceeding 500 cells ml^{-1} decreased from approximately 50–25% (Borkman 2002).

The major features and statistical linkages of *S. costatum*'s altered bloom behaviour in Narragansett Bay—its seasonal timing, duration, magnitude—are consistent with responses to changes in temperature and other parameters modified by climate mediated by the variations in regional scale ocean–atmosphere interactions of the NAO. Temperature, whose change was a prominent feature of the time series, influences numerous biotic processes (Lomas *et al.* 2002), including growth rate, grazing, and nutrient recycling. These processes acting in unison, or progressively sequenced in response to the altered temperature, may have been among the internal drivers contributing to *S. costatum*'s climate-induced changes. The Narragansett time series also shows that concurrent, multiple climate-linked changes may be operative in restructuring phytoplankton behaviour leading to a new equilibrium (i.e. 'baseline'). Not only was there a major

change in the annual dynamics of *S. costatum*, the major species there and regionally, but also reduced abundance and occurrence of the Arctic-Boreal *D. confervacea*, and a significant decrease in chlorophyll biomass. Climate change appears capable of modifying baseline phytoplankton community structure, production, bloom cycles, and successions, thereby establishing new equilibria, to which the local trophic structure must adjust.

4.3 Relationships between phytoplankton and climate in the North Atlantic, the North Sea, Celtic Sea and Skagerrak

Long-term changes in phytoplankton over the North Atlantic have been investigated using results from the continuous plankton recorder (CPR) survey (Colebrook 1982, 1991; Reid *et al.* 1998a). This upper layer plankton monitoring programme has monitored at monthly intervals the presence or abundance of approximately 170 phytoplanktonic and 230 zooplanktonic taxa in the North Atlantic and North Sea since 1948 (Warner and Hays 1994). Recent results using this large data set indicate that year-to-year changes in standing stock, production, and community structure of the plankton might be related to the state of the NAO (Reid and Planque 2000).

All NAO-induced hydroclimatic modifications are not ubiquitous and may even have opposite responses in different regions of the North Atlantic (Reid and Planque 2000; this will also be discussed in the next chapter, on zooplankton). Beaugrand *et al.* (2000) used CPR data from the SA route (i.e. the transect from near Brighton in southern

England to La Coroña in northern Spain). This transect (Fig. 4.5) has the advantage that it traverses different hydrodynamic regions including the Channel, the Celtic Sea, and the Bay of Biscay to examine relationships between plankton, hydrography, and climate. The route was divided into 20 spatial intervals ranging from 20 to 70 km and containing the same number of CPR samples (188 observations for each interval, a total of 3760 samples). Selecting the commonest phytoplanktonic and zooplanktonic species, a three-way table was constructed with the annual mean of taxa abundance for each interval and for each year over the period 1979–95. In oceanography, methods that allow the analysis of such complex tables are rare. A three-mode principal component analysis (PCA; Hohn 1979, 1993; Beffy 1992) was developed and applied in conjunction with cluster analysis. Five different zones corresponding to a distinct year-to-year variability in plankton abundance were identified (Fig. 4.5). The zones were also characterized by distinct physical processes and it was even possible to detect the effects of the Ushant Front corresponding to zone 3 (Fig. 4.5). *T. nitschioides* and *Nitzschia delicatissima* and other zooplanktonic taxa mainly present over the English Channel (Fig. 4.6(a)) showed a year-to-year variability in the abundance of species different to the Bay of Biscay (Fig. 4.6(b), (c)). As the first principal component in

each mode was indicative of plankton abundance, there was an evident decrease in number of the planktonic assemblage especially in the English Channel (Fig. 4.6(b)). This feature corresponded with a very high NAO index as well as the beginning of the 1989/1991 high salinity anomaly (Becker and Dooley 1995). Furthermore, especially for the northeast and central English Channel, a higher abundance was observed for negative or low NAO-index values. During a high positive NAO-index, westerly wind is strengthened through this area. This may lead to a stronger mixing that could delay a relative stabilization of the water column needed for the initiation of the primary production (Dickson *et al.* 1988). From the Ushant Front, no relationships occurred except between air humidity and the principal component in the Celtic Sea. Indeed, species such as *Ceratium fusus* and *Ceratium macroceros* seem to have cyclic and more regular interannual variations in deeper waters (Fig. 6(a and c)). In the period 1983–85 seems to have increased the abundance of these two dinoflagellates and the cyclopoid copepod *Oithona* spp. increased with no indication of a cause.

Reid *et al.* (1998a) investigated the long-term changes of a greenness index derived from the analysis of CPR samples from 1948 to 1995. Over the central North Sea and the central northeastern North Atlantic, a pronounced and significant increasing trend in the concentration of chlorophyll *a* as well as an augmentation in the length of the production season was clearly demonstrated. The analysis has been recently updated by Edwards *et al.* (2002). A stepwise change seems to have appeared after 1985 and it may be expected that this change is linked with a change in community structure in the phytoplankton. This possibly constitutes another adding indication of the regime shift in the ecosystem in this region reported by Reid *et al.* (1998a), Edwards *et al.* (1999) and Reid *et al.* (2001). Since the correlation between the NAO index and these two trends was weak and not significant (Edwards *et al.* 2002) it may be that the effects of this oscillation are the result of the integration of number of hydroclimatic factors (Reid and Planque 2000; Edwards *et al.* 2002). Furthermore, uncertainties remain about the relationship between the period of high NAO of the last 30 years and climatic warming. An opposite trend in the intensity of chlorophyll was also detected in the northern northeast Atlantic and related with the negative temperature anomaly in this area (Reid *et al.* 1998a,b). This

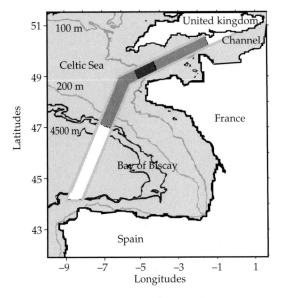

Figure 4.5 Identification of regions based on their year-to-year variations. From Beaugrand *et al.* (2000).

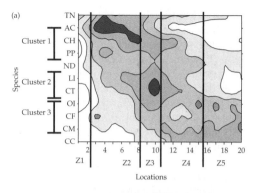

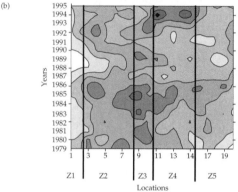

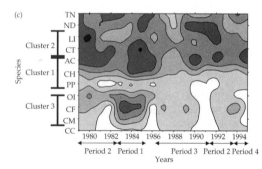

Figure 4.6 Year-to-year variability in first principal components for each species in space and time. As a three-mode PCA was used, the principal component integrates two modes of variability. From Beaugrand *et al.* (2000). (a) Spatial distribution of the first principal component species-location. (b) Year-to-year variability of the first principal component year–location from 1979 to 1995 in each region (see Fig. 4.5). Z1: northern eastern English Channel; Z2: southern western English Channel; Z3: Ushant Front; Z4: Celtic Sea; Z5: Bay of Biscay. (c) Year-to-year variability of the first principal component species-year from 1979 to 1995 and for each species. TN: *Thalassionema nitzschoides*; AC: *Acartia* spp.; CH: *Calanus helgolandicus*; PP: *Para-Pseudocalanus* spp.; ND: *Nitzschia delicatissima*; LI: *Limacina* spp.; CT: *Centropages typicus*; OI: *Oithona* spp.; CF: *Ceratium fusus*; CM: *Ceratium macroceros*; CC: *Clausocalanus* spp.

negative trend may be a direct effect of human induced global warming that might implicate the melting of Arctic ice and permafrost. However, it might also be attributed to a stronger formation of Labrador Sea Intermediate Water and its spread eastwards, which could be induced by the NAO (Drinkwater 2000; Reid and Planque 2000).

A pattern of change in phytoplankton biomass (measured as chlorophyll *a*) was also measured in a monthly time series (1985–96) from a station in the mouth of the Gullmar Fjord in the eastern Skagerrak (Lindahl *et al.* 1998). Belgrano *et al.* (1999) demonstrated that phytoplankton biomass, primary production and counts of three species of the toxic algae genus *Dinophysis* were significantly correlated with the NAO. The links between primary production and the NAO may not be direct as seen in the relationships found between the timing of the spring bloom and the Gulf Stream Index (GSI; Taylor *et al.* 1996). The GSI in turn is correlated with the NAO with an approximately 2-year lag, which is believed to reflect the response time of the ocean to atmospheric forcing (Taylor and Stephens 1998). However, in the Skagerrak a link between primary production in May and the forcing of the NAO (Fig. 4.7) has been reported (Lindahl *et al.* 1998; Belgrano *et al.* 1999). The NAO explained with a lag of 3 months 60% of the variance in the primary production expressed as growth rate (R_t; $R^2 = 0.63$; $p < 0.001$; Belgrano *et al.* 1999; Fig. 4.7). The mechanism relating these correlations need further explanations, however, recent findings seem to relate the NAO with changes in the mixed layer depth and the vertical entrainment of nutrients and ultimately to fluctuations in primary production.

Over the North Atlantic, the phytoplanktonic community structure seems to be largely influenced by both global warming and the NAO (which indeed might be linked; see Chapter 2). However, the hydroclimatic system is complex and the effects of climatic forcing are multiple. Indeed, change in phytoplankton community structure might be direct by climatic forcing (bottom-up control) or more complex by the integrated effects in the food web (top-down). As seen in this chapter, the effects appear to be region-specific. The use of new exploratory statistical analyses such as the three-mode PCA might be able to detect both major and more subtle changes in the community structure for time/space structured data.

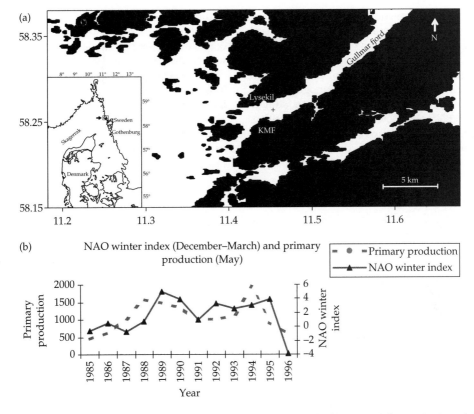

Figure 4.7 (a) Location of the sampling station in the Gullmar Fjord, Sweden (58° N, 11° E) indicating Kristineberg Marine Research Station (KMF). (b) May daily mean primary production (mgC m^{-2} d^{-1}) and NAO winter index 1985–96 ($r = 0.5664$, $p < 0.05$; Belgrano *et al.* 1999).

4.4 Climate and harmful algal blooms

A global expansion in harmful algal blooms (HABs) appears to be in progress (Smayda 1990), the causes of which are unresolved, but may be related to changing climate and concomitant effects on the marine environment (Hayes *et al.* 2001). While it is firmly established that weather plays a significant, short-term role in local red tide outbreaks through its coupled effects on water mass stability, irradiance, and 'water quality' modified by river runoff, larger-scale connections with climate and its changes are obscure. A 'greenhouse' effects scenario has been formulated by Fraga and Bakun (1993) linking HAB events to global climate changes based on decadal changes in wind patterns and bloom dynamics of the toxic dinoflagellate, *Gymnodinium catenatum*, observed off the Galician coast (NW Spain). More convincing evidence (Maclean 1989) that HABs may be elicited by large-scale climatic

phenomena comes from the association observed between ENSO events and western Pacific bloom outbreaks of the toxic dinoflagellate, *Pyrodinium bahamense* var. *compressum*, since its novel outbreak in 1972 off New Guinea. There was exact coincidence of the major bloom events and their regional spread between 1972 and 1987 with ENSO events. Elsewhere, unusual harmful bloom events linked to ENSO have been reported for New Zealand coastal waters (Rhodes *et al.* 1993; Chang *et al.* 1995), and exceptional paralytic shellfish episodes reflective of toxic dinoflagellate blooms occurred during nine ENSO events over a 15-year period in the Pacific northwest coasts of Washington and British Columbia (Erickson and Nishitani 1985).

There is thus some evidence that large-scale climatic perturbations, such as ENSO events of varying intensity, may in some instances promote increases in the abundance of dinoflagellates in both open coastal and inland waters, and lead to their blooms. In oceanic waters, N_2-fixing *Trichodesmium* spp. in the subtropical, North Pacific Ocean increased in abundance along with the change in frequency and duration of ENSO events (Letelier and Karl 1996). However, the bloom-control processes and mechanisms by which the altered oceanographic and atmospheric conditions associated with climate change and ENSO-scale events remain to be clarified.

The relations found between the changes in the NAO, salinity, and SST and the density distribution of three species of *Dinophysis*, suggests the importance of relating climate variability to the occurrence of potentially toxic phytoplankton species in coastal waters (Belgrano *et al.* 1999).

The suggestion that HABs have been increasing in recent years is of considerable concern globally. Evidence to suggest that the occurrence of toxic algae may be related to regional climatic variability as reflected in, for example, the NAO, may enable us to predict the conditions for HABs.

CHAPTER 5

The influences of climate variability on North Atlantic zooplankton populations

Andrew J. Pershing, Charles H. Greene, Benjamin Planque, and Jean-Marc Fromentin

5.1 Introduction

Because of the economic importance of North Atlantic fisheries, there is a long history of research on the variability of not only in the fish stocks, but also other components of the ecosystems in this region (Mills 1989). Time series from this research provide a unique opportunity to study population responses to climate variability. One of the best example of such time series are those derived from the continuous plankton recorder (CPR) surveys conducted by the Sir Alister Hardy Foundation for Ocean Science (SAHFOS—data from which were also used in the preceding chapter). The SAHFOS surveys provide a detailed picture of long-term variability in the ecosystems of the North Atlantic, particularly the eastern Atlantic (Gamble 1994). The SAHFOS surveys, which began in the late 1940s, rely on commercial ships travelling on their usual routes. The CPR sieves plankton on to a silk gauze that is periodically advanced, thereby providing a quantitative record of the abundance of zooplankton and large phytoplankton in discrete segments along the ship's path. Traditional oceanographic surveys complement the CPR surveys, and together, these techniques are beginning to provide a detailed picture of the influence of climate variability on zooplankton populations in the North Atlantic.

Variations in zooplankton populations in regions throughout the North Atlantic have been tied to changes in the Atlantic atmosphere–ocean system (Fig. 5.1). Recent work has tended to focus on the impact on zooplankton of one particular climate mode, the North Atlantic Oscillation (NAO; Hurrell *et al.* 2003; see also Chapter 2). However, it is important to remember that the NAO is only one measure of the climatic conditions in this region. Rather than describing the details of each of these climate–zooplankton associations, we will focus on two particularly well-known regions, the North Sea and the Gulf of Maine. These examples characterize the range of mechanisms by which climate may influence zooplankton. This approach allows us to categorize the important processes linking zooplankton populations and climate.

5.2 Climate variability and zooplankton populations in the northeast Atlantic

Most results linking the NAO to variability in marine ecosystems come from studies in the northeast Atlantic. Starting with the study by Fromentin and Planque (1996) that related variability in the North Sea populations of the copepods *Calanus finmarchicus* and *Calanus helgolandicus* to the NAO,

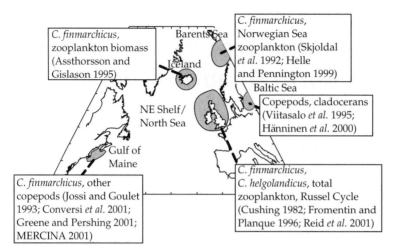

Figure 5.1 Examples of zooplankton populations in the North Atlantic whose variability is associated with regional climate.

researchers have documented NAO-associated changes in marine ecosystems throughout the region (Ottersen *et al.* 2001). However, the importance of the NAO in the eastern Atlantic should not be surprizing, given the strong influence the NAO has on European weather (see Chapter 1).

The atmospheric and circulation changes associated with the NAO are reflected in the correlations between winter sea-surface temperature (SST) and the NAO (Reid and Planque 1999). When the NAO is positive, a large area in the centre of the North Atlantic is cooler, and areas to the south and east are warmer. In the North Sea, SST is positively correlated with the NAO. The correlation is similar to that for air temperature over the North Sea, but is also related to NAO-associated changes in the water flowing into the North Sea from the Atlantic. When the NAO is negative, the inflow increases in the eastern section, bringing cool water from the Norwegian Sea (Stephens *et al.* 1998; Holliday and Reid 2001; Reid *et al.* 2001). When the NAO is positive, the inflow brings warmer water from the south into the region.

In addition to the direct, year-to-year influence of the NAO on conditions in the eastern Atlantic, there is evidence for a cumulative effect of the NAO on this region. The 1980s were the first sustained period of positive NAO-conditions since the 1920s. Reid *et al.* (2001) report the presence of unusually warm water in the North Sea during the late 1980s. This water was associated with a coastal jet that intensified during this period, bringing very warm water and warm-water species into the vicinity of

the British Isles. The observations propose that several years of very positive NAO conditions were required to alter the flow in this jet. See Chapter 1 for a further exposition of this topic.

5.2.1 Response of zooplankton populations

Of all the species in the North Atlantic, few have received more attention than the calanoid copepod, *C. finmarchicus*. *C. finmarchicus* is a large copepod that dominates the spring zooplankton biomass in areas across the North Atlantic (Marshall and Orr 1955; Astthorsson and Gislason 1995; Planque and Batten 2000). To understand many of the associations between this species and climate, it is necessary to understand its seasonal patterns. *C. finmarchicus* is most abundant during late spring and early summer, following the spring bloom. Starting in early summer, some *C. finmarchicus* postpone their development to adults, descend to depths of 400–2000 m, and enter diapause, a state of reduced activity (Hirche 1996; Heath *et al.* 2000). By autumn, most *C. finmarchicus* have entered diapause and their abundance near the surface is greatly reduced. The copepods begin to emerge from diapause and enter the surface waters in winter, in anticipation of the spring bloom (Miller *et al.* 1991). Because their overwintering strategy requires deep water, most areas on the continental shelf are too shallow to support large numbers of diapausing *C. finmarchicus*. Thus, even though *C. finmarchicus* dominates the springtime biomass in shelf regions such as the North Sea (Williams

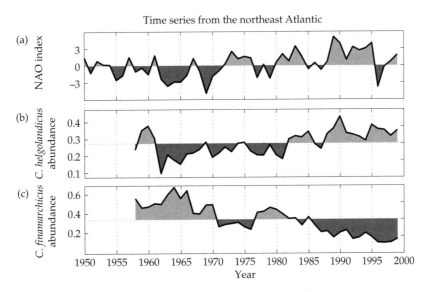

Figure 5.2 The relationship between (a) the NAO and (b) the copepods *C. helgolandicus* and (c) *C. finmarchicus* in the eastern Atlantic.

et al. 1994), the populations of this species in shelf ecosystems should be viewed as expatriates from the open-ocean populations (Greene and Pershing 2001).

Calanus finmarchicus in the northeast Atlantic, together with its cogener *C. helgolandicus*, were the first zooplankton species whose fluctuations in abundance were directly correlated with the NAO (Fromentin and Planque 1996). Starting in the 1960s, the NAO Index exhibited a consistent trend towards more positive conditions (Fig. 5.2(a)). During this time, the abundance of *C. helgolandicus* increased (Fig. 5.2(b)) along with the total abundance zooplankton and phytoplankton as measured by the CPR (Planque and Taylor 1998). The *C. finmarchicus* population exhibited an opposite trend, declining by 37% between 1962 and 1992 (Fig. 5.2(c)). *C. helgolandicus* abundance is positively correlated with the NAO, and exhibits the same pattern as total zooplankton and phytoplankton (Planque and Taylor 1998) (Fig. 5.3(a)). *C. finmarchicus* abundance is negatively correlated with the NAO (Fig. 5.3(b)), and the correlation is stronger than that for *C. helgolandicus*.

Several hypotheses have been advanced to explain the NAO-associated fluctuations in the North Sea between populations of these two copepods. Fromentin and Planque (1996) proposed that changes in temperature associated with the NAO (warmer during positive years,

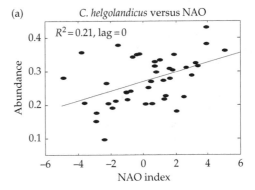

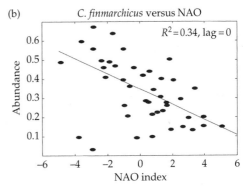

Figure 5.3 Correlations between the NAO and the abundance of (a) *C. helgolandicus* and (b) *C. finmarchicus* in the eastern Atlantic.

colder during negative) drive the species shifts, with *C. helgolandicus*, which is typically found in warmer waters, flourishing during warmer conditions and *C. finmarchicus*, which is typically foun in cold waters, flourishing during colder conditions. They also suggest that the influence of temperature on these species could alter the competitive interaction between them. Stephens *et al.* (1998) suggest that increased flow from the north and west during negative NAO conditions advects large numbers of *C. finmarchicus* to the shelf region from their principle overwintering habitat in the deeper waters. If temperature changes in the North Sea are determined by water mass movements, then these two hypotheses may be related. Specifically, increased flow from the Norwegian Sea during negative NAO conditions would cool the North Sea and bring *C. finmarchicus* on to the shelf while displacing *C. helgolandicus* to the south.

From 1958 to 1995, the phase of the NAO was a good predictor of *C. finmarchicus* abundance in the North Sea (Planque and Reid 1998). However, the dramatic drop in the NAO in the winter of 1996 did not lead to a dramatic increase in *C. finmarchicus* abundance as predicted. The correlation between the NAO and *C. finmarchicus* seems to have broken down in 1996 and 1997 (Planque and Reid 1998), and there are two hypotheses that may explain the anomalous relationship in those years.

The first hypothesis concerns the nature of the NAO index. The NAO index is based on the difference in sea-level pressure (SLP) between Lisbon, Portugal, and Stykkisholmur, Iceland (Hurrell 1995; see also Chapter 2). In most years, these locations are near the centres of the low- and high-pressure cells over the Atlantic. However, in both 1996 and 1997, the subpolar low shifted from Iceland to northern Norway (Ulbrich and Christoph 1999). Thus, even though the NAO Index was low in 1996, it is likely that the conditions in the eastern Atlantic were atypical of negative NAO conditions, possibly explaining the unpredictable response of the North Sea *C. finmarchicus* population.

The second hypothesis is based on the interaction between deep-water circulation patterns and the diapausing strategy of *C. finmarchicus*. In 1994, Backhaus *et al.* (1994) offered a detailed hypothesis to explain the persistence of *C. finmarchicus* in the regions surrounding the British Isles in the face of strong advection. They postulated that the North Sea population of *C. finmarchicus* is formed by animals (and their descendents) that overwinter in the

Faroe–Shetland Channel and are advected into the North Sea via wind-driven surface currents similar to the flows investigated by Stephens *et al.* (1998). These ideas prompted the formation of the ICOS (Investigation of *Calanus finmarchicus* migrations between Oceanic and Shelf seas) project, which investigated the 'Backhaus' hypothesis using both field observations and numerical modelling (Heath 1999). A key result of the modelling efforts was the identification of relative contributions that various deep-water source regions make to the North Sea population under different wind conditions. When the winds are from the southwest or southeast, the important source region is to the west of Scotland, southwest of the Wyville–Thomson Ridge (Gallego *et al.* 1999) (Fig. 5.4, outlined in red). When the winds are from the northwest, the source region shifts to the northeast, extending into the Faroe–Shetland Channel and southern Norwegian Sea

Figure 5.4 Hypothesis explaining interannual and interdecadal variability in *C. finmarchicus* abundance in the North Sea developed by the ICOS study (Heath 1999; Heath *et al.* 1999). Southerly winds associated with negative NAO conditions tend to advect *C. finmarchicus* from an area west of Scotland (red outline). Winds with a more northerly component tend to advect animals from a region further east, in the Faroe–Shetland Channel (blue outline). The concentration of *C. finmarchicus* in the Faroe–Shetland Channel is characteristically higher than that further west; thus, negative NAO conditions should lead to more *C. finmarchicus* in the North Sea. However, this mechanism is complicated by the distribution of NSDW, which is believed to carry *C. finmarchicus* into the Faroe–Shetland Channel. When NSDW production is high, a typical condition during negative NAO winters, the flow of NSDW into the Faroe–Shetland Channel is higher and leads to greater abundance of *C. finmarchicus* in the source region (light blue patch). When the volume of NSDW is reduced, the flow of this water mass and *C. finmarchicus* into the Faroe–Shetland Channel is reduced. (See Plate 4.)

(Fig. 5.4, outlined in blue). The field observations suggest that diapause stocks south of the Wyville–Thompson Ridge are less abundant than in the Faroe–Shetland Channel; and, within the Channel, the concentration of *C. finmarchicus* closely follows the distribution of Norwegian Sea deep water (NSDW; Heath and Jonasdottir 1999). Combining these results provides an explanation for the relationship between *C. finmarchicus* and the NAO that accounts for the low abundance of *C. finmarchicus* in the North Sea in 1996. According to the model, advection of animals from the Faroe–Shetland Channel requires northwest winds, a condition that is associated with negative NAO winters (Dickson 1997). However, the concentration of diapausing *C. finmarchicus* in the Faroe–Shetland Channel is influenced by the flow of NSDW across the Wyville–Thompson Ridge. When NSDW production is high, as it was during the 1960s, this water mass flows into the Faroe–Shetland Channel (Fig. 5.5, light blue), leading to increased abundance of *C. finmarchicus* in the source region and eventually, in the North Sea. The persistent positive NAO conditions during the 1980s and 1990s led to reduced NSDW formation (Fig. 5.4, pink) (Hansen *et al.* 2001) and reduced concentrations of *C. finmarchicus* in the Faroe–Shetland Channel. Thus, even though conditions in 1996 were favourable for advecting *C. finmarchicus* from the chief source region, the actual concentrations there were depressed (Heath *et al.* 1999). The transport of NSDW into the Faroe–Shetland Channel is likely related to the NAO; however, several years of negative NAO conditions may be required for this transport to increase.

The diapause strategy of *C. finmarchicus* implies that shelf populations of this species must be closely tied to oceanic populations. Thus, this species can be considered an indicator of the degree to which a shelf ecosystem reflects an oceanic versus neritic zooplankton community. This idea was first articulated for the North Sea by (Glover 1957), who described a shift in the North Sea zooplankton community from one composed of mostly neritic species such as copepods in the genera *Pseudocalanus*, *Paracalanus*, and *Acartia* as well as *Centropages hamatus*, to a more oceanic community with *Calanus*, *Centropages typicus*, and the pteropod *Clione limacina*. The community change in the early 1950s was associated with decreased biomass in the zooplankton community, a delayed spring bloom, and years with progressively more negative NAO values. The oceanic community dominated through much of 1970s, until the NAO shifted into a consistently positive phase. In the mid-1980s, the North Sea zooplankton shifted to a strongly neritic community (Reid *et al.* 2001). The 'regime shift' observed by Reid *et al.* (2001) included a dramatic increase in the number of horse mackerel in the North Sea as well as an earlier spring bloom, more phytoplankton, and more frequent observations of species typically found much farther south. Reid *et al.* (2001) propose that the ecosystem shift was caused by an extension of the northward-flowing warm-water jet along the shelf break. The changes in this current brought warm water and warm-water species into the region, and this water was then transported into the North Sea. Thus, the interannual abundance patterns of individual species must be interpreted in the light of dramatic shifts in the state of the physical environment and ecosystem of the North Sea.

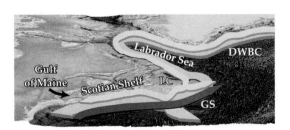

Figure 5.5 The major circulation features of the northwest Atlantic. LC = Labrador Current, GS = Gulf Stream, DWBC = Deep Western Boundary Current. Land is indicated by shades of green, and the sea floor is indicated by shades of grey (light grey on the continental shelf, dark grey in deeper areas). (See Plate 5.)

5.3 Climate variability and zooplankton populations in the northwest Atlantic

The relationship between the NAO and the weather over the Northwest Atlantic Shelf is less clear than in the North Sea. In maps of the correlation between the NAO and atmospheric properties such as wind speed or SLP, the Gulf of Maine/Scotian Shelf region appears as an area of low or zero correlation (Hurrell 1995; Reid and Planque 1999). However, the northwest Atlantic Shelf lies near the intersection of two major current systems (Fig. 5.5), the Labrador Current and the Gulf Stream, both of which are influenced by the NAO (Dickson *et al.*

1996; Taylor and Stephens 1998; see also Chapter 2). Thus, any effects that the NAO has on ecosystems of the northwest Atlantic Shelf are likely to be mediated by oceanographic processes, specifically, the dynamical interaction of the subpolar and subtropical gyres.

The Gulf of Maine/Scotian Shelf region, including Georges Bank is one of the most heavily studied areas in the northwest Atlantic. The US National Marine Fisheries Service and the Canadian Department of Fisheries and Oceans have conducted CPR surveys throughout the region since the 1960s. These studies have followed the same protocols as the SAHFOS surveys, but the temporal and spatial coverage of the western surveys is somewhat limited.

As in the eastern Atlantic, much of the analysis of the CPR data has focused on *C. finmarchicus*. Jossi and Goulet (1993) were the first to note a trend towards increased abundance of *C. finmarchicus* from 1961 to 1989 in the Gulf of Maine (Fig. 5.6(c)). They noted that this trend ran counter to the pattern for this species in the North Sea. After Fromentin and Planque (1996) established the influence that the NAO can have on *C. finmarchicus*, Conversi *et al.* (2001) reanalysed the CPR time-series from the Gulf of Maine. They specifically looked for a connection between the *C. finmarchicus* and NAO index

time-series and found a positive cross-correlation with the NAO at a lag of 4 years. Furthermore, Conversi *et al.* (2001) established cross-correlations between the NAO and SST at a lag of 2 years and between SST and *C. finmarchicus* at a lag of 2 years. These results were supported by Greene and Pershing (2001) who linked changes in *C. finmarchicus* abundance to the temperature in the bottom waters (150–250 m) in the Gulf of Maine/Scotian Shelf region with changes in slope water circulation patterns following the NAO. Both studies suggest a chain of events that starts with a phase shift in the NAO, is followed next by hydrographic changes in the Gulf of Maine/Scotian Shelf region, and ends with a response in the *C. finmarchicus* population.

5.3.1 Environmental conditions and the NAO

Both Conversi *et al.* (2001) and Greene and Pershing (2001) found a correlation between *C. finmarchicus* abundance in the Gulf of Maine and a measure of regional temperature. These results raise two questions: what processes account for the *C. finmarchicus*–temperature correlation, and which temperature measure is a better proxy for these processes? Addressing these questions requires a

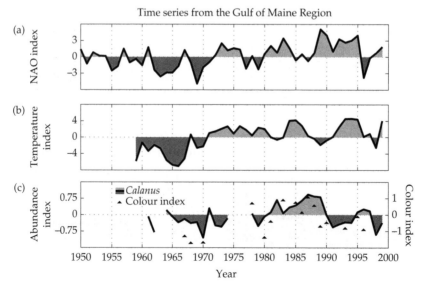

Figure 5.6 The three time series analysed from the Gulf of Maine Region. (a) The winter NAO time-series, (b) The first principal component of the regional temperature, and (c) *C. finmarchicus* abundance index (line and the colour index (triangles). The shading underneath the curve indicates whether the anomalies are positive (light) or negative (dark).

more detailed understanding of the physical and climatic environment of the northwest Atlantic.

Petrie and Drinkwater (1993) established an idea that is key to connecting physical changes in the Gulf of Maine to the NAO. In their analysis of temperature and salinity data from the Gulf of Maine and Scotian Shelf, they found that the amplitude of the interannual temperature oscillations increased with depth and was especially strong below 100 m. When searching for a mechanism, they found little correlation between water temperatures and interannual differences in winter heat flux. Instead, the variability was related to the volume transport of water from the Labrador Sea around the Tail of the Grand Banks. When the transport is high, the waters below 100 m in this region are cooler, and when the transport is reduced, these waters are warmer.

Because of the strong influence of the NAO on winter conditions over the Labrador Sea, it seems reasonable to connect the mechanism proposed by Petrie and Drinkwater (1993) to the NAO. Recent observations following the dramatic drop in the NAO during 1996 have solidified our understanding of the connection between the NAO and physical and biological processes in the Gulf of Maine. Prior to 1996, the bottom water found in the deep basins of the Gulf of Maine and Scotian Shelf was derived from relatively warm and saline Atlantic Temperate Slope Water (ATSW; Greene and Pershing 2001). This water mass extended northeast to the Laurentian Channel during much of the 1980s and 1990s (MERCINA 2001; K. Drinkwater *personal communication*) (Fig. 5.7(a)). Following the drop in the NAO during the winter of 1996, the front separating ATSW from the cooler, fresher Labrador Subarctic Slope Water (LSSW) was pushed to the southwest as the transport of LSSW around the Tail of the Bank increased. By the end of 1998, the LSSW had reached the mid-Atlantic Bight and had intruded into the deep basins on the shelf (MERCINA 2001; K. Drinkwater *personal communication*) (Fig. 5.7(b)). The subsequent years saw a return to positive NAO conditions, and the front between the two water masses returned to the northeast.

The changes in the distribution of the Slope Water masses described above reflect a large-scale reorganization of the distribution of water masses and currents in the northwest Atlantic. As a proxy

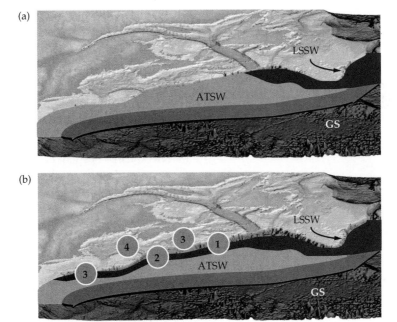

Figure 5.7 The distribution of LSSW, ATSW, and the position of the Gulf Stream (GS) during the (a) warm and (b) cold states of the CSWS. The numbers in (b) show the first observations of LSSW following the 1996 drop in the NAO: (1) = September 1997, (2) = January 1998, (3) = February 1998, (4) = August 1998. Land is indicated by shades of green, and the sea floor is indicated by shades of grey (light grey on the continental shelf, dark grey in deeper areas). (See Plate 6.)

for the modal state of this 'Coupled Slope Water System' (CSWS; MERCINA 2001), we developed the Regional Slope Water Temperature Index (RSWT Index), which is the dominant mode from a principle components analysis of the temperature between 150 and 250 m in eight regions in the Gulf of Maine, Scotian Shelf, and over the continental slope (MERCINA 2001; Pershing 2001). The RSWT Index has its strongest cross-correlation with the NAO at a lag of 1 year, although the correlation at lags of 0 and 2 years are also high (MERCINA 2001; Pershing in press; Greene and Pershing 2003) (Fig. 5.8(a)).

From the description above, the deep-basin temperatures that Greene and Pershing (2001) linked to *C. finmarchicus* variability in the Gulf of Maine should be strongly influenced by the state of the CSWS. Furthermore, the strong winter mixing in the northwest Atlantic implies that the SSTs analysed by Conversi *et al.* (2001) are also influenced by the

CSWS. The cold, continental air masses that pass over the northwest Atlantic lead to large heat fluxes out of the surface waters, creating cold, dense waters, and deep winter mixed layers (Brown and Beardsley 1978; Dickson *et al.* 1996). The formation of Labrador Sea Water is an extreme example of this phenomenon. A consequence of this winter mixing is that variability in winter temperature, even in the surface waters, is strongly dependent on the temperature at the base of the mixed layer (Petrie and Drinkwater 1993).

An analysis of the relationship between *C. finmarchicus* abundance in the Gulf of Maine and the state of the CSWS confirmed the results of Conversi *et al.* (2001) and Greene and Pershing (2001). From 1961 to 1999, *C. finmarchicus* abundance in the Gulf of Maine is strongly correlated with the state of the CSWS, as indicated by the RSWT Index. A cross-correlation analysis suggests that the copepod population increases after a shift in the CSWS from cold to warm, with the strongest correlation at a lag of 3 years (Fig. 5.8(b)). However, *C. finmarchicus* abundance declined in the same year as the dramatic change in the state of the CSWS that occurred in 1997.

As in the eastern Atlantic, the NAO's value in predicting changes in the abundance of *C. finmarchicus* in the Gulf of Maine decreased in the 1990s. There is no significant relationship between the NAO time series and *C. finmarchicus* abundance in the Gulf of Maine, if the data from the 1990s are incorporated. However, the relationship between copepods and the state of the CSWS still holds. The increasing trend in the NAO values over the time period studied by Conversi *et al.* (2001) and Greene and Pershing (2001) likely made the effect of the NAO on the copepods easier to detect. The NAO was predominantly positive during the 1990s, although not as high as in the late 1980s. Adding data from the 1990s to the analysis decreases the sensitivity of the correlation analysis to the large changes (1960s versus 1980s) and increases the focus on smaller, year-to-year fluctuations. Because the NAO's influence on the conditions in the Gulf of Maine is indirect—mediated by the CSWS, the Gulf of Maine should be less sensitive to subtle changes in the NAO.

As in the eastern Atlantic, the mechanisms that could account for the climate-induced fluctuations in *C. finmarchicus* fall into two categories: those involving changes in local biological processes and those involving circulation changes. There is

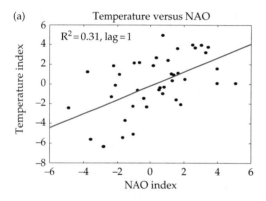

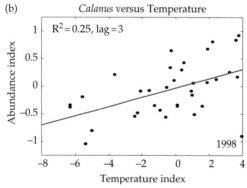

Figure 5.8 The significant relationships between (a) NAO and temperature and (b) *C. finmarchicus* identified by the cross-correlation analysis. All of the best-fit lines have slopes significantly different from zero.

evidence to support both possibilities. In addition to their different temperature and salinity signatures, the two water masses associated with the CSWS have distinct nutrient concentrations (Petrie and Yeats 2000). LSSW is a relatively young water mass that is formed by cooling of winter surface waters in the Labrador Sea; thus, this water mass has relatively low concentrations of nitrate and silicate (Petrie and Yeats 2000). ATSW forms by mixing between LSSW, Gulf Stream, and coastal water masses (Gatien 1976), and consequently, it has higher nutrient concentrations. The CPR Color Index data from the Gulf of Maine indicate higher phytoplankton standing stocks when ATSW is present (Fig. 5.6(c)), and this could lead to higher growth rates for the *C. finmarchicus* population. This hypothesis assumes that *C. finmarchicus* is always abundant on the northwest Atlantic Shelf; however, this species' preference for deep water during winter calls this assumption into question.

Variability in *C. finmarchicus* transport into the Gulf of Maine from different source areas is another hypothesis. Unlike the North Sea, the Gulf of Maine and Scotian Shelf have a few areas deeper than 250 m that provide habitat for diapausing *C. finmarchicus* (Sameoto and Herman 1990). Simulations of *C. finmarchicus* in climatological flow fields indicate that the summer population in the Gulf of Maine—the population to which the CPR is most sensitive—is composed of animals that were on the Scotian Shelf in late-winter and spring (Miller *et al.* 1998). Thus, the abundance observed by the CPR in the Gulf of Maine in a given year is determined by a combination of the size of the initial late-winter population on the Scotian Shelf and the growing conditions during the spring. Sameoto and Herman (1990) found high concentrations of diapausing *C. finmarchicus* in the Emerald Basin on the Scotian Shelf in 1985 and 1986 and hypothesized that these stocks seed the Scotian Shelf population. However, a longer study in this region suggests that conditions in 1985–86 were exceptional, and that in most years, the initial population on the Scotian Shelf is derived from the Slope Water (Head *et al.* 1999).

The scenario described above suggests two transport pathways that could affect the Gulf of Maine population: the transport of *C. finmarchicus* onto the Scotian Shelf, and the transport of *C. finmarchicus* from the Scotian Shelf to the Gulf of Maine. Modelling studies comparing the circulation over of the northwest Atlantic shelf between positive and negative NAO conditions found little change

in the shelf circulation (Loder *et al.* 2001). This suggests that any transport mechanism linking *C. finmarchicus* abundance in the Gulf of Maine and the state of the CSWS likely affects the coupling between the Scotian Shelf and Slope Waters. To get from the continental slope region on to the Scotian Shelf, *C. finmarchicus* must be advected across the shelf-break front. When LSSW is present in the region, the density gradient across this front is stronger (Pickart *et al.* 1999) and may be located further off the shelf (Head *et al.* 1999) thereby decreasing cross-shelf exchange.

In the North Sea, *C. finmarchicus*'s negative association with the NAO sets it apart from most other zooplankton populations in the region. In the Gulf of Maine, the abundance time-series of several other zooplankton species, including *Metridia lucens* and *Pseudocalanus* spp., had a similar pattern to that of *C. finmarchicus* with highs in the early 1960s and throughout most of the 1980s and lows during the late 1960s, early 1970s and a brief period in the early 1980s (Jossi and Goulet 1993). However, *C. finmarchicus* was the only species Jossi and Goulet (1993) found that had a significant trend; and despite the similarities among the time series, the *C. finmarchicus* time-series stood apart from the other species. Their work suggests a common mechanism that allows for increased zooplankton production in the Gulf of Maine during positive NAO/ATSW-dominated conditions. The differences in nutrient concentrations between the two Slope Water masses and the resulting changes in phytoplankton abundance could influence the abundance of a wide range of zooplankton species. This could account for the commonalities among the copepod time-series. If the modal shifts in the CSWS are associated with circulation changes, especially changes in the coupling between the shelf and slopewater regions, then *C. finmarchicus*'s sensitivity to advective changes would enhance this species' association with the CSWS.

5.4 Discussion

The examples presented above suggest that the effects of Atlantic climate variability on zooplankton populations fall into two categories. *C. finmarchicus* exemplifies the first class of effects, which we call *translations* (Pershing 2001). Translations involve movements of organisms from one place to another such as the advection of *C. finmarchicus* from the continental slope on to the shelf. These changes are based entirely on the physical changes produced by climate variability. We will refer to the other class as *indirect* effects. Indirect effects link a population to climate only through its effect on another species, such as climate-induced changes in its prey or predators.

Indirect effects have been recognized in many ecosystems and by many authors (Ottersen *et al.* 2001), however, translations are a unique class of climate effects that have not been distinguished previously. Zooplankton populations are very sensitive to translational effects because their horizontal movements are, by definition, determined by advection. *C. finmarchicus*, in particular, is especially susceptible to circulation changes because of its requirement for deep water during the fall and early winter diapause period. In both the Gulf of Maine and North Sea, a significant portion of the interannual variability of this species—and its association with the NAO—is associated with physical changes that could alter the coupling between shelf and deep water populations. The variability in *C. finmarchicus* abundance is also linked to circulation in other Atlantic regions. In the Barents Sea, the abundance of *C. finmarchicus* is closely tied to the supply of warm Atlantic Water flowing from the Norwegian Sea (Skjoldal *et al.* 1992; Helle and Pennington 1999; Dalpadado *et al.* 2003). When the flow increases, it advects *C. finmarchicus* and other Norwegian Sea zooplankton into the Barents Sea. Although the abundance of *C. finmarchicus* in the Barents Sea has not been formally correlated with the NAO, the transport of Norwegian Sea water is associated with positive NAO conditions (Dickson *et al.* 2000; Ottersen and Stenseth 2001). The abundance of *C. finmarchicus* around Iceland has been linked to changes in the distribution of water masses; however, it is unclear if the association is related to direct advection of *C. finmarchicus* in the water masses or differing patterns of primary productivity associated with them (Astthorsson and Gislason 1995).

Translations associated with frontal shifts have an especially strong effect on species with distributions tied to a specific geographic feature, provided the shift passes over the feature. The nesting grounds of marine birds and breeding grounds of seals are limited to a few locations. During frontal shifts such as those occurring during El Niño in the Pacific, these species are often the most heavily impacted (Graybill and Hodder 1985). Many commercially important fish species have similar restrictions, spawning only at a few locations,

principally submarine banks (Mann and Lazier 1996). The dramatic nature of these indirect effects results from the species sampling their environment from a fixed point. Shelf ecosystems are similarly fixed relative to adjacent circulation features; thus, they should be particularly sensitive to changes in circulation. This may explain the strong response of shelf populations such as those in the North Sea to climate variability, and the predominance of translational effects linking climate variability to population changes in the ocean.

The changes in the Gulf of Maine suggest that translations and indirect effects are independent and not exclusive of each other. Conditions favourable for the transport of *C. finmarchicus* into the Gulf occur concurrently with increased primary production, and both likely play a role in the connection between the variability in the *C. finmarchicus* population and the NAO. Distinguishing the relative influence of physical processes such as climate-induced translations versus biological processes is a fundamental problem in biological oceanography, one that increases the complexity of uncovering the mechanisms linking population changes to climate. Because translational effects are due entirely to physical processes, recognizing the importance of translational effects offers the potential to simplify these investigations.

A third class of effects known as *direct effects* was not obviously present in the examples discussed above, although this is a key class in other categorizations of population responses to climate (Ottersen *et al.* 2001; see also Chapter 1). Direct effects include the influence of physical factors controlled by climate, temperature being the most obvious, on a population's vital rates. An example of this class is the association between the growth rate of larval cod and temperature in the Barents Sea (Ottersen *et al.* 2001). In the Baltic, salinity changes associated with the NAO appear to determine the proportion of oceanic versus freshwater zooplankton (Viitasalo *et al.* 1995; Hänninen *et al.* 2000); however, the community changes may result from an alteration in the inflow of Atlantic water (a potential translation) or differences in phytoplankton dynamics caused by buoyancy changes (an indirect effect) rather than the direct influence of salinity of zooplankton growth and survival.

5.4.1 Interannual versus longer timescales

The North Sea responds to the NAO on two main timescales. The longest scale involves the large regime shifts reported by Glover (1957) and Reid *et al.* (2001). These shifts are similar to the changes associated with the Russell Cycle (Cushing and Dickson 1976; Cushing 1982; see also Chapter 10). Although these interdecadal shifts can occur abruptly, at least for some species (Reid *et al.* 2001), many species exhibit a more gradual rise and fall between the extreme states. The long-term changes are likely driven by a combination of physical and biological mechanisms. Physical features such as the extension of the coastal jet described by Reid *et al.* (2001) and the decline in *C. finmarchicus*'s overwintering habitat described by Heath *et al.* (1999) likely require several years of consistent environmental conditions for their effects to be detected. Superimposed on the interdecadal changes in the North Sea are interannual fluctuations. These changes involve year-to-year differences in the circulation field or the timing or magnitude of spring stratification. The physical changes lead to changes in the abundance of various species, but are rarely strong enough to lead to regime shifts in any 1 year.

5.5 Conclusion

Climate variability, especially that associated with the NAO, is important in determining the distribution and abundance of zooplankton populations around the North Atlantic. These examples suggest that climate influences zooplankton populations in two main ways. Through its effect on ecosystem properties such as the timing or magnitude of the spring bloom, climate variability can cause changes in zooplankton populations. Effects of this kind, known as indirect effects, are likely present in all of the ecosystems reviewed, but the complexity of marine ecosystems can mask their influence. Climate variability can have a significant effect on ocean circulation patterns, and thus, the distribution of zooplankton. Effects of this kind, known as translations, depend solely on physical processes and thus, are easier to detect. Translations are especially important in shelf ecosystems and for the copepod *C. finmarchicus*.

There are still relatively few examples of the influence of climate on zooplankton populations. Partly, this is due to the rarity of biological time-series long enough to compare with the NAO pattern, but it is also influenced by our incomplete knowledge of the physical aspects of climate variability. Enhanced knowledge of ecosystem–climate interactions is a prerequisite for sustainable management of the marine resources in the North Atlantic.

PART III

Fish and seabird populations

Fish and seabirds are in many obvious ways very different vertebrate groups. While ectothermic fish are confined to the sea, marine birds are endotherms and spend much of their time in the air or on land as well as on or in the sea. The two chapters in this part, Chapter 6 dealing with fish and Chapter 7 with seabirds, thus reflect both differences and similarities with regards to impacts of climate. Climate influences both marine birds and fish directly through physiology as well as indirectly through affecting interactions with predators, prey, and competitors in addition to regulating suitable habitat (see Chapter 2 for background information on climate). Marine, especially coastal, ecosystems, typically include both birds and fish; birds often as top predators, fish at intermediate levels. Fluctuations in the atmospheric and oceanographic climate as well as changes introduced in the ecosystem by other causes may thus be expected to impact both fish and birds, but the responses may be quite different. The relation between seabirds and fish may be that of predator and prey (seabirds are typically piscivorous) or that of competitors.

On a more practical level—but nevertheless scientifically very important—the inclusion of these two chapters within the same section bring together two quite different traditions within the field of ecology—the marine based (fish ecology) and the more terrestrial based (seabird ecology) approaches: although the study of seabirds has obvious links to the field of marine biology, it brings with it the terrestrially-based ornithological scientific tradition.

In Chapter 6, Ottersen and co-workers discuss how climate fluctuations may influence North Atlantic fish populations through affecting growth patterns, spawning and reproduction, distribution and migration, abundance and recruitment, catchability and mortality (Community ecology of fish will be dealt with in Chapter 10, while Chapters 11 and 12 look at population dynamics of fish from a Pacific point of view). Partly due to their commercial importance, the literature on response of North Atlantic fish to climate fluctuations is extensive. The authors are thus able to draw upon examples covering many different species and populations. On the other hand, to keep within the book's space-limit, a more in-depth inspection is confined to a few different, but important fish species in the region; the demersal cod, the pelagic herring, and sardines and the anadromous salmon. (Some general principles and examples with regards to modelling marine ecosystems, and fish in particular, were given by Werner and co-authors in Chapter 3.)

When assessing the impacts of climate fluctuations on marine birds in the North Atlantic in Chapter 7, Durant and co-workers have far less material to build upon than what is available for fish. However, by linking in with examples from other parts of the world a comprehensive view of complex patterns is developed. Durant and co-workers review responses to climate fluctuations according to direct influences on reproduction, mortality, and energetics as well as more indirect impacts through temperature, currents, and fronts in the upper parts of the sea. (In Chapter 16 Lima and Jacsic review effects of El Niño on different groups of animals including seabirds.)

Many seabirds feed on fish, particularly small pelagics. Thus, climate conditions favourable for the fish may also favour the birds. However, not only fish abundance but also their availability to the birds as prey, especially their distance from nesting areas, is decisive. An example is the survival of first-year herring, which responds dramatically and quickly to changes in ocean climate (Chapter 7). This leads to alternating periods of high and low abundance, respectively, which again could explain similar trends in the Atlantic puffin's population. If the herring availability does not match the puffin requirement at the time of the rearing, this may easily cause a reduction of the chick survival.

Furthermore, for birds, which are located towards the top of the food chain, the relationship with climate is complicated because the biology of members of the lower levels of the food web also must be taken into account. For example, ambient temperatures favourable for the physiology of a bird and its main prey could at the same time be unfavourable for the food resources of the prey. Consequently, such a temperature level might become unfavourable for the bird if it implies a reduction of the prey's food resource.

Plate 4 Hypothesis explaining interannual and interdecadal variability in *C. finmarchicus* abundance in the North Sea developed by the ICOS study (Heath 1999; Heath *et al*. 1999). Southerly winds associated with negative NAO conditions tend to advect *C. finmarchicus* from an area west of Scotland (red outline). Winds with a more northerly component tend to advect animals from a region further east, in the Faroe–Shetland Channel (blue outline). The concentration of *C. finmarchicus* in the Faroe–Shetland Channel is characteristically higher than that further west; thus, negative NAO conditions should lead to more *C. finmarchicus* in the North Sea. However, this mechanism is complicated by the distribution of NSDW, which is believed to carry *C. finmarchicus* into the Faroe–Shetland Channel. When NSDW production is high, a typical condition during negative NAO winters, the flow of NSDW into the Faroe–Shetland Channel is higher and leads to greater abundance of *C. finmarchicus* in the source region (light blue patch). When the volume of NSDW is reduced, the flow of this water mass and *C. finmarchicus* into the Faroe–Shetland Channel is reduced.

Plate 5 The major circulation features of the northwest Atlantic. LC = Labrador Current, GS = Gulf Stream, DWBC = Deep Western Boundary Current. Land is indicated by shades of green, and the sea floor is indicated by shades of grey (light grey on the continental shelf, dark grey in deeper areas).

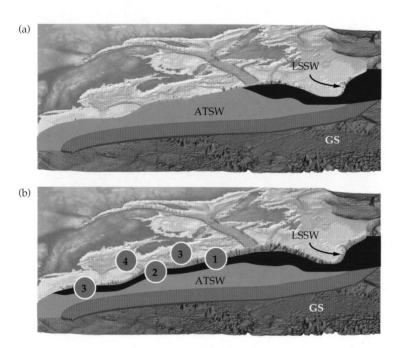

Plate 6 The distribution of LSSW, ATSW, and the position of the Gulf Stream (GS) during the (a) warm and (b) cold states of the CSWS. The numbers in (b) show the first observations of LSSW following the 1996 drop in the NAO: (1) = September 1997, (2) = January 1998, (3) = February 1998, (4) = August 1998. Land is indicated by shades of green, and the sea floor is indicated by shades of grey (light grey on the continental shelf, dark grey in deeper areas).

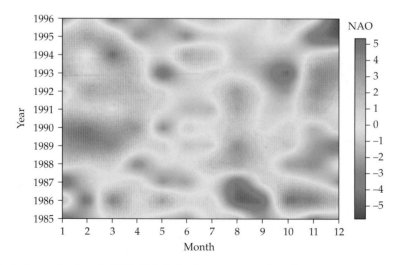

Plate 7 (a) NAO distribution covering the period 1985–96, indicating the strong positive phase during winters in the late 1980.

CHAPTER 6

The responses of fish populations to ocean climate fluctuations

Geir Ottersen, Jürgen Alheit, Ken Drinkwater, Kevin Friedland, Eberhard Hagen, and Nils Chr. Stenseth

6.1 Introduction

As early as the 1870s Spencer Fullerton Baird, a fisheries biologist and the first US Fish Commissioner, recognized the importance of the environment for fluctuations in fish stocks. His original programme for fisheries research therefore included oceanographic and meteorological investigations (Kendall and Duker 1998). However, for several decades it was still widely held that the abundance of sea fish was boundless, and fluctuations in catches essentially were caused by variations in migratory patterns. The innovative thinking of the Norwegian fisheries biologist Johan Hjort contributed to a shift of focus in fishery research. Hjort (1914) realized that most fish populations consist of several year classes, and their abundance was by and large determined early in life. Differences in the sizes of year classes were further considered responsible for most variations in abundance of fish populations (Kendall and Duker 1998).

Around this time two other pioneering Norwegian scientists, Bjørn Helland-Hansen (meteorologist) and Fridtjof Nansen (physical oceanographer and zoologist, although primarily known for his Arctic expeditions), initiated a more systematic search for relations between physical conditions and different aspects of fish stocks. In 1909 they wrote:

'It is to be expected that variations in the physical conditions of the sea have great influence upon the biological conditions of the various species of fishes living in the sea, and it might therefore also be expected that such variations are the primary cause of the great and hitherto unaccountable fluctuations in the fisheries. It is therefore obvious that it would be of very great importance, not only scientifically but also practically, if the relation between the variations in the physical conditions of the sea and the variations in the biological conditions of the various food fishes could be discovered' (Helland-Hansen and Nansen 1909).

How does the physical environment affect fish and shellfish? This occurs both as a direct response to ocean–climate fluctuations, as well as to changes in their biological environment (including predators, prey, species interactions and disease) and fishing pressures. While this multiple forcing sometimes makes it difficult to establish unequivocal linkages between changes in the physical environment and the response of fish or shellfish stocks, some climate effects are clear (e.g. see reviews by Cushing and Dickson 1976; Bakun et al. 1982; Cushing 1982; Sissenwine 1984; Shepherd et al. 1984; Sharp 1987; Drinkwater et al. 2003).

Direct physiological effects include metabolic and reproductive processes. Fish often appear to seek specific temperatures and salinities. Climate variability may produce sub-optimal conditions, especially for those individuals residing at the boundaries of their species distribution. As these fish move towards their preferred hydrographic conditions, this may lead to distribution changes of

DIRECT PHYSIOLOGICAL IMPACTS - suboptimal conditions for performance especially at the margin of the distribution feeding rate, digestion, reproduction

the species. If fish remain in the sub-optimal conditions, performance is reduced and the fish become weakened. This in turn can result in increased mortality rate by being less able to avoid predation or through direct starvation. Reproduction is typically temperature dependent with gonad development occurring more quickly under warm conditions. Thus, temperature determines the time of spawning with higher temperatures resulting in earlier spawning.

Climate variability may influence the abundance of catchable fish, principally through effects on recruitment (how many young survive sufficiently long to enter the segment of the stock being exposed to the fishery). The physical environment also affects feeding rates and competition through favouring one or another species, as well as the abundance, quality, size, timing, spatial distribution, and concentration of food. Climate also affects predation through its influences on the abundance and distribution of predators. Fish diseases leading to either a weakened state or death may also be environmentally triggered. Particular temperature ranges may, for instance, be more conductive to allowing disease outbreaks.

Besides temperature and salinity conditions, fish are also influenced by mixing and transport processes in the ocean. For example, mixing can affect the timing and the size of the spring bloom of phytoplankton through promoting nutrient replenishment of the surface layers and influencing vertical stratification of the water column. Mixing also influences the encounter rate between larvae and their prey organisms. (Chapters 4 and 5 dealt with phytoplankton and zooplankton, respectively.) Ichthyoplankton (fish eggs and larvae) can be dispersed by the currents, which may carry them into or away from either nursery areas, regions of good food production, optimal temperature or salinity conditions, or a combination thereof. Transport processes may ultimately determine whether individuals are lost to the original population.

Individual fish and populations undoubtedly experience climate locally and regionally, through temperature, wind, and currents as discussed above. Nevertheless, such meteorological and oceanographical features are often governed by phenomena extending over much larger areas. Interaction between the ocean and atmosphere may form dynamic systems (see, e.g. Chapter 2), exhibiting complex patterns of variation, which may profoundly influence ecological processes in many ways (see, e.g. Chapter 1).

Understanding how climate variability affects fish and fisheries and how the effects differ between species is of paramount importance when trying to predict the potential impacts of anthropogenically introduced climate change. In the following, we first show, through examples, how climate variability may influence North Atlantic fish populations through affecting spawning and reproduction, abundance and recruitment, individual growth rates, distribution and migration, natural mortality, catchability and availability to fisheries. Although, they are principally derived from the northwest Atlantic Ocean, the responses are representative of many species around the world. Building upon this summary account, we provide a more thorough overview of the impact of climate on some of the most important fish species in the North Atlantic; the demersal cod, the pelagic herring and sardines, and the anadromous salmon.

6.2 Main processes by which climate may influence fish

6.2.1 Spawning and reproduction

The physical environment affects the reproductive cycle of fish. For example, the age of sexual maturity of certain fish species is influenced by ambient temperatures. Atlantic cod off Labrador and the northern Grand Bank mature at age 7 and in the northern Gulf of St Lawrence and the eastern Scotian Shelf at age 6, while in the warmer waters off southwest Nova Scotia and on Georges Bank they mature at 3.5 and 2 years, respectively (Drinkwater 1999). On the eastern side of the Atlantic the pattern is similar. In the Barents Sea, cod start to spawn at ages 6–7, while coastal cod in the warmer fjords of southern Norway reach maturity at age 2–3 years.

Spawning times may also be influenced by temperature. Low temperatures typically result in delayed spawning through slow gonad development as has been observed in Atlantic cod on the northern Grand Bank (Hutchings and Myers 1994). While high temperatures promote gonad development resulting in earlier spawning, the relationship between temperature at the spawning site and time of spawning depends on local hydrography and fish distribution. In contrast to the positive relationship between local temperatures and time of spawning on the northern Grand Banks, cold temperatures lead to earlier spawning of cod off southern

Newfoundland (Hutchings and Myers 1994). However, these fish reside in warm offshore waters and move onto St Pierre Bank prior to spawning. In very cold years on the Bank, they appear to delay migration onto the Bank thereby remaining in the warm offshore waters longer, resulting in faster gonad development and an earlier readiness to spawn. The time of peak spawning of Georges Bank cod also has been found to vary with temperature (Serchuk *et al.* 1994). On the other hand, the timing of Arcto-Norwegian cod spawning is not significantly affected by temperature variability (Cushing 1969; Ellertsen *et al.* 1989).

Temperature-dependent spawning is not limited to cod. In the early 1990s, extremely low temperatures during spring off Newfoundland delayed capelin spawning by over a month, which lead to slow growth rates and poor condition (Nakashima 1996). Marak and Livingston (1970) found that a 1.5–2 °C temperature change produced a difference in spawning time of haddock on Georges Bank by a month with earlier spawning and a longer duration in warm years. From studies in the Baie des Chaleurs within the Gulf of St Lawrence, spawning of the giant scallop (*Placopecten magellanicus*) has been found to be associated with rapid temperature changes caused by wind-induced upwelling (Bonardelli *et al.* 1996).

Miller *et al.* (1995) found that 52% and 70% of the seasonal variance of egg and larval size at hatch, respectively, of Atlantic cod on the Scotian Shelf, are temperature dependent over the range 2–14 °C with size decreasing as temperature increases. A similar dependence of egg size on temperature was found by Ware (1977) for Atlantic mackerel (*Scomber scombrus*) in the Gulf of St Lawrence. This is believed to concur in part an ecological advantage in order to match available prey size at the time of hatching, as the latter is also temperature dependent (Ware 1977).

Incubation times of cod eggs are also temperature dependent. Page and Frank (1989) found they varied from 8 days at 14 °C to 42 days at 1 °C for Atlantic cod on the Scotian Shelf. Thus, eggs in colder water are more vulnerable to predation due to longer exposure time and may therefore experience lower survival.

Turbulence in the rearing environment furthermore plays a role in feeding throughout the larval stage. Various studies have shown that moderate turbulence enables increased feeding rates (e.g. Sundby and Fossum 1990; MacKenzie and Leggett 1991; MacKenzie *et al.* 1994; Muelbert *et al.*

1994; Fiksen *et al.* 1998; Werner *et al.* 2001). Muelbert *et al.* (1994) modelled Atlantic herring feeding with and without turbulence, and found high incidence of starvation in stratified, non-turbulent cases. In turbulent conditions, zooplankton and icthyoplankton are moved past each other, and the occurrence of predator–prey encounter is increased (Rothschild and Osborn 1988). The threshold food concentration for a larva to survive thus decreases with increased contact rates (Muelbert *et al.* 1994). MacKenzie *et al.* (1994) found a dome-shaped relationship between turbulence and feeding success, while prey encounter rates increase with turbulence, the probability of successful pursuit is limited in highly turbulent conditions. The turbulence generated by tidal forces in most coastal areas tends to be of the correct magnitude to enhance feeding (Muelbert *et al.* 1994). Large storms or high river run-off may, however, periodically result in detrimental levels of turbulence into the system (MacKenzie *et al.* 1994).

6.2.2 Abundance and recruitment

Trying to understand what regulates recruitment variability has been the number one issue in fisheries science since the early twentieth century. Evidence of changes in fish abundance in the absence of fishing suggests the likelihood of possible environmental causes (O'Connell and Tunnicliffe 2001; see also Chapter 3). Since the advent of intensive fishing, it has become increasingly difficult to sort out the relative importance of fishing versus environment as the cause of recruitment variability. Still, recruitment levels have frequently been associated with variations in temperature during the first years of life of the fish (Drinkwater and Myers 1987; Ellertsen *et al.* 1989; Ottersen and Stenseth 2001).

The Norwegian spring-spawning stock of herring has undergone dramatic fluctuations during the twentieth century. Spawning stock biomass increased during the first 30 years, from a low level of about 2 million tons to more than 15 million tons in 1945. From about 1950 it decreased steadily until its collapse in the late 1960s. The stock has increased in biomass during the 1990s and in light of history still seems to be in a rebuilding state (Toresen and Østvedt 2000). These long-term fluctuations in spawning stock biomass seem to be caused by variation in survival of recruits. Toresen and Østvedt (2000) furthermore, showed that the long-term changes in spawning stock abundance are highly correlated with the long-term variations in the mean annual temperature of the Atlantic water masses flowing into the Barents Sea from the south. Recruitment is positively correlated with mean winter temperatures at the Kola section in the south central Barents Sea, suggesting that environmental factors govern the large-scale fluctuations in production for this herring stock.

Further south in the northeast Atlantic, the large pelagic tuna also exhibit recruitment patterns related to sea temperature fluctuations. The abundance of bluefin tuna in the northeast Atlantic increased in the period of warming between the 1920s and 1950s. During this period bluefin tuna, together with swordfish, appeared as far north as off the coasts of the Faroe Islands and Iceland (Cushing 1982). Santiago (1997) and Borja and Santiago (2001) examined the relationship between the North Atlantic Oscillation (NAO; see also Chapter 2) index and eastern bluefin and northern albacore tuna for the period 1969–95. For northern albacore, recruitment during high NAO years was approximately half that of low NAO years. The opposite occurred in the case of eastern bluefin, mean recruitment during high NAO situations being near twice that during low NAO conditions. Mejuto and de la Serna (1997) found a statistically significant relationship ($p < 0.05$) between North Atlantic swordfish year-class strength and the NAO. As in the case of northern albacore, high NAO years were associated with low recruitment; and low NAO years with high recruitment levels. (Chapter 11 will return to tuna, looking into the relationship between El Niño-Southern Oscillation (ENSO) and Pacific tuna fluctuations.)

American lobster landings in Canada and the United States increased steadily during the 1980s and into the 1990s to all-time historic highs in most regions. This is due primarily to higher recruitment rather than increased fishing effort (Drinkwater et al. 1996). Relationships between temperature and lobster landings had been established in several areas (i.e. from the Gulf of Maine to the Gulf of St Lawrence) prior to the large increase in landings showing higher landings during warm temperatures. This suggested that perhaps the recent high landings may have been due to a large-scale warming trend. However, examination of the data showed no such warming and using recent temperature data, the temperature–landing relationships were unable to predict any significant rise in landings during the 1980s and 1990s (Drinkwater et al.

1996). This is an example of a 'failed' relationship, one in which a linear regression between an environmental variable and fish or shellfish abundance was established only to find that it broke down when observations from later years became available. These can arise because the abundance is influenced by more than one factor or in some case the original correlations may have been spurious. In the former, one factor may be a dominant forcing mechanism for a time, only to be replaced later by another factor or factors.

Most of the Atlantic salmon that spawn in the rivers of eastern North America in summer, subsequently migrate to the Labrador Sea where they overwinter (Reddin and Shearer 1987). The young salmon—or 'smolts'—also travel to the Labrador Sea, where they reside until ready to return to the rivers. There is large variability in the numbers of salmon returning to the rivers of eastern Canada each year. The similarity in the interannual variability from different rivers over widely separated regions suggests that the numbers of returning salmon are most likely determined in the marine environment. A wintertime index of the areal extent of sea-surface temperatures (SSTs) (4–8 °C) in the Labrador Sea has been developed that shows a high positive correlation with the number of salmon returning to North America during the following spring and summer (Friedland et al. 1993; Reddin and Friedland 1993) (we will return to salmon later on in this chapter).

While most of the examples mentioned above have involved changes in temperature, there is also a transport-related effect on recruitment. Eggs and larvae are affected by currents. The 1987 haddock year class on eastern Georges Bank, which appeared to have been spawned normally in early spring, was located almost entirely in the Middle Atlantic Bight by June. This unusually large southwestward displacement was the result of an enhanced transport of water from the bank (Polochek et al. 1992). Recent improvements in numerical models of the currents over the continental shelves have allowed scientists to study the potential drift patterns of eggs and larvae (e.g. Hinrichsen et al. (2001) in the Baltic Sea and Ommundsen (2002) off Norway; a comprehensive list of biological transport models is provided in Chapter 3). Advection into unfavourable sites leads to reduced recruitment if the fish die or if they cannot make it back to the parent stock to reproduce (Sinclair 1988). One example of transport-related effects on recruitment involves

Gulf Stream rings off northeastern United States and eastern Canada. Large meanders in the Gulf Stream will sometimes pinch off and separate from the stream to form Gulf Stream rings or eddies. Eddies on the north side of the stream rotate clockwise and tend to trap warm Sargasso Sea water in their centre giving rise to the terminology 'warm-core' rings. The rings that approach the shelf often entrain large amounts of shelf water, transporting them off the shelf into the adjacent deeper slope water region. Greater numbers of Gulf Stream rings close to the continental shelf during the spawning or larval periods has been shown to lead to reduced recruitment in 15 of 17 groundfish stocks, including those of Atlantic cod, redfish, haddock, pollock, and yellowtail flounder (Myers and Drinkwater 1989). The leading hypothesis is that rings entrain shelf waters laden with eggs and larvae, transporting them off the shelf where conditions are less favourable. Indeed, Drinkwater *et al.* (2000) found redfish transported off the shelf were in poorer condition than those remaining on the shelf. Many die because of either lack of appropriate food or being more susceptible to predation, or in the case of those that make it to metamorphosis, because they cannot find the appropriate habitat. Death can also occur if they encounter temperatures that are too high, as observed in the case of cod larvae transported off Georges Bank (Colton 1959).

Interannual variability in sardine (*Sardina pilchardus*) recruitment off the northwest coast of the Iberian peninsula has been attributed to upwelling linked to wind conditions (Guisande *et al.* 2001). Moderate upwelling intensity prior to spawning seems to be more favourable than both lower and higher intensities. In both cases, the upper layer is strongly mixed, which might negatively affect primary production due to light limitation. Furthermore, Guisande *et al.* (2001) showed that sardine recruitment success was enhanced during low NAO index years with higher than normal winter–spring temperatures and onshore transport anomalies.

6.2.3 Growth

Environmental conditions have a marked effect on individual growth rates of many fish species. Since they are ectotherms, temperature is a key environmental factor for fish (Fry 1971). Individual growth in fish is the integrated result of a series of physiological processes (feeding, assimilation, metabolism,

transformation, and excretion) where rates are all controlled by temperature (Brett 1979; Michalsen *et al.* 1998). In addition to such direct effects, temperature affects growth of fish indirectly through influencing ecosystem productivity, changes in the length of the feeding season, effects on the distribution of the fish themselves, their prey, predators, and competitors.

Interactions between the factors affecting growth complicate attempts to understand the causes of interannual variation in growth in the wild. The interaction between the effect of temperature and food ration poses a particularly difficult problem. An increase in temperature will accelerate growth if food is not limiting but reduce growth if it is. Further, whether food is 'limiting' may depend on temperature. There is an optimum temperature for growth at any particular food ration. If temperature is below this optimum, then food is not limiting and an increase in temperature will accelerate growth. If temperature is above this optimum, then food is limiting and an increase in temperature will decelerate growth. That temperature preferences may depend on food availability further complicates the issue (Swain and Kramer 1995).

A workshop on the dynamics of growth in Atlantic cod (*Gadus morhua*), organized by the ICES/GLOBEC Cod and Climate Change Working Group, covered many aspects of cod growth, including the role of ambient temperatures for both inter- and intra-stock variability (Andersen *et al.* 2002). The fact that mean bottom temperatures account for 90% of the observed (10-fold) difference in growth rates between different Atlantic cod stocks in the North Atlantic (Brander 1994, 1995) shows how important environmental conditions are for growth of fish in the wild. Higher temperatures lead to faster growth rates.

Regional studies have shown similar results (Fleming 1960; Shackell *et al.* 1995). The fastest growing cod is found in the Irish Sea where a 4-year old fish is, on average, five times larger than one off Labrador and Newfoundland. Temperature not only accounts for differences in growth rates between cod stocks but also year-to-year changes in growth rates within a stock. Thus, sea temperature declines were responsible for approximately 50% of the observed decrease in size-at-age of Atlantic cod on the northeastern Scotian Shelf from the mid-1980s to the mid-1990s (Campana *et al.* 1995) and off Newfoundland (de Cárdenas 1996; Shelton *et al.* 1996). This is particularly important given that

50–75% of the declines in the spawning stock biomass of the Newfoundland, Gulf of St Lawrence and northeastern Scotian Shelf cod stocks during this period were due to reduced weight-at-age (Sinclair 1996). Fifty percent of the recent decline in weight-at-age of 4–8 year old northern cod can be accounted for by the weight at age 3, a finding that indicates how size at an early age may determine future fish production (Krohn and Kerr 1996). Similar conclusions were drawn by Michalsen *et al.* (1998) and Brander (2000).

Furthermore, the effect of temperature on growth changes with age. The maximum growth rate of cod decreases with age, and temperature has a greater influence on growth potential of juvenile cod compared to adults (Bjørnsson and Steinarsson 2000).

The effect of temperature on individual growth of cod is evident in, for example, the Arcto-Norwegian stock in the Barents Sea, where warm conditions lead to higher growth rates (Loeng *et al.* 1995; Michalsen *et al.* 1998; Ottersen and Loeng 2000). The same is true in lower latitudes and is observed, for example, in the case of North Sea cod (Brander 1995). However, the regions inhabited by cod generally have temperatures well below what is optimal for growth and reproduction when fed to saturation in a laboratory setting (Jobling 1988). Conditions other than optimum temperature must consequently be the determinant of the distribution of cods stocks throughout the North Atlantic (Sundby 1994). The fact that cod in the wild choose temperatures well below the optimal temperature for growth and preferred temperature during laboratory conditions is explained in ICES (1991) as a result of interaction with other species (both prey, competitors, and predators). Thermal conditions may be a correlate of a more immediate signal coming from other biotic or abiotic factors (Rose *et al.* 1994). A prime candidate is food availability (Lilly 1994). The high abundance of copepods and euphausids, which is the food for the early stages of cod, and the abundance of capelin (*Mallotus villosus*), being the main food for the juveniles and adults, may explain the high abundance of cod in the 'hostile' arctic-boreal part of the North Atlantic (Sundby 1994).

Low bottom temperatures resulted in decreased growth rates of adult American plaice (*Hippoglossoides platessoides*) on the Grand Banks during the 1980s (Brodie 1987; Brodie *et al.* 1993). Fifty percent of the interannual variations of growth of herring (*Clupea harengus harengus*) ages 3–7 in two separate Newfoundland bays were accounted for by the March to December water temperatures (Winters *et al.* 1986). Reduced length-at-age and weight-at-age for capelin of ages 3 and 4 off Newfoundland during the 1990s have been shown to be a direct response to low ocean temperatures (Nakashima 1996). For 2- and 3-year-old Barents Sea capelin, Gjøsæter and Loeng (1987) found a positive correlation between temperature and length-at-age. Both when comparing between geographical regions and between years, higher growth rates corresponded to higher temperatures. Recently, examining the mechanisms behind the above link more thoroughly (Orlova *et al.* 2002), found links between ambient temperatures, the abundance of zooplankton prey available, and capelin feeding patterns and condition.

Length of 0-group (5-month old) fish fluctuates in synchrony among several Barents sea species (such as herring, haddock, and cod; Loeng *et al.* 1995). Furthermore, length of all three species is positively correlated with sea temperature. This indicates that these species, having similar spawning and nursery grounds, respond in a similar manner to large-scale climate fluctuations. Also for many species in the Northwest Atlantic, including Atlantic cod, haddock (*Melanogrammus aeglefinus*) and winter flounder (*Pseudopleuronectes americanus*), growth rates of larvae have been found to be temperature dependent (Morse 1989).

Reduced growth rates at lower temperatures are, in part, due to changes in feeding rates. Laboratory experiments reported by McKenzie (1934, 1938) showed that Atlantic cod from the Bay of Fundy and Scotian Shelf ate well at temperatures within their normal tolerance range but ceased feeding at very high (>17°C) or very low temperatures (<0°C). Reduced feeding with subsequent weight loss at low temperatures for adult American plaice from the Grand Banks has also been measured in the laboratory by Morgan (1992). Besides lower feeding rates, reduced growth may also arise through delayed spawning (as discussed earlier in this chapter), initially causing a short growing season, and subsequently smaller size later in life. At too high a temperature the metabolic costs outweigh the energy gained through feeding.

Large-scale atmospheric climate variability, acting through both sea temperature and other regional climate variables is expected to influence fish growth. This is the case for growth of Canadian northern cod. Growth, as defined by the weight

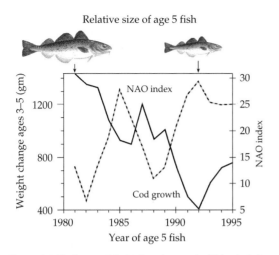

Figure 6.1 The impact of the NAO on the growth of Atlantic Cod off Labrador and Newfoundland, Northern cod. Time series of weight gain between age 3 and 5 (darker, full line) and 3-year average of the NAO index for the equivalent years (Drinkwater 2002; Drinkwater *et al.* 2003). The pictures of the cod show the relative size of 5-year old fish under best and worst NAO conditions.

gain between ages 3 and 5, is inversely related to the NAO index averaged over the 3 years the cod is growing. A high index, which produces cold ocean temperatures, strong winds, and much ice off Labrador and Newfoundland, results in slow growth (Fig. 6.1; Drinkwater 2002; Drinkwater *et al.* 2003).

6.2.4 Distribution and migration

Temperature is one of the primary factors, together with food availability and suitable spawning grounds, in determining the large-scale distribution pattern of fish. Because most fish species (and stocks) tend to prefer a specific temperature range (Coutant 1977; Scott 1982), long-term changes in temperature may lead to expansion or contraction of the distribution range of certain species. These changes are generally most evident near the northern or southern boundaries of the species range; warming results in a geographic shift northward and cooling draws species southward.

Several studies have documented distributional changes off West Greenland in response to fluctuations in sea temperatures. Jensen and Hansen (1931) and Hansen (1949) reviewed the response of West Greenland waters to their warming in the 1920s. It included (1) the introduction of new species that came from Iceland via the Irminger Current, (2) the

extension northward of warm-water species such as cod, herring, and halibut, (3) fish such as herring and redfish being found in areas outside their tradition grounds, and (4) overwintering in northerly regions of some species. In addition, the southward migration of white whales in autumn was delayed and the abundance of cold-water species such as capelin increased in the north while decreasing in the south.

Capelin off Newfoundland and Labrador, spread southward as far as the Bay of Fundy when temperatures declined south of Newfoundland in the mid-1960s and retracted northward as temperatures rose in the 1970s (Tibbo and Humphreys 1966; Colton 1972; Frank *et al.* 1996). During cooling in the later half of the 1980s and into the 1990s, capelin again extended their range, eastward to Flemish Cap and southward onto the northeastern Scotian Shelf off Nova Scotia (Frank *et al.* 1996; Nakashima 1996). For example, small quantities of capelin began to appear in the groundfish trawl surveys on the Scotian Shelf in the mid-1980s and since then numbers have increased profoundly (Frank *et al.* 1996). Initially only adult capelin were caught but later juveniles appeared, suggesting capelin were successfully spawning.

This shift appears to have been part of a larger-scale ecosystem change. While capelin were spreading onto the Scotian Shelf, Arctic cod (*Boreogadus saida*) were moving southward. The latter is a small cold-water pelagic fish, whose primary grounds have traditionally been the Labrador Shelf stretching southward to northern Newfoundland. In the late 1980s and early 1990s as the temperatures decreased, Arctic cod pushed southward onto the Grand Banks and into the Gulf of St Lawrence in large numbers. This southward movement was suggested by Gomes *et al.* (1995) and substantiated through annual autumn groundfish surveys off Newfoundland through the 1990s (Lilly *et al.* 1994). Since the mid-1990s this region has experienced warming and the Arctic cod abundance declined and appears to have retreated northward (Dalley *et al.* 2000). Southward shifts in the distribution of groundfish species off Newfoundland and Labrador during the late 1980s and 1990s have also been documented; for Atlantic cod by deYoung and Rose (1993), Taggart *et al.* (1994), and Rose *et al.* (1994), and for fish assemblages consisting of both commercial (e.g. Greenland halibut (*Reinhardtius hippoglossoides*) and American plaice) as well as non-commercial species by Gomes *et al.* (1995).

Changes in distribution were also observed during the warming trend in the 1940s in the Gulf of

Maine, which produced a northward shift in abundance and distribution of Atlantic mackerel, American lobster (*Homarus americanus*), yellowtail flounder (*Limanda ferruginea*), Atlantic menhaden (*Brevoortia tyrannus*) and whiting (*Merluccius bilinearis*) as well as the range extension of more southern species such as the green crab (*Carcinus maenas*) (Taylor *et al.* 1957). In contrast, during the cooling trend in the Gulf of Maine from 1953 to 1967, Colton (1972) noted that several species retracted southward, including American plaice and butterfish (*Peprilus triancanthus*). Mountain and Murawski (1992) have documented north–south shifts in distribution as a function of temperature within the Gulf of Maine. The weighted-mean catch for 14 out of 30 stocks investigated from groundfish surveys conducted during 1968–89 was found to increase northward with increasing temperature. This relationship was found to be strongest for Atlantic mackerel, Atlantic herring, and silver hake (*M. bilinearis*).

Temperature-related displacement of Arcto-Norwegian cod has been reported on an interannual timescale as well as on both small and large spatial scales. In periods of warm climate, the cod distribution is extended towards the lower-temperature eastern and northern parts of the Barents Sea as compared to colder periods when the fish tend to concentrate in the south-western region (Ottersen *et al.* 1998).

Many species that migrate appear to use environmental conditions as cues. For example, Atlantic mackerel migrate from their overwintering grounds off the Middle Atlantic Bight across the Gulf of Maine along the Atlantic coast of Nova Scotia and into the Gulf of St Lawrence. Their arrival at any location along their route requires temperatures warmer than 7–8 °C (Sette 1950). Similarly, the north–south migrations of American shad along the Atlantic coast of North America are regulated by the seasonal movement of waters in the 13–18 °C range (Leggett and Whitney 1972). April SSTs and ice conditions in the southern Gulf of St Lawrence determine the average arrival time of Atlantic herring on their spawning grounds (Lauzier and Tibbo 1965; Messieh 1986). Ice conditions also appear to control the arrival time in spring of Atlantic cod onto the Magdalen Shallows in the Gulf of St Lawrence (Sinclair and Currie 1994). This is in contrast to their return migration in the autumn to the deep waters in the Laurentian Channel south of Cabot Strait that appears to be unrelated to environmental

conditions. The timing and geographical distribution of Atlantic salmon (*Salma salar*) along the Newfoundland and Labrador coasts have been shown to depend upon the appearance of the 4° C water (Narayanan *et al.* 1995). Salmon arrive earlier during warmer years.

6.2.5 Natural mortality

Direct mortality of adult fish caused by environmental conditions is rare, however, some documented cases suggest that it does occur. Drastic declines in the spawning stock biomass of sole (*Solea vulgaris*) in the North Sea in 1964 was caused by a high natural mortality in the winter of 1963–64 when the water temperature was very low (ICES 1979; 2000). A similar event occurred in 1995/96 (ICES 1997).

Dead cod observed off Newfoundland have been attributed to very low temperatures (Templeman 1965; pers. comm. George Lilly, NW Atl. Fish. Centre, St. John's, Newfoundland). While cod normally can survive quite low temperatures, and indeed have antifreeze proteins in their blood (Goddard *et al.* 1999), rapid exposure to very cold temperatures or the presence of ice crystals in the water may lead to death. Another example of expected environmental induced mortality was that of tilefish (*Lopholatilus chamaeleonticeps*) that live on the continental slope off New York. They burrow into the ocean floor and do not move in response to temperature variability. In the early 1880s, millions of dead tilefish were found floating in the vicinity of the shelfbreak. Marsh *et al.* (1999) suggest that increased transport of the cold Labrador current leading to a rapid temperature change of 4–6°C caused the death of the tilefish.

6.2.6 Catchability and availability

Climate can affect fisheries through influences upon availability and catchability. Availability is how many fish there are for the fishermen to catch, and catchability is how difficult it is for the fishermen to catch them. Availability and catchability depend upon the total abundance of fish as well as when and how they are distributed. For example, if migrating fish such as herring are abundant, but do not arrive on the fishing grounds during the time the fishery is permitted to fish, then the availability is low. Also, if fish are abundant but widely distributed such that concentrations are low, then catchability is also likely to be low. Similarly, if the fish are not very abundant but highly concentrated, then catch rates in those areas containing fish are good. The environment can affect catchability, for example, when temperatures are low, lobster are known to move slowly, reducing the potential for encountering lobster traps, and hence, cause reduced catchability (McLeese and Wilder 1958). The landings in cod traps off Newfoundland have also been shown to depend upon temperature variability (Templeman 1966). If the traps are located in waters that are too cold, catches are low. Only when the temperatures are warm enough do catches increase. Similar results were observed in cod traps from Quebec off the north shore of the Gulf of St Lawrence (Rose and Leggett 1988).

Catch rates in the groundfish surveys conducted by the Canadian Government have also been shown to be influenced by environment conditions. For example, the abundance of age 4 Atlantic cod on the eastern Scotian Shelf caught in the annual spring surveys is greater during years when a larger proportion of ocean bottom is covered by so-called Cold Intermediate Layer (CIL) waters, that is, temperatures less the 5 °C and salinities of 32–33.5 percentile (Smith *et al.* 1991). This may result from cod seeking preferred conditions or an inability to avoid the trawls due to reduced swimming speeds at lower temperatures (Smith and Page 1996).

The combination of politically restricted fisheries patterns and environmental variability may have pronounced impacts on availability. For instance, most of the Barents Sea is under either Russian or Norwegian jurisdiction, but there is a relatively small region of international waters in the centre. This area is aptly named the Loophole and at times there is an extensive fishing activity there. Most of the international fisheries in the Loophole take place in the southern part, where in warmer years fish of a variety of species and all sizes are found throughout the year. On the other hand, in colder years there may be hardly any fish in the area for prolonged periods, the fish then having moved further south and west. The reason for this pattern lies in the oceanography of the region. The western Barents Sea is always slightly warmer than the eastern part due to Atlantic water entering from the southwest. While the shift towards colder water masses is gradual as one moves eastwards, the north–south temperature gradient is much sharper in the Polar Front region, which separates Atlantic and Arctic water masses. The southern part of the Loophole lies near the Polar Front so relatively small movements of the Front may greatly impact water temperatures in the region. These movements of water masses are most pronounced between warm and cold years in the Barents Sea as a whole, but also occur on a timescale of weeks. The consequence of colder water masses entering the Loophole is that most of the fish normally in the area become unavailable to the international fishing fleets.

6.3 Impact of climate on cod recruitment and abundance

Having described how the ocean environment may affect a broad range of fish, in this and the following two sections we discuss in more detail the role of ocean climate on three of the major commercial species in the North Atlantic.

Cod (*Gadus morhua*) fisheries can be traced back to the Middle Ages and even before (see Kurlansky 1997). Cod is found all around the North Atlantic, from the Bay of Biscay to the Barents Sea on the eastern side of the Atlantic and from Cape Hatteras, North Carolina, to the southern part of Baffin Island and the west coast of Greenland on the western side (Fig. 6.2). Stocks have been heavily exploited during the last three decades leading to a general decline in numbers. However, already at the beginning of the last century the profound fluctuations in the yearly landings of cod were well known and caused considerable puzzlement and problems for fishermen. Many (imaginative) causes for this great variability were suggested, but little was known at this time about the reasons for fluctuations in fish stock abundance (Øiestad 1994).

A large body of literature now exists relating ocean climate to cod recruitment and a number of different mechanisms have been suggested. Table 6.1 provides an overview of some of the mechanisms acting during early lifestages that may affect interannual recruitment variability. Following spawning, cod eggs, and later young stages are generally distributed in the upper mixed layer before they settle towards the bottom as half-year olds. The main part of the variability in year-class strength of cod is determined during these early life stages (Myers and Cadigan 1993; Sundby *et al.* 1989), stages during which climatic conditions have a

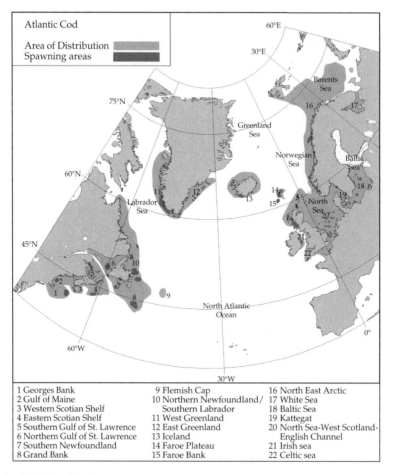

Figure 6.2 Distribution of Atlantic cod stocks.

1 Georges Bank	9 Flemish Cap	16 North East Arctic
2 Gulf of Maine	10 Northern Newfoundland/	17 White Sea
3 Western Scotian Shelf	Southern Labrador	18 Baltic Sea
4 Eastern Scotian Shelf	11 West Greenland	19 Kattegat
5 Southern Gulf of St. Lawrence	12 East Greenland	20 North Sea-West Scotland-
6 Northern Gulf of St. Lawrence	13 Iceland	English Channel
7 Southern Newfoundland	14 Faroe Plateau	21 Irish sea
8 Grand Bank	15 Faroe Bank	22 Celtic sea

great effect (Cushing 1966; Ellertsen *et al.* 1989; deYoung and Rose 1993; Dickson and Brander 1993; Ottersen and Sundby 1995; Drinkwater *et al.* 2003).

While a number of environmental factors have been demonstrated to influence recruitment related processes of one stock or the other, the most thoroughly studied is sea temperature. Temperature affects cod recruitment through many different processes. However, the relative importance of the processes and the role of temperature varies from region to region. This leads to both the degree and the direction in which the accumulated effects of temperature influence recruitment differing highly between the various cod stocks. This is not surprising, considering that the distribution of the species encompasses a wide temperature range, with stocks distributed in waters with annual mean temperatures (at 100 m depth) varying from about 0 to 12 °C.

It should be noted that many of the initially statistically significant correlations between environment and fish recruitment have been shown to break down when tested against new data (Myers 1998). This could be taken as evidence for climate fluctuations having no influence on recruitment, or it may also be that the nature of the relationships in many cases varies over time and is not fully understood. However, for populations close to the northern limit of a species range temperature–recruitment relations seem to be consistently positive for fish in general (Myers 1998).

The closer temperature–recruitment linkage observed in cold-water stocks may well be an effect of temperature-growth-survival links that

Table 6.1 Mechanisms during early life stages considered to affect interannual recruitment variability of North Atlantic cod stocks (updated from Ottersen 1996)

Stock	Physical process/feature	Biological feature	Reference	Hypothesis
Arcto-Norwegian	Temperature	Larval feeding success linked to syncrony of cod larvae and *Calanus finmarchicus* nauplii production	Ellertsen *et al.* (1989) Ottersen and Sundby (1995) Sundby (2000) Sirabella *et al.* (2001)	Match–mismatch (Cushing 1969, 1974, 1982, 1990)
	Temperature	Early survival related to growth rate	Ottersen and Loeng (2000)	'Bigger is better' (Gulland 1965; Ware 1975)
	Summer turbulent mixing by wind and tide	Larval feeding success related to contact rate between cod larvae and *C. finmarchicus* nauplii	Sundby and Fossum (1990) Sundby *et al.* (1994)	Encounter rate (Rothschild and Osborn 1988)
	Wind direction, residual currents, advection	Larval feeding success	Ottersen and Sundby (1995) Ottersen and Stenseth (2001) Sundby (2000)	Member–vagrant (Sinclair and Iles 1989)
	Wind direction, residual currents, advection	Reduced predation risk connected to spreading of eggs and larvae	Ottersen and Sundby (1995)	Member–vagrant (Sinclair and Iles 1989)
North Sea	Spring vertical stratification	Larval feeding success linked to timing and abundance of *C. finmarchicus* nauplii	Cushing (1984)	Match–mismatch (Cushing 1969, 1974, 1982, 1990)
	Summer heat content/ Spring turbulence	Early larval feeding success related to increased primary production	Svendsen *et al.* (1995)	—
	Sea temperature (in general)		Dippner (1997) Sirabella *et al.* (2001)	
Irish	Sea temperature (in general)		Planque and Fox (1998) Planque and Fredou (1999)	
Baltic	Volume of water with temperature, oxygen, and salinity conditions allowing successful egg development ('reproductive volume')	Survival of eggs and larvae	Jarre-Teichmann *et al.* (2000) MacKenzie *et al.* (2000) Köster *et al.* (2001)	—
	Salinity and oxygen levels through inflow and river runoff impacted by large-scale climate fluctuations	Survival of eggs and larvae	Zorita and Laine (2000) Haenninen *et al.* (2000)	—
	Transport of larvae and pelagic 0-group	Feeding success, survival	Köster *et al.* (2001)	Match–mismatch (Cushing 1969, 1974, 1982, 1990) Member–vagrant (Sinclair and Iles 1989)
Faeroe Plateau	Wind direction, horizontal circulation, advection	Larval retention on shelf	Hansen *et al.* (1994)	Member–vagrant (Sinclair and Iles 1989)
	Wind direction, advection	Larval feeding success linked to spatial synchrony of cod larvae and *C. finmarchicus*	Hansen *et al.* (1994)	Match–mismatch (Cushing 1969, 1974, 1982, 1990)

(Continued)

Table 6.1 (*Continued*)

Stock	Physical process/feature	Biological feature	Reference	Hypothesis
Icelandic	Spring vertical stratification	Larval feeding success through availability of *Calanus* eggs	Jakobsson (1992) Malmberg and Blindheim (1994) Brander *et al.* (2001)	Match–mismatch (Cushing 1969, 1974, 1982, 1990)
West Greenland	Upper layer sea temperature	—	Hermann (1953) Buch *et al.* (1994) Horsted (2000)	—
	Increased advection[1] due to change in windfield[2]	Import of larvae from Iceland	[1]Hansen and Buch (1986) [2]Dickson and Brander (1993)	—
Northern	Salinity level	—	Myers *et al.* (1993)	—
	Annual temperature signal	Distribution and spawning ground shift	deYoung and Rose (1993) Rose *et al.* (2000)	'Right site' (deYoung and Rose 1993)
	Wind	Larval retention on shelf	deYoung and Davidson (1994)	Match–mismatch (Cushing 1969, 1974, 1982, 1990)
Georges Bank	Wind, horizontal circulation, advection	Larval retention on shelf	Werner *et al.* (1993) Serchuk *et al.* (1994) Lough *et al.* (1994)	Member–vagrant (Sinclair and Iles 1989)
	Temperature	Plankton production	Martin and Kohler (1965)	—

are not necessarily present in populations in other environments. This makes sense in light of the hypothesis stating that the larvae of species inhabiting a cold environment should be more exposed to environmental fluctuations than species that inhabit warmer waters as a consequence of longer development times (Houde 1989, 1990).

Temperature influences growth and survival at the early life stages in at least two distinct ways: it affects the development rate of the fish larvae directly, but also indirectly through regulating the production of prey. In cold-water regions one may expect that zooplankton biomass is generally larger when temperature is high (as demonstrated for the Norwegian and Barents Seas by Nesterova (1990)). High food availability for larval and juvenile fish results in higher growth rates and greater survival through the vulnerable stages when year-class strength is determined. Temperature also affects the development rate of the fish larvae directly and, consequently, the duration of the high-mortality and vulnerable stages decreases with higher temperature (Ottersen and Sundby 1995).

The effects of temperature on recruitment of Atlantic cod across its entire distributional range was examined by Ottersen (1996) and Planque and Fredou (1999). Stocks inhabiting areas at the lower part of the overall temperature range of the species (Canadian northern (Labrador and Newfoundland), West Greenland and Arcto-Norwegian cod) were found to benefit from positive temperature anomalies while stocks occupying the warmer areas seem to profit from negative temperature anomalies (e.g. Irish and North Sea). For stocks in mid range temperatures the results are not decisive (Ottersen 1996; Planque and Fredou 1999). This is consistent with the hypothesis of Templeman (1972), which states that during periods of cooling cod recruitment is successful towards the southern end of the species range and weak in the north and vice versa during warmer periods.

Another example illustrates how two of the main cold-water stocks have adjusted differently to regional conditions. One way in which temperature variation has been suggested to affect the Canadian Northern cod is through altering the spawning locations, a northerly spawning distribution connected to warm ocean conditions being important for strong recruitment (deYoung and Rose 1993; Helbig *et al.* 1992). When northern cod migrate upstream to spawn, they have to move northwards into colder water masses with temperatures that can be very low. In cold years the northern cod may therefore be forced to spawn further south were larval retention and hence survival is poor (deYoung and Rose 1993).

On the other hand, Ellertsen *et al.* (1989) concluded that Arcto-Norwegian cod spawn at the same sites every year. The main reason for this difference between the Arcto-Norwegian and the Canadian northern cod stock may be as follows. Arcto-Norwegian cod occupy water masses that have a mean annual bottom temperature of 4 °C as opposed to the 2 °C of northern cod. When Arcto-Norwegian cod move upstream from their feeding grounds to spawn, the migration is, contrary to that of the northern stock, southwards towards warmer water. The spawning takes place in the thermocline between cold coastal water and warmer Atlantic water at temperatures between 4 °C and 6 °C (Ellertsen *et al.* 1989). The vertical temperature differences are greater than the year-to-year differences (Brander 1993). Hence, although the depth of the thermocline may vary considerably from year-to-year (Eggvin 1932), the desired temperature is found at roughly the same latitude every year.

The cod stock at West Greenland is probably more dependent on climatic variations than most other fish stocks (Jakobsson 1992). In general periods with positive temperature anomalies may be connected with high abundance of cod and vice versa (Buch *et al.* 1994). More specifically recruitment in the area has been found to be better in warm years (e.g. Hermann *et al.* 1965). Dickson *et al.* (1994) state that 'the climatic control appears the dominant one, regulating the supply of warmth and larvae on which the stock depends'. Jakobsson (1992) notes that the almost complete failure in recruitment in the late 1960s and early 1970s coincided with the presence of the 'Great Salinity Anomaly' and generally colder temperatures. However, interpreting the effect of temperature variability on cod recruitment at West Greenland is somewhat hampered by the complex cod stock structure in that area.

In addition to an inshore spawning and a local offshore-spawning stock component, it has long been known that part of the recruitment for West Greenland is a result of inflow of larvae spawned off Iceland (Tanning 1937). Such a transport provides an interesting example of how variability in wind and residual currents determines the distribution of cod eggs and larvae, and heavily influences survival rate and thus recruitment. This partly explains why the recruitment variability for West Greenland cod as a whole is higher than for any other Atlantic cod stock. However, the full story of these relationships, is certainly more complex.

Buch *et al.* (1994) state that the interannual variability in larval drift to Greenland from the spawning grounds off southwestern Iceland is determined by the strength and direction of the Irminger Current. This is supported by Dickson and Brander (1993) who attribute the variability to changes in the winter–spring windfield. The hypothesis is strengthened further by evidence of changes from cold to warm periods coinciding with increased easterly airflow. Buch *et al.* (1994) pointed to this being the cause of the simultaneous increase in the two offshore stock components at West Greenland during the middle of this century. During periods of intensified easterly airflow the component of Icelandic origin will be enhanced by increased larval advection and simultaneously the local offshore component will greatly benefit from the strengthened influence of warmer currents on West Greenland waters.

The economical consequences of such climate-related events are major and not confined to the effect on the cod catch at West Greenland. It is now apparent that cod surviving the drift to Greenland as larvae later may return as adults to Iceland (Dickson and Brander 1993). Schopka (1991) showed that the return of the 1945 year class alone represented an unpredicted increment of over 700,000 tonnes of 8-year-old fish, calculated by Dickson and Brander (1993) to be worth over £1 billion at 1993 prices.

The contrast between year classes of this size and the situation towards the end of the 20th century is striking. Adverse environmental factors (for instance relatively low temperature of upper water layers since about 1968) seem to be the main reason for the recent drastic decline of the stock size. Low larval productions at Greenland and relatively low transport of larvae from Iceland to Greenland waters resulted in a period with lower recruitment (Horsted 2000).

However, the latest results suggest that although cod have not been observed in Greenland waters in considerable numbers yet (Anon 2002; Wieland and Hovgård 2002), several of the factors crucial for Greenland cod seem to be in place. In the recent years, temperature conditions have been favourable at east and southwest Greenland (Borovkov and Stein 2001; Buch *et al.* 2002) and spawning stock biomass of Icelandic cod as well as 0-group abundance in Icelandic waters has been at or above the levels reported for the 1980s. Although changes in survival related to other factors than temperature might be at play, larval drift across the Denmark Strait appears to be crucial for a short-term recovery of the cod stock in Greenland offshore waters (Wieland and Hovgård 2002).

In the Baltic Sea the mechanisms controlling cod recruitment are different from most other places. The Baltic is the largest brackish water body in the world. Recruitment of central/eastern Baltic cod critically depends on favourable oceanographic conditions in the deeper basins of the Baltic Sea that determine the survival rates of eggs, larvae, and juveniles (Bagge and Thurow 1994; Jarre-Teichmann et al. 2000). The most important factors presently acknowledged to influence cod recruitment include: (1) The volume of water with temperature, oxygen, and salinity conditions that meet the minimum requirements for successful egg development ('reproductive volume'); (2) Potential egg production by the spawning stock (linked to the spawning stock age-structure); (3) the timing of spawning; (4) predation mortality on eggs due to sprat (*Sprattus sprattus*) and herring (*Clupea harengus*); (5) transport of cod larvae, and (6) cod cannibalism (Jarre Teichmann et al. 2000; Köster et al. 2001).

Furthermore, by means of exploratory statistical analysis such factors were incorporated into stock-recruitment models. The best model was able to explain 69% of the variation in 0-group recruitment in the most important spawning area, the Bornholm Basin. Hence, relatively simple models proved sufficient to predict recruitment of 0-group cod in these areas, suggesting that key biotic and abiotic processes may be successfully incorporated into recruitment models (Köster et al. 2001).

Dickson and Brander (1993) summarized the factors that had been suggested to promote effective inflow of saline and oxygenated water and concluded, citing a number of studies, that the main cause is persistent westerly winds. Recent publications support the importance of fluctuations in the large-scale atmospheric circulation. Hänninen et al. (1999) emphasised that major pulses of saline water most often occur during winter. They pointed to the combined effect of diminished river runoff and increased intensity of westerly winds, driven by the NAO during this season (the NAO is thoroughly described in Chapter 2).

A stronger meridional sea-level-pressure (SLP) gradient over the North Atlantic (this implies a high/positive NAO phase), and therefore stronger westerly winds, causes positive rainfall anomalies in the Baltic Sea catchment area and increased run-off giving rise to decreased salinities at all depths, while the oxygen concentration seems to increase. The mechanisms by which a stronger zonal atmospheric circulation enhances the oxygen concentrations may be related either to a weakened stratification through the reduced salinity (at long timescales), or by stronger or more frequent inflows of North Sea waters (at short timescales) (Zorita and Laine 2000). Thus, the actual effect of a stronger meridional SLP gradient (positive NAO) on the 'reproductive volume' of the cod depends on additional factors.

Since the NAO is a dominant pattern of atmospheric behaviour in the North Atlantic sector, its fluctuations have also been linked to variability in other cod stocks. Out-of-phase fluctuations in year-class strength of cod between the northeast and northwest Atlantic has been hypothesized (Izhevskii 1964; Templeman 1972) and suggested to be related to the NAO (Rodionov 1995). Links between strongly positive (negative) NAO events and poor (good) recruitment and growth for the Canadian northern cod stock are shown by Mann and Drinkwater (1994) while positive NAO anomalies have been linked to favourable conditions for Arcto-Norwegian cod (Dippner and Ottersen 2001; Ottersen and Stenseth 2001; see Fig. 6.3).

This anti-phase pattern is reasonable considering current knowledge. Both regions are characterized by sea temperatures towards the lower end of the overall range inhabited by cod. Cod recruitment tends in both areas to be higher in warmer years than in colder (deYoung and Rose 1993; Ellertsen et al. 1989). Inverse fluctuation in Barents and Labrador Sea temperatures has been pointed out earlier (Izhevskii 1964). The impact of the NAO on this 'seesaw' pattern has also been demonstrated (van Loon and Rogers 1978). During the last decades the NAO has accounted for approximately 50% of the interannual climate variability both in the Labrador Sea Region (Drinkwater and Myers 1997) and the Barents Sea (Ottersen and Stenseth 2001), but the signs of the correlations are opposite.

This east–west anti-synchrony in the effect of the NAO must, however, not be extended too far geographically. In regions like the North Sea, where below-average sea temperatures have been demonstrated to be favourable for cod recruitment (Dickson et al. 1973), one cannot expect cod recruitment to benefit from a positive NAO phase (Fig. 6.3).

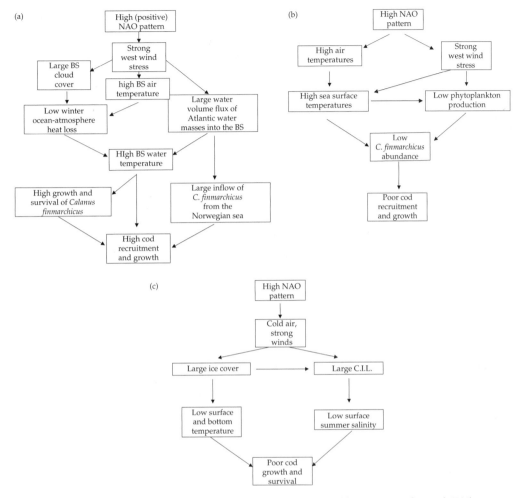

Figure 6.3 Mechanisms relating the NAO to cod recruitment for (a) The Barents Sea (adapted from Ottersen and Stenseth 2001), (b) the North Sea region (adapted from Fromentin and Planque 1996), and (c) Canadian northern cod (adapted from Mann and Drinkwater 1994).

6.4 Atlantic salmon and climate variation in the North Atlantic

Few fish species in the North Atlantic are as intensely affected by climate variation as the Atlantic salmon. The ocean migrations of Atlantic salmon rival those of the large pelagic species such as tuna, with documented returns of North American salmon from the eastern side of the Atlantic and European fish from the western side (Tucker *et al.* 1999; Hansen and Jacobsen 2000). Transoceanic migration affects growth, maturity schedules, availability to fisheries, and eventually

recruitment of salmon populations (Narayanan *et al.* 1995; Friedland 1998). The migrations themselves are known to vary in response to currents and temperature distributions, among other factors (Thorpe 1988). However, the environment's effect on salmon is not limited to the conditions adults experience at sea. As anadromous fish, they return to freshwater as adults and their progeny leave freshwater as juveniles. Freshwater mortality, hitherto considered the less important mortality factor compared to marine mortality (Chadwick 1987), may be amplified by changing climate patterns associated with global climate change (Cunjak *et al.* 1998). During their first year at sea the migration cues, first feeding opportunities, and ocean nursery conditions affecting juveniles are to a large extent climate controlled. In order to understand the relationship between salmon and climate variation, it is important to deal with three key life history stages of salmon: juveniles in freshwater, juveniles during their first year at sea, and maturing adults.

The essence of anadromy is the use of freshwater habitats for reproduction and juvenile rearing, thus improving the survival of early life history stages by protecting them from marine predators. Obviously there are tradeoffs, though marine predators cannot prey upon salmon parr (the name given to salmon juvenile in freshwater), there are many other predators that make the freshwater habitat less than benign. Predation in freshwater is reasonably well understood and can be related to the density of predators (Mills 1989). However, it was recognized in early studies of juvenile salmon (Huntsman 1938) that precipitation, and thus flow and water depth, might affect predation rates and thus juvenile production (Ghent and Hanna 1999).

Climate can also affect the dynamics of juvenile salmon populations in freshwater nurseries through modulation of growth rates, principally by the effect of temperature on growth. Habitat availability is a limiting factor in the production of juvenile salmon and the factors that impact the pace at which cohorts move through rearing habitat, impact the overall production of pre-recruits to the stocks (Bardonnet and Baglinière 2000). Juvenile rearing in freshwater may last for as many as 7 years in northern streams to as few as 1 year in southern habitats (Power 1981). Since migration from freshwater is growth mediated, climate forcing that affects growth will determine the pace at which cohorts leave nursery streams (Saunders *et al.* 1994). It is predicted that smolt ages (smolt is the name given to the juvenile stage adapted to life in the ocean) will decrease and precocious maturation will increase across much of the rearing habitat in North America if the anticipated increases in temperature resulting from global climate change are realized (Minns *et al.* 1995; Juanes *et al.* 2000).

The accumulated mortality in freshwater is thought to be relatively predictable, which is reflected in the general adherence in many salmonid populations to spawner-recruit relationships during the freshwater phase (Solomon 1985). However, this perception may change due to concerns about mortality on the parr the winter before

they leave freshwater. This seasonal mortality effect is referred to as overwinter mortality and has its own set of dynamics linked to climate and the structure of the rearing habitat.

Overwinter mortality is associated with the relationship between pre-migrant parr and their rearing habitat, the margins of which appear to provide an unstable refuge during their last winter in freshwater. Parr that attain a sufficient size to migrate to the ocean will metamorphose into smolts during the spring of migration. During their last winter in freshwater, pre-migrant parr are relatively large for their habitat where they often have to live beneath frozen stream water (Cunjak *et al.* 1998). There is growing concern that the mortality of pre-migrant parr may be quite large for some populations and subject to climate variations, which affect the stability of the ice cover (Whalen *et al.* 1999). Ironically, it also appears that the smaller members of the nursery population may be better adapted to surviving these shifting conditions since their smaller size makes more specialized refuges available (Cunjak 1988). The anticipated dynamics of climate change over the mid to high latitudes include scenarios of higher storm frequency and thermal variation, all of which strike at the heart of this delicate balance salmon have developed in freshwater (Monahan *et al.* 2000). These climate conditions may destabilize ice cover and cause pre-migrant parr mortality to increase.

The next transition for salmon is the movement of smolts into the ocean, which is affected by climate conditions in several ways. At the outset, smolt migrations are cued by environmental signals such as temperature conditions in freshwater rearing areas (Solomon 1978; Jonsson and Ruud-Hansen 1985). In theory, the cues are intended to guide salmon smolts to specific 'migration windows' in the coastal ocean, where the fish try to take advantage of specific prey, avoid specific predators, and find suitable habitat conditions. The fish are already under a physiologic stress since they are challenged by the transition of moving from fresh to saline water; optimizing the ecological aspects of the migration would seem important (Friedland and Haas 1988). If adaptations to initiate the migration to sea are not robust to climate variability, the consequence for regional stock groups may be profound, especially for stocks at the margins of salmon distribution. This is illustrated in the behaviour of North versus West Coast Icelandic salmon stocks where yield variability is associated with the variability of air and sea temperature (Scarnecchia *et al.* 1989b). The North Coast stocks, which are at latitudes at the northern end of the range for salmon, have much greater yield variation associated with variability of air temperatures over the rivers (likely affecting smolt migration cues) and coastal ocean areas (likely affecting thermal conditions for juvenile first entering the sea).

The largest component of natural mortality affecting Atlantic salmon populations in the marine environment occurs during the first year at sea. The juveniles are referred to as post-smolts during this period, which begins after their migration to the ocean and up until the first winter. The post-smolt period has emerged as a critical period for salmonids worldwide because of the magnitude and variability of post-smolt mortality (Pearcy 1992). Long-term patterns of stock abundance for regional and continental stock complexes are defined by post-smolt survival, but perhaps more importantly these patterns are often associated with climate forcing (Friedland 1998). The stock complexes for Europe and North America appear to have different climate controlling mechanisms reflecting the theorized differences in post-smolt distribution and ecology.

The nursery zone for European post-smolts is located in the open ocean whereas North American post-smolts appear to utilize inshore habitats. Holm *et al.* (2000) described the distribution of European post-smolts from surface trawling operations in the Northeast Atlantic. The nursery is confined to a region within the Norwegian Sea, the northern extent of which would appear to be defined by current transport. The post-smolts co-occur with surface shoals of herring and mackerel and occupy a similar ecological niche (Jacobsen and Hansen 2000). In North America, post-smolts may be found in high numbers in the Labrador Sea during the fall of the year (Reddin and Short 1991). However, during the earlier part of the post-smolt period (i.e. through the spring and summer months), fish are also found in the Gulf of St Lawrence, the coast of Nova Scotia, and elsewhere (Dutil and Coutu 1988; Ritter 1989; Friedland *et al.* 1999). Furthermore, North American stocks may not physically mix for many months after entering the ocean while it appears that European stocks are concentrated in a single, albeit large, ocean area (Friedland and Reddin 2000).

If in some years the North American post-smolt nursery is distributed along the coast, it would function in fundamentally different ways than the European post-smolt nursery, thus affecting the climate response of the respective stock complexes. Predation losses for the North American stocks may be the result of more avian and mammal predators emanating from specific rookery locations on the coast (Montevecchi *et al.* 1988); whereas European fish would be expected to face growth-mediated predation pressure associated with the dynamics of an oceanic ecosystem (Sogard 1997). Growth data provides a practical test for this idea. Inter-annual post-smolt growth variation for Scottish and Norwegian stocks correlated with survival supports the idea that mortality of these stocks is controlled by their growth (Friedland *et al.* 2000). Post-smolt growth increments of North Esk (in Scotland) salmon reflect the growth of juvenile fish during their first season at sea, which correlates to the cohort specific survival of 1SW returns the next year (Fig. 6.4). However, similar growth signatures for North American stocks are uncorrelated with survival (Friedland and Haas 1996). It is important to qualify this observation by stating that the analysis was done on hatchery fish and that survival differences among stocks may, at times, be attributed to growth regardless of the pattern of interannual growth variation (Friedland *et al.* 1996).

A fundamentally different response to climate forcing due to the nature of the post-smolt nursery is reflected in our current state of knowledge concerning climate forcing and recruitment of Atlantic salmon. From analyses of SST temperature distribution, spring thermal conditions have emerged as

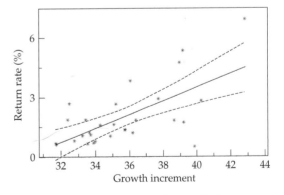

Figure 6.4 Scattergram and regression between 1SW return rate and post-smolt growth increment matched by smolt years for the North Esk stock ($r = 0.65$; $p < 0.001$). Dashed line represents 95% confidence interval (from Friedland *et al.* 2000).

an important forcing in the recruitment of European salmon stocks. The relationship was first quantified for trends in the entire stock complex (i.e. for all sea age fish), using time series subjected to temporal averaging (Friedland *et al.* 1993). Subsequently, when examined for a distinct set of stocks from central Europe using a time step of one month, the relationship became much clearer (Friedland *et al.* 1998a). Warm-water conditions during the first months at sea were found to be critical for the survival of one seawinter salmon (1SW, or those fish at sea for one winter) suggesting a positive correlation between temperature and survival. The importance of thermal conditions during the first few months at sea is also supported with data for regional stocks in Iceland and the Baltic Sea (Scarnecchia 1984; Salminen *et al.* 1995). However, a different relationship has emerged in respect to the North American stock complex. The first climate signal that appeared to be related to the North American stock complex was characterized as a relationship between 2SW salmon and winter thermal habitat in the northwest Atlantic (Ritter 1989; Friedland *et al.* 1993). However, it is not clear whether this relationship is related to the survival of post-smolts given the season of the climate effect; instead, it has been suggested that this relationship may be related to age of maturation (Friedland *et al.* 1998b). Taking into consideration the distribution of post-smolt nursery habitats and emigrational corridors in North America, researchers have found that spring thermal conditions in inshore areas co-vary with recruitment time series for North American stocks (Friedland *et al.* 2003a,b). It would appear that European and North American salmon stocks are responding to climate forcing, such as temperature and factors likely to co-vary with temperature, affecting the early marine stages. But interestingly, the relationship between temperature and survival appears to be negative in North America, suggesting there may be an optimal window for ocean entry and survival in salmon. These studies provide a framework of testable hypotheses related to the effect of climate and growth on post-smolt survival and salmon recruitment.

Age at maturation has important consequences on the total complement of eggs deposited during spawning. The 1SW spawners are referred to as grilse and may contribute varying proportions of the gametes during the annual spawning run. When a high proportion of a cohort matures as grilse, egg deposition is low per spawning because the younger fish do not produce the same quantity of eggs as multi-seawinter salmon. Though the decision to mature has a strong genetic component (Gjerde 1984), environment can also play a significant role (Saunders *et al.* 1983). Climate may be acting on this process in a number of different ways. Conditions that affect the growth of salmon may determine whether an individual is physiologically ready to return to the river to spawn. There is evidence that growth at various times during the post-smolt year may be important to achieving maturity (Scarnecchia *et al.* 1989a, 1991; Gudjonsson *et al.* 1995; Duston and Saunders 1999), suggesting hypotheses regarding genetic and temporally fixed mechanisms lack the plasticity realized by wild stocks (Thorpe 1986). Alternatively, some investigators have suggested that climate variations that extend migrations beyond a reasonable return distance contribute to a variable proportion of grilse in the return (Martin and Mitchell 1985). And in fact, some animals on migration routes away from home rivers are actually approaching maturation and likely regress when they fail to find their home rivers (Friedland *et al.* 1998b).

The NAO has been associated with the climate forcing of Atlantic salmon in respect to survival and maturation. The spring thermal habitat areas associated with post-smolt survival of central European salmon stocks are derived from large ocean areas where the distribution of SST is affected by the NAO (Dickson and Turrell 2000). Likewise, the winter thermal habitat associated with the abundance of specific age components in the northwest Atlantic are also derived from areas where SST distribution is correlated with the NAO (Friedland *et al.* 1993). However, it would be premature to suggest that the NAO is the only source of climate force affecting salmon or that the NAO is the only way to visualize the relevant impact of climate variation on salmon. For example, high frequency current fluctuation in the Barents Sea appears to create a lagged linkage between Icelandic and Russian salmon stocks (Antonsson *et al.* 1996). And other atmospheric indices, such as the atmospheric circulation index (ACI), have been useful in developing hypotheses on transoceanic and global stock synchrony (Klyashtorin 1998) and may be of use also in studies of Atlantic salmon.

The unprecedented decline in Atlantic salmon abundances over the past few decades raises concern over the future impact climate change may have on Atlantic salmon. With climate at the core of many of the factors contributing to the decline of stocks, the effect of further shifts, beyond the reactive norms to which salmon populations have adapted, now pose the threat of range shift for the species. If these climate changes are being accelerated by anthropogenic factors, the speed at which the changes occur may be beyond the time scales salmon need to develop requisite survival adaptations, thus posing the threat of widespread extinctions.

6.5 Climate forcing of herring and sardine populations

6.5.1 Background

Small pelagic fishes such as sardine, anchovy, herring, and others represent about 20–25% of the total annual world fisheries catch. They are widespread and occur in all oceans. They support important fisheries all over the world and the economies of many countries depend on those fisheries. They respond dramatically and quickly to changes in ocean climate. Most are highly mobile; have short, plankton-based food chains and some even feed directly on phytoplankton. They are short-lived (3–7 years, except herring), highly fecund and some can spawn all year-round. These biological characteristics make them highly sensitive to environmental forcing and extremely variable in their abundance. Thousandfold changes in abundance over a few decades are characteristic for small pelagics and well-known examples include the Japanese sardine, sardines in the California Current, anchovies in the Humboldt Current, sardines in the Benguela Current, and herring in European waters. Their drastic stock fluctuations often caused dramatic consequences for fishing communities, entire regions and even whole countries. Their dynamics have important ecological consequences as well as economic ones. They are the forage for larger fish, seabirds, and marine mammals. The collapse of small pelagic fish populations is often accompanied by sharp declines in marine bird and mammal populations that depend on them for food (Hunter and Alheit 1995). Major changes in abundance of small pelagic fishes may be accompanied by marked changes in ecosystem structure. The great plasticity in growth, survival and other life-history characteristics of small pelagic fishes is the key to their dynamics and makes them ideal targets for testing the impact of

climate variability on marine ecosystems and fish populations (Hunter and Alheit 1995; Alheit and Hagen 2001, 2002).

6.5.2 The stocks

The Bohuslän region of Sweden lies adjacent to the Skagerrak, north of Gothenburg. It is exposed to a very variable hydrographic regime as the Skagerrak connects the North Sea and the Baltic Sea. Periodically, large amounts of spent herring from the North Sea migrated to the Bohuslän coast in autumn and overwintered in the skerries and fjords close to the shore. These migrations occurred for periods of 20–50 years and supported considerable fisheries. They were interrupted by even longer time spans of 50–70 years when North Sea herring did not migrate in large masses to Bohuslän (Devold 1963; Höglund 1978; Cushing 1982; Lindquist 1983; Sahrhage and Lundbeck 1992; Jahnke 1997; Alheit and Hagen 1997, 2001, 2002). These Bohuslän herring periods can be traced back for about 1000 years and nine such periods have been identified. Relatively good data on the Bohuslän fishery are available because whenever the herring return to Bohuslän a new fisheries period began, the whole economy of the region changed completely. These remarkable changes were remembered for a long time and written down in sagas, chronicles, tax reports, etc. The catches were quite considerable, particularly when considering that the herring were not caught by fishing vessels, but mainly from the shore using beach seines. During the fishing season 1895/96 more than 200,000 MT (metric tonnes) were caught (Lindquist 1983) and in one particular season in the eighteenth century 270,000 MT were reported (Höglund 1978).

The Norwegian spring-spawning herring is the largest stock of the Atlanto-Scandian herring. Its total biomass may have ranged between 15 and 29 million MT in a virgin state (Dragesund et al. 1980). Its main habitat is the Norwegian Sea. Historical records indicate large fluctuations of the fishery during the last 500 years (Devold 1963; Beverton and Lee 1965; Skjoldal et al. 1993; Toresen and Østvedt 2000). Periods of large catches have alternated with periods of extreme scarcity. Beverton and Lee (1965) and Røttingen (1992) point out that it is not clear how much of this periodicity is due to real changes in abundance or to changes in migration routes preventing the fish from coming within the limited range of coastal fishing fleets. Already Boeck (1871) and Petterson (1926) suggested that the migration patterns of herring were profoundly changed over secular periods and seem to be linked to stock size (Dragesund *et al.* 1980). The fishing periods of the Norwegian spring-spawners and the Bohuslän periods seem to alternate (Devold 1963; Beverton and Lee 1965; Skjoldal *et al.* 1993; Alheit and Hagen 1997).

The distributions of herring and sardine (*S. pilchardus*, also called pilchard) overlap in the English Channel. Fishing for both species off the south-western tip of England, off Cornwall and Devon, has been reported since the sixteenth century (Southward *et al.* 1988). The geographical boundary between the two species seems to shift on a decadal scale and, consequently, periods of the herring fishery have alternated with those of the sardine fishery and were in phase with those of the Bohuslän herring. Also, they seem to be linked to the Russell Cycle, which describes the synchronous alternation of appearance and disappearance of a large number of pelagic species, zooplankton and fish, including fish eggs and larvae, recorded since 1924 in the western Channel (Cushing 1982). Unfortunately, when the Russell Cycle (see Chapter 10) changed from a warm-water to a cold-water system between 1965 and 1979, the herring failed to re-appear.

There are less data on herring and sardine fisheries of French fishermen in the English Channel since the eighteenth century. However, the herring periods of this fishery were from about 1750–1810 and from 1880–1910 (Binet 1988). Peak catches reached about 50,000 MT in 1905. Most sardines caught in the French fishery in the English Channel were taken off southern Britanny and the Vendée region. Occasionally, a sardine fishery developed north of Brest, off northern Brittany. There are various reports from the period between 1726 and 1764. A notable decrease of sardine abundance was reported in the first years of the nineteenth century, but the fishery resumed in the 1860s and 1870s (Binet 1988).

A small herring population that was first mentioned in 1728 exists in the Bay of Biscay off southern Brittany. Due to its relative insignificance, reports are rather sporadic, but indicate, nevertheless, periods of presence and absence (Binet 1988).

6.5.3 Fluctuations in climate and fisheries

The 'fisheries periods' are reported historically as persistent presence or absence of fish and fisheries at certain locations. The dramatic socio-economic changes between fish-rich and fish-poor periods were the cause for the historic records of these fisheries. The fluctuations of these fisheries could be the result of real fluctuations of biomass or of decadal changes in migration routes moving the fish to areas not accessible to the limited range of the fishing methods employed in historic times. However, fluctuations in biomass and migration routes are likely to be concomitant phenomena (Alheit and Hagen 1997).

The fisheries described above form two groups with alternating periods of occurrence. The first group consists of the Bohuslän herring, the herring of southwestern England, the herring caught by French fishermen in the eastern English Channel and the Bay of Biscay herring. The second group comprises the Norwegian spring-spawning herring, the sardines of southwestern England and the sardines caught by the French fleet in the English Channel (Fig. 6.5).

Herring and sardine differ in their adaptation to temperature (Southward *et al.* 1988). The distribution of both species overlap in the English Channel. The herring is an Arcto-boreal species and the English Channel is usually the southern limit of its distribution. In contrast, the Channel usually represents the northern limit of the distribution of the sardine. Therefore, cooling of the waters in the Channel should favour a more southern distribution and increasing biomass of herring; warming should have the opposite effect. The same mechanism should work for the sardine with opposing reactions. Indeed, a strong correlation between temperature and occurrence and abundance of both species has been recorded for the last 400 years along the south coast of Devon and Cornwall (Southward *et al.* 1988). The herring fishery was favoured during cold periods and extended further west, whereas the sardine fishery was then restricted to western Cornwall. During the extremely cold period called the 'Little Ice Age' (during the second half of the seventeenth century), sardines were very scarce while herrings were abundant. The sardine fishery was more prevalent in warm periods. The southward displacement of the French herring fishery in the English Channel coincided with the climatic deterioration at the end of the eighteenth century. Also, the extension and decline of the French herring fishery in the English Channel corresponded to herring catch records off the south coast of Cornwall and Devon. Clearly, both fisheries were linked to long-term changes in the

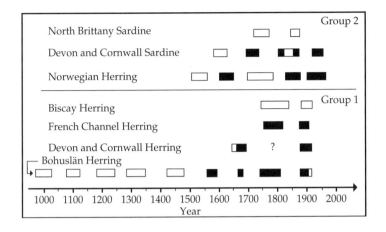

Figure 6.5 Historical periods of European herring and sardine fisheries. Open rectangles depict periods where the extensions are not precisely known. Modified from Alheit and Hagen 1997.

temperature regime (Alheit and Hagen 1997). The fisheries for the Bohuslän herring and the Norwegian spring herring are linked in a similar manner to opposing temperature regimes. For example, the last period of the Bohuslän fishery (1878–1906) was during a cool period whereas the Norwegian herring period from 1920 to 1950 occurred during a period of elevated temperatures. Also, the high abundance of the Norwegian spring-spawning herring started when Icelandic ice cover decreased and ceased when ice cover increased (Beverton and Lee 1965). Consequently, there is a tendency for Bohuslän periods to appear during periods of Icelandic ice cover (Cushing 1982; Alheit and Hagen 1997). In general, a drastic decrease of temperature over NW Europe followed by a cold period seems to be the prerequisite for an abundant Bohuslän period (Fig. 6.6). Also, since 1350, the Bohuslän periods coincide with times when the frequency of southwesterly surface winds over England had their minimum values (Alheit and Hagen 1997).

In summary, the fluctuations of herring and sardine fisheries in the Northeast Atlantic are related to temperature regimes and fluctuate on a decadal scale. Therefore, when searching for factors forcing the herring and sardine populations, we clearly have to take into account climatic variability in response to changes in the oceanic current system on the basin-scale of the North Atlantic. Cold and warm periods on the decadal scale over the Northeast Atlantic are reflected in the strongest climate signal of the Northern Hemisphere, the NAO (see Chapter 2). Consequently, herring and sardine periods on the decadal scale may be related to the variability of the NAO (Alheit and Hagen

1997). The first group of the fish stocks under investigation, the Bohuslän herring, the herring of southwestern England, the herring caught by French fishermen in the eastern English Channel, and the Bay of Biscay herring, exhibited strong fisheries when the NAO index was in a negative phase, that is, when there were cold temperatures at mid- and higher latitudes of the eastern North Atlantic, negative sea surface temperature anomalies (SSTAs), increased ice cover off Iceland and in the northern Baltic Sea, reduced westerly winds, minimum frequency of southwesterly winds over England and cold water in the North Sea, English Channel, and in the Skagerrak. In contrast, the second group of stocks, the Norwegian spring-spawning herring, the sardines of southwestern England, and the sardines caught by the French fleet in the English Channel, were thriving when the NAO was in a positive phase, that is, when there were warm temperatures at mid- and higher latitudes of the eastern North Atlantic, positive SSTAs, reduced ice cover off Iceland, intensified westerly winds, increasing frequency of southwesterly winds over England and relatively warm water in the North Sea, English Channel and in the Skagerrak (Alheit and Hagen 1997).

The conclusion that climate variability governs the dynamics of herring and sardine populations is, however, not only derived from the coincident occurrence of positive and negative NAO periods and herring and sardine fishing periods on the decadal scale. The teleconnection patterns between the dynamics of the herring and sardine populations are strong evidence for climate dependence. In spite of the distances of several thousand kilometres between the herring and sardine populations, they react to the same basin-wide forcing

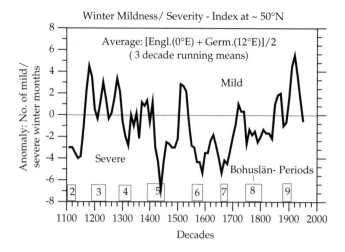

Figure 6.6 Winter mildness/severity index and Bohuslän herring periods (rectangles). Modified from Alheit and Hagen (1997).

(Fig. 6.6; Alheit and Hagen 1997). Now, more than 90 years after the last Bohuslän period, it remains to be explained why we have not observed any more recent Bohuslän herring period. For example, in the 1960s the NAO had another negative period. However, there was a prolonged recruitment failure of North Sea herring from the late 1960s to the early 1980s, partly due to heavy fishing pressure (Corten 1990; Corten and van de Kamp 1992): thus one would not necessarily have expected to encounter large quantities of North Sea herring off the Bohuslän coast.

6.6 Conclusions

In this chapter, we have reviewed several examples of the impacts of environmental variability on fish stocks. As stated by Frank *et al.* (1990), taking the next step, to predict the response of local marine organisms to possible climate variability or climate change scenarios, becomes a highly speculative exercise. However, observations of changes to the fish stocks due to climate variability in the past allow us to predict some general responses. Climate change (and here we will not deal with causes, nor

the amplitude or sign of the change) may be expected to result in distributional shifts in species with the most obvious changes occurring near the northern or southern boundaries of their range. Migration patterns will shift and thus result in changes in arrival times along the migration route. Growth rates are expected to vary with the amplitude and direction being species dependent. Recruitment success could be affected due to changes in time of spawning, fecundity rates, survival rate of larvae, and food availability.

Qualitative predictions of the consequences of climate change on fish resources will require reliable regional atmospheric and oceanic models of the response of the ocean to climate change. Improved knowledge of the life histories of those species for which predictions are required and further understanding of the role environment, species interactions, and fishing play in generating the variability of growth, reproduction, distribution, and abundance of fish stocks is also needed. The multi-forcing and numerous past examples of 'failed' environment–fish relationships indicate the difficulty fisheries scientists face in providing reliable predictions of the fish response to climate change.

CHAPTER 7

Marine birds and climate fluctuation in the North Atlantic

Joël M. Durant, Nils Chr. Stenseth, Tycho Anker-Nilssen, Michael P. Harris, Paul M. Thompson, and Sarah Wanless

7.1 Introduction: seabirds and their food web

Very few studies have directly assessed the relationship between climate and population performance in seabirds in general, and North Atlantic seabirds in particular. Nevertheless, there are many data from diverse sources that provide insights into the likely impact of climate variations upon these species. Most of these studies have been conducted over relatively short, recent timescales, being based on direct observations of bird populations over the last century or so. Nevertheless, paleoecological studies of penguin populations have illustrated that longer-term changes in abundance, occurring over a period of 3000 years, may also be related to climate variation (Sun *et al*. 2000). In this chapter we summarize—and synthesize—what currently is known about the ecological effects of climate fluctuation on seabirds in the North Atlantic region.

Any effects of climate on these species are likely to occur through two main processes: either directly through physiological effects or indirectly through an influence on prey availability (see Chapter 1). Direct physiological effects include metabolic processes during key stages of the life cycle such as reproduction and moult. Variations in the physical environment may also affect feeding rate or competition for food resources through changes in either energetic requirements or food availability. Such effects may be seen as a modification by the climate of the threshold level in energy necessary to carry out particular life-history activities (cf. Stearns 1992). For example, there is an important energy trade-off between reproductive investment and maintenance (cf. Williams 1966; Stearns 1992).

The fact that their prey includes a wide variety of organisms, each with populations that may fluctuate in response to climatic change, means that we must also consider indirect effects through changes in food availability. Because seabirds are found at higher trophic levels and may take prey from various levels, their relationship with climate becomes even more complex. For example, an ambient temperature that is favourable for both seabirds and their main prey, might at the same time be unfavourable for the prey's primary food resources. Consequently, such a temperature might be globally unfavourable for the bird because of the reduction in the availability of its own prey. Therefore, the relationship between seabirds and climate will be complicated by the biology of the lower members of the food web. In this context, oceanographic factors (e.g. water temperature and currents) and large-scale climatic and hydrographic processes (e.g. the North Atlantic Oscillation (NAO)) generate variation in the production, distribution and abundance of organisms upon which birds feed (Chapters 4–6).

In the following, we show how climate might influence seabirds directly through variations in temperature and wind. We also provide an overview of the potential indirect impact of climate variability on North Atlantic seabird populations. First, however, we briefly describe key features of seabird biology that are relevant for an improved understanding of the nature of such climatic effects.

7.2 Seabird biology: breeding on land, feeding at sea

Several bird species depend upon marine food resources at some point in their annual cycle. These include species from orders such as Gaviiformes, Podicipediformes, and Anseriformes. However, here our main focus will be on birds that are completely dependent on the marine environment; the seabirds. Seabirds are represented by only four orders (as compared to around 28 orders for terrestrial birds). In the Northern Hemisphere, only three orders of seabirds are found: the Charadriiformes, Procellariiformes, and Pelecaniformes. Each of these orders has specific adaptations, but all depend upon the sea for their food resources.

Seabirds have in general low fertility. Many species lay only one egg per year (Jouventin and Mougin 1981) and in some species reproduction does not occur every year (Jouventin and Mougin 1981; Weimerskirch 2001). On the other hand, this low birth rate is compensated for by high longevity (review in Weimerskirch 2001). Hence, since most seabirds cannot adjust clutch size in response to food supply, they have less flexibility in their breeding response to environmental fluctuation than terrestrial birds. Long-lived species tend to adjust their expenditure on parental care to balance benefits to the offspring against costs to the parents, thus maximizing individual fitness (Williams 1966; Erikstad et al. 1998; Weimerskirch et al. 2000b). Therefore, even if climatic variation does not have a profound effect on adult survival rate, it might have important effects on fledging success. However, since many of these species typically delay reproduction until they are between 2 and 9 years old (Jouventin and Mougin 1981), and in albatrosses the age of first breeding can even reach extreme values of 13 years (Marchant and Higgins 1990), it may take several years for

such effects of climate variation on population size to become apparent (Thompson and Ollason 2001).

Seabirds are typically not constrained to a central place (Lack 1968; Ashmole 1971; Prince et al. 1992; Weimerskirch et al. 1993), and they can therefore often overcome to a large extent the problem of environmental variability. Their great mobility allows them to exploit locally and ephemerally favourable conditions and resources over great distances. However, during the breeding season, seabirds are typical central-place foragers, tied to a breeding site on land and foraging for marine resources. During foraging trips many seabirds regularly traverse hundreds or thousands of kilometres within a period of days (Harrington 1977; Stahl et al. 1985; Jouventin and Weimerskirch 1990; Flint 1991). A major constraint on breeding for seabirds is the distance between the breeding grounds on land and the feeding zones at sea (Weimerskirch and Cherel 1998). The distance of foraging is limited by the need to incubate egg(s) or to rear chick(s), neither of which can usually be left alone for long periods. For many species, suitable breeding sites are limited, and the dependence of birds to foraging areas around these sites increases the effects of temporal variation in environmental conditions within flying distance around the nest-site. During the chick(s) growth period, adults must make frequent visits to the nest in order to feed the young, even if both parents normally share the task of incubation and rearing the young. Making foraging trips at a frequent rhythm to feed their chick(s) adds an additional energy constraint upon the parents through fasting. Some species have even evolved a dual strategy for these feeding trips, within which adults alternate short trips to feed their chick(s) with a long trip during which they increase their body mass (Fig. 7.1; Weimerskirch et al. 1997a,b, 1999; Weimerskirch and Cherel 1998; Catard et al. 2000; Dearborn 2001; Watanuki et al. 2001). On average, birds conduct one long foraging trip followed by two short trips, with the duration of the long trip depending on the need to increase body mass and replenish the body reserves during the chick-rearing period (Catard et al. 2000; Dearborn 2001).

In the Northern Hemisphere, foraging distances from breeding colonies are typically smaller than in the Southern Hemisphere (Hunt et al. 1999). Northern seabirds usually forage within 200 km

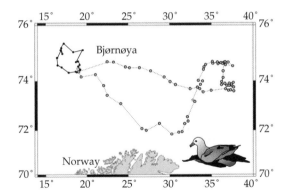

Figure 7.1 Two foraging trips (short and long) during the late brooding period of a female northern fulmar (*F. glacialis*) breeding at Bjørnøya (adapted from Weimerskirch *et al.* 2001).

from their colonies. This is, first, because available breeding sites are spread more evenly through suitable near-shore foraging areas in the north. Second, prey availability differs between the two hemispheres, with more invertebrate prey in the southern oceans. Consequently, the long-distance forager species (mainly Procellariiformes) are more represented in the Southern Hemisphere. For example, the short-tailed shearwater *Puffinus tenuirostris* may feed more than 2000 km away from their breeding colony during the chick-rearing period (Klomp and Schultz 1998; Nicholls *et al.* 1998). Despite its extreme foraging range, this shearwater is still able to provision its chicks at a sufficient rate by using the two-fold strategy that alternates long and short feeding trips.

The long foraging trips of these seabirds appear to result from the need to obtain prey from patchy oceanic resources. Seabirds take a wide variety of prey, but they typically favour small pelagic schooling fishes, moderately sized pelagic crustaceans, and squid from the upper- and mid-water column (Montevecchi and Myers 1996; Garthe 1997). Oceanographic features (such as fronts, pycnoclines) may concentrate these prey species and provide for seabirds a spatially and temporally predictable food supply (Hunt 1990; Schneider 1990; Begg and Reid 1997; Mehlum *et al.* 1998) explaining why they forage preferentially at such physical conditions (Hunt and Schneider 1987; Begg and Reid 1997; Hunt *et al.* 1999; Hoefer 2000; Skov and Durinck 2000).

7.3 Direct influences of climate on seabirds

Reproductive characteristics, such as clutch size or timing of breeding, are typically related to latitude (Olsen and Marples 1993; Sanz 1999)—partly through climatic conditions. In the North Atlantic, few studies have been conducted on the influence of climate on seabird biology (Table 7.1). However, data from more general studies of seabird ecology together with findings from the more detailed studies conducted in the Pacific (Schreiber 2001) provide valuable insight on how climate variability may influence energetic costs, reproductive output, and mortality rates in these species.

7.3.1 Reproduction

Birds require much resources to produce eggs, and the quality of the produced eggs may affect the survival of chicks (Carey 1996). Obtaining resources for egg production may be particularly difficult when other factors constrain the timing of the breeding season (Perrins 1996), implying that birds must obtain the necessary resources by a certain date. Consequently, the timing of breeding itself is often dependent upon food availability, meaning that laying date is (by-and-large) correlated to the natural changes in food resources (Meijer and Drent 1999).

During incubation, the adult uses part of its energy reserves to maintain the egg(s) at the optimal temperature for embryonic development, a temperature usually ranging from 36 °C to 38 °C (see Stoleson and Beissinger 1999); if the egg's temperature drops below 24–27 °C (physiological zero), embryo development is halted. Excessive exposure to temperatures between this physiological zero and normal incubation temperature (i.e. on average 24–36 °C) can lead to abnormal development or the death of the embryo. Consequently, decreases in ambient temperature may lead either to an increase in the transfer of heat between the adult and the egg, resulting in higher energy costs of incubation, and/or a decrease in hatching success (Williams 1996). During incubation, adults must support both the cost of incubation and their own metabolic needs, either by leaving the egg to go foraging, or by drawing upon their body reserves with fasting that could last several months in the extreme cases of the emperor penguin (*Aptenodytes forsteri*:

Table 7.1 Relationships between climate variability and some seabirds species in the North Atlantic

Species	Climate variable(s)	Population parameter(s)	Observed effect	Ref.
Arctic tern (*Sterna paradisaea*)	Salinity	Sea distribution	−	1
Atlantic puffin (*F. arctica*)	SST	Hatch + Fldg + Brd. S.	none	2
		Laying date	+	3
	Sea temperature	Fldg. S.	+	4
Black-baked gull (*L. marinus*)	SST	Hatch + Fldg + Brd. S.	+	2
Black-headed gull (*L. ridibundus*)	Salinity	Sea distribution	−	1
Common guillemot (*U. aalga*)	SST, salinity	Sea distribution	+	1, 5
	Stormy conditions	Foraging cost	+	6
	SST	Hatch + Fldg + Brd. S.	none	2
	SST	Laying date	−	7
	Air temperature	Fledging date	−	8
Common gull (*Larus canus*)	Salinity	Sea distribution	−	1
Common tern (*Sterna hirundo*)	Salinity	Sea distribution	−	1
Herring gull (*L. argentatus*)	Salinity	Sea distribution	−	1
	SST	Hatch + Fldg + Brd. S.	+	2
Black-legged kittiwake	SST, salinity	Sea distribution	+/none	1, 5
(*R. tridactyla*)	SST	Hatch + Fldg + Brd. S.	+	2
Leach's storm petrel	SST	Hatch + Fldg + Brd. S.	none	2
(*O. leucorhoa*)				
Manx shearwater (*P. puffinus*)	SST, salinity	Sea distribution	+	5
Northern fulmar (*F. glacialis*)	SST, salinity	Sea distribution	+	1, 5
	Wind speed	FMR	−	9
	NAO, air temp	Hatch + Fldg. S.	−/+	10
Northern gannet (*Sula bassana*)	SST	Breeding density	+	11
Razorbill (*Alca torda*)	SST	Sea distribution	+	5
	SST	Laying date	−	12
European shag (*P. aristotelis*)	Wind	Laying date	+	13

SST = sea surface temperature; Hatch + Fldg + Brd. S. = hatching success, fledging success and breeding success; FMR = Field metabolic rate; NAO = North Atlantic Oscillation.
+ means that an increase in the value of the climate variable is correlated to an increase of the population parameter.
References: 1. Garthe 1997; 2. Regehr and Rodway 1999; 3. Harris *et al.* 1998; 4. Durant *et al.* 2003, 5. Begg and Reid 1997, 6. Finney *et al.* 1999, 7. Harris and Wanless 1988, 8. Hedgren 1979, 9. Furness and Bryant 1996, 10. Thompson and Ollason 2001, 11. Montevecchi and Myers 1997, 12. Harris and Wanless 1989, 13. Aebischer and Wanless 1992.

Le Maho 1977). In either case, reductions in ambient temperature will result in an increased energy cost for re-warming the egg (Williams 1996).

Chick growth rate and fledging mass are sometimes directly correlated with survival up to breeding age (Chastel *et al.* 1993; Croxall *et al.* 1988). Brief periods of low food availability can be overcome by using stored fat (Ricklefs and Schew 1994), but longer periods may result in slower growth. In a long-term study of Atlantic puffin (*Fratercula arctica*), the general sigmoidal avian growth curve (Ricklefs 1968, 1973) differed between years (Anker-Nilssen and Aarvak 2002). Higher asymptote was generally observed in years when food delivery to the chick was greatest, and chicks, in turn, had higher fledging success (Fig. 7.2); the influence of food supply on chick growth rates was later verified experimentally (Øyan and Anker-Nilssen 1996; Cook and Hamer 1997). However, several factors may explain differences in chick provisioning rates. The availability of the birds' primary prey is the more likely factor, but the presence of alternative prey and the distance to foraging areas may also influence parental effort (Erikstad *et al.* 1998).

A central issue in life history is how animals balance their investment in young against their

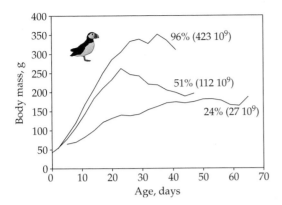

Figure 7.2 Mean body mass in relation to age of Atlantic puffin chicks (*F. arctica*) in Røst, North Norway during three different years. For each year, their fledging success (in percent) and the abundance (numbers in parentheses) of their main prey, the first-year herring, are indicated (adapted from Anker-Nilssen and Aarvak 2002).

own chances to survive and reproduce in the future (Stearns 1992). The long-lived seabirds are presumably less likely to increase their effort when raising young to ensure that they do not jeopardize their own survival. It has even been suggested that these long-lived species have evolved a fixed level of investment in their young in order to maximize their own survival (Sæther *et al.* 1993). Thus, they may be adjusting their feeding effort in relation to both their own body condition and to the short-term needs of the chicks (Erikstad *et al.* 1997). In years of poor feeding conditions, birds can reduce their parental effort so as to sustain themselves, resulting in a lower food supply for the chick, delayed growth (Øyan and Anker-Nilssen 1996) and potentially a lower chick survival. Indeed, the chances of a chick surviving to breed appears to be maximized if the chick reaches a high asymptotic mass during growth (Weimerskirch *et al.* 2000b). This suggests that climatic variation may have a stronger influence on breeding success and recruitment rate than on adult survival, particularly as these climatic effects are likely to be primarily linked to food availability (Cairns 1987, 1992). Thus, only the more drastic climatic events are likely to have clear effects on adults (see below), although the impact of those events influencing adult survival are expected to have the stronger influence on population dynamics.

7.3.2 Mortality

Direct evidence of increased adult mortality caused by environmental conditions is rare for seabirds. This may largely be due to the difficulty in determining the weather conditions that seabirds experience while they are at sea (assessments of the influence of weather on adult mortality are restricted to periods when they are on land). Consequently, in the absence of any obvious pathology, adult mass mortality is typically attributed to starvation. However, the cause of the starvation could be either the absence of prey or the inaccessibility of prey due to bad weather. For example, in 1983, 30,000 auks washed ashore from the North Sea following a series of storms (Harris and Wanless 1984). Conversely, in the Gulf of Alaska large numbers of common guillemots (*Uria aalge*) were found dead in 1993, apparently having died from starvation most probably due to the offshore unavailability of food (Piatt and van Pelt 1997). In the southeast Bering Sea, hundreds of thousands of emaciated short-tailed shearwaters died in 1997—a phenomenon quite likely due to long-term climatic changes (Baduini *et al.* 2001). These climatic effects could either be severe weather that hampered foraging, or anomalous oceanographic conditions that change the distribution and abundance of prey (Harris and Wanless 1996; Piatt and van Pelt 1997). For example, the highest numbers of seabird carcasses found along the central California coast between 1980–86 occurred during years of strong El Niño (Bodkin and Jameson 1991).

For chick mortality the relationship with weather conditions is more easily observed. The tendency for more extreme storms, or variations in prevailing wind conditions, may also have direct effect on the populations by increasing egg loss or chick mortality. For instance, heavy rain during chick period resulted in chicks dying of exposure when birds' feeding is disrupted in European shag (*Phalacrocorax aristotelis*) in the Cies Islands (NW Spain; Velando *et al.* 1999). Similarly, a severe gale at Isle of May, Scotland destroyed 49% of exposed European shag nests. This event forced the adults to rebuild their nest and lay a replacement clutch (Aebischer 1993). Burrow-nesting species, on the other hand, typically suffer nest loss during heavy rain after flooding and subsequent erosion (Warham 1990; Rodway *et al.* 1998). In general, such extreme weather will primarily affect birds nesting in the lower-quality nest-sites, and hence mostly affect the less-experienced birds (Coulson 1968).

7.3.3 Energetics

Birds are endothermic animals that maintain a constant core temperature by utilizing ingested or stored energy reserves. The thermoneutral zone (TNZ) is the range of temperatures between which the metabolism is not affected by temperature changes (Schmidt-Nielssen 1997). Whenever ambient temperature is outside a bird's TNZ, the bird experiences a thermoregulatory response that results in an increase in energy use. Ambient temperatures can vary over a very wide range in the North Atlantic, and seabirds are therefore often faced with environmental temperatures outside their TNZ (Dawson and O'Connor 1996). This is particularly so, for species making deep dives to catch their prey as sea temperature decrease with depth (Fig. 7.3; Koudil *et al.* 2000). However, over a typical annual cycle, sea-surface temperature (SST) generally varies much less than air temperature. Consequently, birds may respond behaviourally to extreme bouts of very hot or very cold air temperatures by remaining in contact with seawater, thus reducing the cost of thermoregulation.

The thermoregulatory response to a decrease in temperature has been demonstrated by measuring an increase in the field metabolic rate (FMR); the organism's daily energy expenditure measured in the field (Schmidt-Nielssen 1997; Ellis and Gabrielsen 2001). The maintenance of core temperatures through thermogenesis requires energy substrate deriving either directly from an increased food intake or from the utilization of body reserves. Opportunities to increase foraging effort may therefore be crucial

during periods of cold or wet weather. Incubation causes additional energetic costs due to the exchange of heat between their brood patch and the egg(s). When ambient temperature falls below the TNZ, incubation behaviour may increase adult metabolic rates by 19–50% compared to nonincubating birds. In seabirds, the metabolic rate during incubation is 1.2 times the basal metabolic rate (BMR) (Williams 1996). Furthermore, if eggs are left unattended, or when parents exchange incubation duties in very cold conditions, the subsequent re-warming of egg has additional energetic costs (Williams 1996; Schmidt-Nielssen 1997). During the chick-rearing period there may also be a marked increase in field metabolic rate, with the FMR/BMR-ratio varying from 1.8 to 4.8 (Ellis and Gabrielsen 2001). All stages of reproduction are certainly energetically stressful for seabirds, and poor weather conditions may further increase these costs. Obviously the availability of suitable prey within range of breeding colonies is very important during these critical periods.

Seabird chicks are highly dependent upon adults for the delivery of food required for their development. Young chicks may also be brooded by their parents to reduce thermoregulatory costs. More typically, however, the long distances between breeding colonies and foraging areas mean that chicks are left alone while adults forage at sea. The length of time that chicks are left alone will differ between species and sites, but may also vary in relation to climate-driven variation in the location of prey or of the cost of travel (see below). Chicks left alone in this way must confront the problem of heat loss which, again, may vary in extent due to

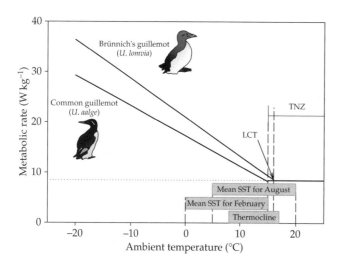

Figure 7.3 The influence of ambient temperature on the metabolic rate (MR) in two seabirds. The lower critical temperature (LCT) is the temperature under which the bird has to increase its metabolism to maintain its body temperature. LCT is the lower limit of the TNZ. Both seabirds are diving in water that changes in function of the season and latitude in North Atlantic (SST = sea surface temperature). Moreover, they are able to dive as deep as 200 m and are confronted to a thermocline; here average range for temperate water (adapted from Croll and McLaren 1993).

Note: TNZ could be very broad for polar species such as the Ivory gull (*Pagophila eburnea*) in which MR does not increase until −30 °C.

variations in ambient temperature, wind speed, and precipitation (Konarzewski and Taylor 1989). Nestlings may maintain high body temperatures in cold environments through plumage insulation, particular behaviours such as huddling, and thermoregulation. However, since the clutch size in seabirds is typically small, benefits from huddling does not occur except for penguins forming crèches. Consequently, chicks often have to rely on their own capacity to thermoregulate, which therefore may drain crucial energy that could otherwise be used for growth. Compared to adults, chicks also have a higher surface-to-volume ratio, which is less favourable for heat conservation (Visser 1998), and their underdeveloped muscles contribute little to heat production by thermogenesis (Hohtola and Visser 1998). Hence, changes of ambient temperature may have a strong impact on chick's energy budget. The type of nest site may moderate this problem, with the protection afforded by cavity-nesting reducing the chick's energetic costs compared to open-nesting (Martin and Li 1992). In addition, the cavity-nesters' chicks tend to grow slower than the open-nesters' and thus require relatively less food per day, which in turn can decrease the daily parental effort in foraging and nest protection (Martin and Li 1992).

Studies on several species (e.g. Manx shearwater, *Puffinus puffinus*: Harris 1966; Atlantic puffins: Anker-Nilssen 1987; Øyan and Anker-Nilssen 1996; and yellow-eyed penguin, *Megadytes antipodes*: van Heezik 1990) suggest that developing chicks faced with food shortages allocate resources preferentially to certain body parts. For instance, the rapid development of thermogenic tissue is especially important for chicks. Periods of bad weather preventing adults from foraging may then have an important effect on the chick's development and survival (Fig. 7.4(a); see also below). In Atlantic puffins, this relationship between the chick's development and climate is illustrated by a threshold value for mean sea temperature (Fig. 7.4(b)), below which there is no fledging. However, this relationship may be explained as a result of an indirect effect on food supply as well as by the expected direct effect of temperature on the chick's metabolism (Williams 1996; Anker-Nilssen and Aarvak 2002).

Climatic fluctuation could also directly influence seabirds by influencing the cost of flight which, in turn, leads to variations in the cost of foraging. There are considerable differences in the style of flight between different groups of seabirds, with

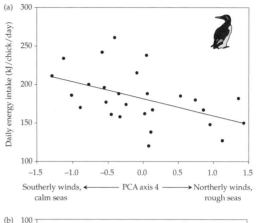

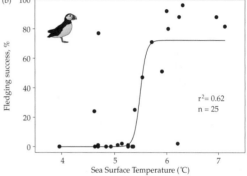

Figure 7.4 Relationship between reproduction and climate: (a) Relationship between mean daily energy intake of young common guillemots (*U. aalge*) on the Isle of May and the most important explanatory weather covariate (data from all-day watches carried out between 1983 and 1997). During stormy weather, the mean energy value of loads and the proportion of chicks attended are reduced indicating a decrease in the foraging efficiency (Finney *et al.* 1999). (b) Relationship between fledging success of Atlantic puffin chicks (*F. arctica*) in Røst in 1975–2001 (Anker-Nilssen and Aarvak 2002) and mean sea temperature at 0–75 m depth from March to July (G. Ottersen, IMR, Bergen, personal communication). A logistic regression curve is fitted to the data set ($F_{2,24} = 21.40$, $P < 0.0001$, Durant *et al.* 2003).

the two extremes being gliding and flapping. Variations in wind speed may profoundly affect the cost of flight (Furness and Bryant 1996) but the extent of this influence depends upon the flight style (Spear and Ainley 1997). For example, birds relying on flapping will be disadvantaged when the wind is strong, whereas the effect will be the opposite for gliding species (Furness and Bryant 1996; Finney *et al.* 1999). For example, the northern fulmar (*Fulmarus glacialis*) has a high at-sea FMR during low wind speeds because it uses gliding flight extensively during foraging (Furness and Bryant 1996). As a consequence, the lack of wind might limit the breeding range of this and other Procellariiformes species. In contrast, flapping species such as black-legged kittiwake (*Rissa tridactyla*) and the little auk (*Alle alle*) have been shown to have higher FMR during periods of strong wind (Gabrielsen *et al.* 1987). Similarly, Hodum *et al.* (1998) found that high FMR typically is due to the high cost of flight and pursuit diving in the pelagic feeding Cassin's auklet (*Ptychramphus aleuticus*).

As one of the major features of climatic fluctuation is the variation of wind speeds and direction, suggesting an important influence upon foraging energetics. Furthermore, such influences may affect different members of the seabird community in different ways. As such, changes in climatic conditions could affect the strength of both inter- and intra-specific competition. Gliding species with low flight costs can forage at great distances from breeding colonies (Weimerskirch *et al.* 2000a) and in areas of low productivity (Ballance *et al.* 1997). During periods of low productivity, flight proficiency becomes increasingly important because only species with relatively low flight costs may be able to move between prey patches (Ballance *et al.* 1997). Changes of climatic conditions may therefore influence both population distributions and competitive interactions between different seabirds.

7.4 Indirect influences of climate on seabirds

Seabird populations are typically more likely to be affected by climate variation indirectly rather than directly, through changes in the availability of key habitats or prey (cf. Schreiber 2001). For instance, climate change may create new, or redistribute existing, feeding areas for Arctic seabirds by melting the high-Arctic ice pack (Brown 1991). Alternatively, there may be changes in breeding site availability or quality through sea-level change or variations in the frequency of extreme storm events. Variations in temperature may also affect the extent of sea ice, which has been shown to influence the mode and cost of travelling between breeding and foraging areas in incubating emperor penguins (Williams 1995; Barbrand and Weinershirch 2001; Croxall *et al.* 2002).

Changes in prey availability have been shown to influence several key demographic parameters, even

if most studies have focussed upon variations in reproductive success (e.g. Martin 1987; Barrett and Krasnov 1996). For example, successful reproduction in several seabirds in the northwest Atlantic is related to the availability and timing of the inshore movements of the capelin (Montevecchi and Myers 1996). One of the earliest studies to link variations in climate to such a relationship between prey availability and reproductive success in North Atlantic seabirds was carried out by Aebischer *et al.* (1990) and documenting parallel long-term trends in weather conditions, prey abundance, and breeding performance of North Sea kittiwakes. This indirect role of climate variation has later been suggested through several studies. For example, in years when the arrival of capelin (*Mallotus villosus*) in Newfoundland is delayed, hatching, fledging, and breeding success of kittiwake, herring gull (*Larus argenteus*) and great black-baked gull (*Larus marinus*) are reduced (Regehr and Rodway 1999). Such delayed capelin arrival was explained by an anomalously cold SST resulting in a delay of one month of the spawning migration of the capelin (Nakashima 1996). Data on the Atlantic puffins of Røst, North Norway show a threshold relationship between food resources (first-year herring, *Clupea harengus*) and fledging success (Anker-Nilssen 1992; Anker-Nilssen and Aarvak 2002; Durant *et al.* 2003) such that there is complete breeding failure when prey abundance is below a certain level. The Norwegian spring-spawning stock of herring has experimented great fluctuations during the twentieth century (Toresen and Østvedt 2000, Chapter 6); presumably to a large extent as a response to changes in ocean climate.

These examples may be explained by the match/mismatch of food availability and requirement (Cushing 1990; see also Chapter 1). If herring availability does not match the Atlantic puffin's requirements at the time of rearing, it produces a dramatic reduction in chick survival (Anker-Nilssen 1992; Anker-Nilssen and Aarvak 2002; Durant *et al.* 2003). Even during years of high herring productivity, a too early puffin's breeding relative to the growth and migration of its main prey, would render the prey unavailable for the chick rearing. This mismatch can thus be considered both in terms of timing and abundance. Changes in climatic conditions between the period when the birds assess the environmental quality prior to laying and the actual time of chick rearing could modify food availability creating a mismatch.

As discussed above, foraging seabirds select habitats where prey are more predictably concentrated and more easily captured. For seabirds, the choice of such foraging habitats is especially important since prey densities are low in many oceanic areas, and prey may remain at inaccessible depths. In both the horizontal and vertical dimensions, then, seabirds must focus their foraging activities in areas where prey interact with different processes to produce predictably located concentrations in near-surface waters. In some cases, such concentrations may be a result of the interactions with other predators (e.g. where the foraging activities of sub-surface predators enhance the surface availability of prey for shallow diving seabirds; Ballance and Pitman 1999). In other cases, however, prey concentrations occur where physical processes produce either areas of high productivity, or aggregations of prey.

Obviously, changes in ocean climate may thus influence seabird prey availability by affecting timing, location or strength of these oceanographic features. Recent attempts to understand the potential indirect impacts of climate variation on seabirds have therefore explored relationships between these oceanographic features and the birds' distribution and demographic parameters. In the following, we address the potential effects on seabirds by changes in SSTs, frontal systems and larger-scale proxies of ocean climate.

7.4.1 The influence of SST on seabirds

The abundances and assemblages of seabirds are influenced by short- and long-term changes in SST (Veit *et al*. 1996, 1997; Guinet *et al*. 1998). For example, in the Antarctic, the blue-petrel (*Halobaena caerulea*) breeding performance is reduced if their body condition is lowered as a result of a high SST during the preceding winter. Similarly, in the Pacific Ocean there is a coupling between seabird reproduction and ocean temperature (Ainley *et al*. 1994, 1996; Veit *et al*. 1996) leading to poorer reproductive performance during warm-water years for inshore species such as the sooty shearwater (*Puffinus griseus*) and increase for offshore species such as the Leach's storm petrel (*Oceanodroma leucorhoa*; Veit *et al*. 1996). This implies that a long-term increase of the SST certainly could result in a decrease in the abundance of some seabirds (Veit *et al*. 1996). Although such links are now being described for more systems, the causal relationships between SST and reproductive success are less clear in many of the sea-bird

systems. Nevertheless, some work does point to potential links through known effects of temperature on key prey populations. During the early 1990s, cold-water events in the northwest Atlantic appear to have inhibited migratory pelagic species such as mackerel (*Scomber scombrus*) and squid (*Illex illecebrosus*) from moving into the region (Montevecchi and Myers 1997). As the distance to food supply is a main factor influencing seabird reproduction, this in turn created a major shift in the pelagic food webs (Montevecchi and Myers 1997). As a consequence, there were profound negative effects on the reproductive success of surface-feeding birds such as black-legged kittiwakes (Regehr and Montevecchi 1997). This highlights that a slight change in oceanographic conditions, possibly associated with climate change, might have a large-scale and profound effect on seabird population. This indirect effect of SST on seabirds through changes in their prey resources is also seen in the Atlantic puffin. At the puffin colonies in Røst, fledging success is related to both sea temperature and food availability, both factors being correlated (Durant *et al*. 2003). Here, the lower sea temperatures affect the population of the main prey for the seabird, creating a mismatch between the Atlantic puffins' energy requirements and their food availability. However, this influence is quite complex since changes in SST do affect different species in different ways (Fig. 7.5). Warm-sea temperatures tend to decrease the plankton productivity, but low temperatures may also negatively affect fish growth (Chapter 6), potentially having effects thousand of kilometres from the seabirds' breeding colonies (Montevecchi and Myers 1997). Depending upon the birds' feeding biology (planktivorous or piscivorous) changes in the SST may have a variety of effects. For example, the reproductive success of planktivorous auklets in northwest Pacific (*Aethia cristatella* and *Cyclorhinchus psittacula*) is negatively correlated with SST, whereas for piscivorous puffins (*Lunda cirrhata* and *Fratercula corniculata*) it is positively correlated (Kitaysky and Golubova 2000).

7.4.2 Fronts, currents and seabirds

Oceanographic features such as fronts at the boundaries of water masses, ice edges, and currents that interact with bathymetry may all concentrate prey. The mixing of water masses at these features creates conditions that support all the members of the food web. Frontal systems support enhanced stocks of phytoplankton, zooplankton, fish, and seabirds

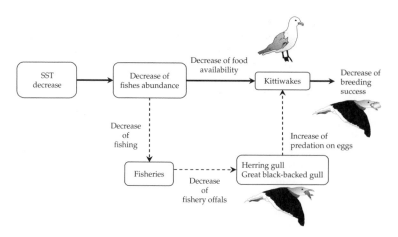

Figure 7.5 Interactive and synergic effect. The decrease of the SST has a double negative effect on kittiwakes (*R. tridactyla*) trough the decrease of the food supply and the increase of nest predation by gulls (*L. argentatus* and *L. marinus*; adapted from Regehr and Montevecchi 1997).

(Kinder *et al*. 1983; Coyle and Cooney 1993; Decker and Hunt 1996). As a consequence, seabird populations are associated with these physical features (Hunt 1990; Schneider 1990); for example, the distribution of Antarctic and sub-Antarctic seabirds is closely linked to the polar front and sub-Antarctic front (Hunt 1991; Guinet *et al*. 1997; Ainley *et al*. 1998; Charrassin and Bost 2001). The abundance and the success of the reproduction of these seabirds are intimately linked to the extent of the sea-ice (Hunt 1991; Barbraud *et al*. 2000; Barbraud and Weimerskirch 2001; Croxall *et al*. 2002) and to the changes of winter temperature that influence ice formation (Barbraud and Weimerskirch 2001). In the North Atlantic we may expect similar phenomena, and several frontal systems within the Irish Sea and North Sea appear to provide predictable resources for seabirds (Begg and Reid 1997; Hunt *et al*. 1999; Skov and Durinck 2000). However, such fronts may only be formed seasonally and can be subject to variations in response to wind-induced mixing (Allen *et al*. 1980). Usually, the influence of hydrography on seabird distribution is through variations in surface salinity, transparency and thermal stratification. For example, fulmar (and to some extent common guillemot) occurrence is correlated with highly saline, thermally stratified water with high-water clarity—all characteristics of their main prey habitat (Garthe 1997). In the northwest Atlantic, seabird populations off Newfoundland are linked to sea currents, and variations in the

strength of the Gulf Stream may influence the migration of the pelagic prey (Montevecchi and Myers 1995).

7.4.3 Large-scale influences on seabirds

In the Southern Hemisphere and the Pacific, changes in seabird populations have been well-studied in relationship to large-scale climatic phenomena (reviewed in Schreiber 2001) where El-Niño-Southern Oscillation (ENSO) influences wind and sea currents which may lead to important changes in temperature, precipitation, and food resources. This phenomenon can affect birds all around the world, as illustrated by the black-throated blue warbler (*Dendroica caerulescens*), a North American migratory passerine whose demographic rates varied in relation to the ENSO (Sillett *et al*. 2000). For seabirds, ENSO may reduce both breeding success and adult survival. During the most severe cases of ENSO, many adult seabirds can die due to the disappearance of their food sources, affecting both population size and population structure (Schreiber and Schreiber 1984; Duffy 1990; Piatt and van Pelt 1997; Bertram *et al*. 2000). ENSO seems to affect conditions during the breeding season in many seabirds species (Croxall 1992; McGowan *et al*. 1998). In the Galapagos penguin (*Spheniscus mendiculus*), it has been reported that body condition is related to the ENSO events, with a deterioration in body condition during ENSO leading to a reduced

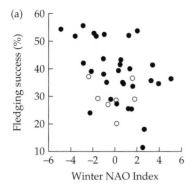

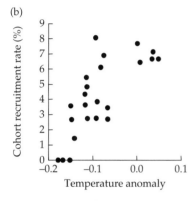

Figure 7.6 Variation in (a) the percentage of northern fulmars (*F. glacialis*) nests producing fledglings in relation to the NAO index for the winter before the breeding season (1952–95) and (b) the effect of the temperature on the percentage of each cohort of chicks (1958–80) that recruits to the colony. The recruitment rate for different cohorts of chicks is significantly related to anomalies in Northern Hemisphere growing season temperatures (Thompson and Ollason 2001).

breeding success (Boersma 1978) and a decline of the population (Boersma 1998).

In the Atlantic Ocean, an equivalent phenomenon to the ENSO is the NAO. The impacts of the NAO appear less extreme (and less clear) than the mass mortalities associated with ENSO, and it is only since the mid-1990s that temporal patterns in the NAO have been related to variability in biological populations (Ottersen *et al.* 2001; Stenseth *et al.* 2002). Reported effects of the NAO on the abundance of zooplankton (Chapters 4 and 5) and key fish prey (Chapter 6) suggest that the NAO may influence the dynamics of seabird populations, but it is only recently that studies have started to explore these relationships. Nevertheless, there is evidence that the winter NAO influences both the probability of breeding and subsequent reproductive success in the northern fulmar (Thompson and Ollason 2001). Furthermore, cohort recruitment rates at this Scottish colony were related to temperature anomalies in the birds' first year (Fig. 7.6), highlighting the potential for further research that explores the relationships between these different large-scale proxies of climate variation and seabird population dynamics.

7.5 Conclusion

Effect of climate change on seabird populations may take many years to become apparent (Thompson and Ollason 2001). Its effect is complex and involves a large number of physical and biological processes. To understand the true mechanisms it is often necessary to conduct a deep, thorough ecological study of the food web and its many different relationships with the environment. However, an overall pattern of the response of seabirds to climate begins to appear and very interesting interdisciplinary studies are becoming more and more common. Seabirds are sensible to climate change either positively as shown by the extension of the fulmar population or negatively as shown by the Atlantic puffins. Thanks to their position as top predators, their response to climate change is a good index of its effect on the whole food web. In order to improve our scientific understanding of what might happen under various scenarios of global change, the study of seabird populations could be of great value.

PART IV

Community ecology

While previous chapters have primarily focused on the response of *populations* to climate variability, this part deals mainly with climate impact on *communities*. As pointed out by Lekve and Stenseth in Chapter 10, to understand the influence of climate variability on community ecology we need to consider the responses of each species as well as how the interactions between species within the community are influenced. This is so whether we are dealing with communities of phytoplankton (Chapter 8), benthos (Chapter 9) or fish (Chapter 10).

In Chapter 8, Belgrano and coworkers study the potential connections between climate and phytoplankton community composition. They first describe a general statistical approach for analyzing population dynamics of phytoplankton species within the community setting, considering both density-dependence and exogenous factors like climate. Through an example from the Gullmar fjord on the Swedish west coast, they show how the North Atlantic Oscillation (NAO) may influence phytoplankton species diversity. In this case, the positive phase of the NAO corresponded to an increase in diversity. Furthermore, large phytoplankton blooms are linked to nutrient transport associated with stronger southwesterly winds from the Kattegat, typical for positive NAO phases. In general, Belgrano and coworkers emphasis that the phytoplankton species have a unique covariance relationship, resulting in a transfer function that indicates that each species can react to the same climate signal in a different manner.

We have no chapter explicitly dealing with climate impact on zooplankton community ecology. However, spatial variability in sea-surface temperature (SST) has been suggested to explain geographical distribution patterns in Atlantic zooplankton diversity. Analyses have shown that SST measured by satellite explained nearly 90% of the geographic variation in planktic foraminiferal species richness throughout the Atlantic Ocean (Rutherford, S., D'Hondt, S. and Prell, W. 1999. *Nature*, **400**; 749–753). The close relation was explained by zooplankton diversity being primarily controlled by vertical niche availability dictated by the thermal structure of the near-surface ocean. Dynamics within the zooplankton communities are also dealt with in Chapter 4.

Most of the available literature on benthic communities considers only aspects related to species diversity, species composition, and succession. The reason for this may be partly due to climate variability has until a few years ago generally not been considered an important regulating factor for benthic community dynamics. On the other hand, the lack of long-term, regularly sampled benthic data has restricted the feasibility of statistical analyses of climate–benthos relations. However, recent studies have pointed towards the importance of considering the effects of large-scale climatic fluctuations in relation to changes in the dynamic of benthic populations across both temporal and spatial scales. In Chapter 9, Hagberg and coworkers presents a variety of studies on climate influence on marine benthos communities in the North Sea, Skagerrak, and Baltic. Their main conclusion is that the NAO affects single species, as well as whole communities and functional groups, but that the impact may differ geographically and depend on the species composition of the communities.

In Chapter 1 we described how a population's response to a climate signal may be linear or non-linear. However, Lekve and Stenseth in Chapter 10 point to the concept of linearity having a different meaning at the community level. Here linearity is to be understood as linear among species, that is, most species are affected in a qualitatively similar fashion, without changing interaction effects among species. Lekve and Stenseth present several examples of how fish communities may respond to climate variability. The worldwide out-of-phase fluctuations of sardines and anchovies is one, the Russel cycle in the English Channel another. The latter is a classic example of how the geographic range of groups of species may shift as a response to warming or cooling. The main example of Chapter 10 is based on the unique community data regularly sampled along the Norwegian Skagerrak coast since 1919. A model for studying the impact on species richness of the NAO as well as regional wind conditions and sea temperature is introduced and discussed.

CHAPTER 8

Responses of phytoplankton communities to climate variability

Andrea Belgrano, Mauricio Lima, Nils Chr. Stenseth, and Odd Lindahl

8.1 Introduction

The potential connections between climate and phytoplankton community composition dates back to the work by Hutchinson (1967). More recently, studies by Lehman (1992, 2000) and Jassby *et al.* (1996) suggest that climate variability may be linked directly and indirectly to phytoplankton in relation to streamflow as well as changes in biomass and chlorophyll-*a* (Lehman 1996). Shifts in phytoplankton species composition in relation to climate forcing have been reported for the California coast by Tont (1989) observing that dinoflagellates abundance was higher than diatoms in relation to a reduced upwelling during the El-Niño years.

The 1977 climate regime shift along the California coast was also related to a decrease in diatom density (Lange *et al.* 1990). Barber and Chavez (1983) reported a change in the phytoplankton species composition from diatoms to dinoflagellates in relation to changes in the upwelling region off the Peruvian coast linked to El Niño. In the North Sea and neighbouring areas, changes in phytoplankton density and species composition have been observed in relation to the climate shift around 1977 (Dickson and Reid 1983; Colebrook 1986; Aebischer *et al.* 1990) as well as in the late 1980s. Furthermore, long-term fluctuations have been related to the North Atlantic Oscillation (NAO), as reported by Reid *et al.* (1998), Beaugrand *et al.* (2000), Edwards *et al.* (2001), and Edwards *et al.* (2002). Data from the continuous plankton recorder (CPR) survey were fundamental for these studies (see Chapter 5 for description of the CPR data). The anomalous periods in the late 1970s and late 1980s seem to be largely synchronous with unusual ocean climate conditions that have occurred

episodically over a timescale of decades. The unusual ocean climate conditions prevailing at these two time periods appear to contain important hydrographical elements that involve oceanic incursions into the North Sea (Edwards *et al.* 2002).

When analysing the effects of climate on community dynamics we need to disentangle the impact of abiotic and exogenous biotic factors, as well as factors internal to the population. In this chapter, we first present a general statistical approach to studying changes in phytoplankton species abundance in relation to climate variability. Thereafter we apply this method to an example where changes in phytoplankton species diversity, the dynamics of three phytoplankton species in a Swedish fjord, are related to the NAO and other abiotic factors. Finally, we conclude with a general outlook underlining the importance of further studies of changes in phytoplankton biovolume.

8.2 A statistical approach

Population dynamics of phytoplankton species are considered the result of feedback and exogenous structure (see also Chapter 1). We can represent these ecological relationships using a very general model in terms of reproduction and survival of individuals (Berryman 1999), which represent a variant of the Ricker (1954) discrete-time logistic model influenced by climate and stochastic forces:

$$N_t = N_{t-1} \cdot e^{[a_N + f_1(N_{t-1}) + f_2(C_t^1) + K + f_i(C_t^{i+1}) + \varepsilon_t]} \qquad (8.1)$$

Here N_t is the phytoplankton abundance at time t and C_t^{i+1} is the *a* variable representing the exogenous factors. The functions $f_i(\bullet)$ represents the effects of density-dependence and exogenous factors on phytoplankton population dynamics and ε_t represent normal distributed stochastic perturbations. An alternative way to express Eq. (8.1) is in terms of the realized per capita population growth rates, which represent the processes of individual survival and reproduction that drive population dynamics, this is the R-function (*sensu* Berryman 1999). Defining $R_t = \log(N_t) - \log(N_{t-1})$ Eq. (8.1) can be expressed as a R-function (*sensu* Berryman 1999):

$$R_t = a_N + f_1(N_{t-1}) + f_2(C_t^1) + K + f_i(C_t^{i+1}) + \varepsilon_t. \tag{8.2}$$

This model represents the basic feedback structure and integrates the exogenous and stochastic forces that drive population dynamics in nature. The basic idea for population analysis is to choose a family of functional forms for f to fit time-series data. This model formulates Eq. (8.2) as an additive non-linear model (see Bjørnstad *et al.* 1998 for an ecological example). Model (8.2) represents a Generalized Additive Model (GAM; Hastie and Tibshirani 1990). The choice of the functional form of the f_i functions can be approached using natural cubic splines (Green and Silvermann 1994; see Stenseth *et al.* (1997) and Bjørnstad *et al.* (1998, 1999) for ecological examples). The most parsimonious

model can be selected by criteria like the Akaike Information Criterion (AIC) or Schwartz's Bayesian Criterion (SBC).

8.3 Phytoplankton diversity and dynamics, an example

The data used in this example consist of abundance estimates of 40 phytoplankton species (cell counts), with a focus on the three species *Skeletonema costatum* (diatom), *Ceratium tripos*, and *Ceratium furca* (dinoflagellate), and abiotic factors including the NAO and local winds (Belgrano *et al.* 1999, 2001). Measurements were made on a monthly basis from 1986–96 at a station located at the mouth of the Gullmar Fjord on the Swedish West Coast (58° 15 N 11° 26 E; Fig. 8.1).

The NAO fluctuations (Fig. 8.2(a)) shows a switch in the late 1980s and early 1990s from a negative to a positive phase concomitant with warmer winter sea surface temperature (SST) as reported by Belgrano *et al.* (1999). This is a good indication that during the end of the 1980s, as shown by Reid *et al.* (1998), the seasonal pattern of phytoplankton has been extended in the North Sea and adjacent area beyond the usual pattern of spring/summer/autumn due to the warmer SST condition that possibly forced by the shift in the NAO from a negative to a positive

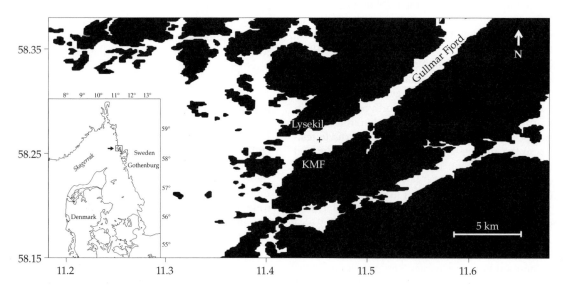

Figure 8.1 Location of the sampling station in the Gullmar Fjord, Sweden (58°N, 11°E) indicating Kristineberg Marine Research Station (KMF).

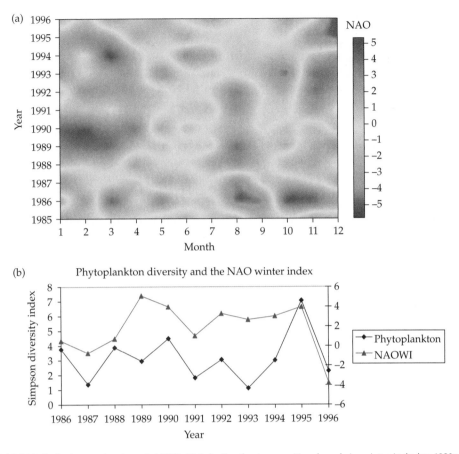

Figure 8.2 (a) NAO distribution covering the period 1985–96, indicating the strong positive phase during winters in the late 1980s and (b) Simpson's diversity index together with the NAO winter index (called NAOWI in panel b). (See Plate 7.)

phase in 1988 (Rodwell *et al.* 1999). The Simpson's diversity index (*D*; Simpson 1949) (Fig. 8.2(b)), was correlated ($r = 0.41$; $p < 0.01$) with the NAO winter index. This indicates that higher diversity in phytoplankton species composition may be related to the positive phase of the NAO. A closer look at the changes in species composition showed that phytoplankton diversity was low during 1987 due to a large dominance of *S. costatum*. The increased number of cells of the dinoflagellate species *Ceratium furca* in 1992 coincided with the absence of two diatoms species *Chaetoceros socialis*, and *Thalassiosira nordenskioeldii* between 1990 and 1993. *Thalassiosira nordenskioeldii* can be regarded as an important species in terms of biomass for the spring bloom due to its size (Tiselius and Kuylenstierna 1996). *Chaetoceros socialis* can be regarded as an important food source for

copepods and its absence might have caused a shift in their grazing habits. These temporal changes in the species compositions during the spring bloom may result in an increase in the grazing activity by heterotrophic dinoflagellates. This may in turn prevent an increase in the diatom concentrations and as reported by Tiselius and Kuylenstierna (1996) disrupt the phytoplankton spring bloom. A plausible explanation (Lindahl *et al.* 1998) is that stronger winds during the high NAO years lead to more vertical mixing, which in turn enhanced nutrient concentrations in the surface layers, thus favouring an increase in dinoflagellate abundance. The relationship observed between the NAO, diatoms, and dinoflagellates on the Swedish west coast seems to be a general pattern very similar to the one observed for the northeast Atlantic by Edwards *et al.* (2001).

Phytoplankton diversity related to a number of limiting resources may be generated by non-linear dynamics (Huisman and Weissing 1999). A first order density dependence was detected in the three species of phytoplankton (Figs 8.3(a), 8.4(a) and 8.5(a)). The results presented in Fig. 8.3 shows the response of that changes in the abundance of the diatom *S. costatum* where strongly related to the variability in nitrates NO$_3$, salinity and density associated with the southwest winds. These changes in the abiotic condition along the Swedish west coast have earlier been observed by Lindahl *et al.* (1998). They suggested that a transport of nutrients associated with stronger SW winds from the Kattegat area directly linked to a typical positive NAO scenario, could be related to large phytoplankton blooms as the one observed for *S. costatum* in 1987. The model for *S. costatum* including the covariates in (SC, Fig. 8.3) explained 95% of the observed variance, however, no direct relation with NAO was found for this species. The models obtained for the dinoflagellate *C. furca* (CF, Fig. 8.4) showed that this species was directly

related to changes in the NAO, SW winds and density. *Ceratium tripos* (CT, Fig. 8.5), showed a direct relation to changes in temperature, nitrate concentrations, southeast wind and northwest wind. The model that included nitrates explained 75% of the variance, reflecting the direct dependence of this species for nitrates availability. The abiotic conditions for the formation of dinoflagellates blooms along the Swedish west coast (Belgrano *et al.* 1999) suggest that the covariates selected by the models for CF and CT may be regarded as the explanatory variables underlying the changes in the abundance of these species. The use of GAMs models in the analysis of ecological time-series provided important information on the effect of both exogenous and stochastic forcing of the dynamics of marine phytoplankton.

8.4 Outlook

Species abundance models have been discussed in great detail by May (1975). However, the results

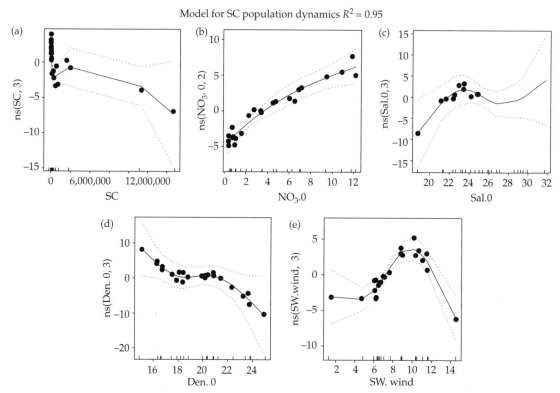

Figure 8.3 GAM Model for *S. costatum* population N_{t-1} (a), nitrates NO$_3$ (b), salinity (c), density (d), and southwest wind (e).

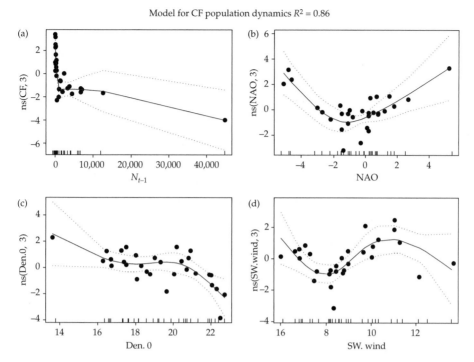

Figure 8.4 GAM Model for *C. furca* population N_{t-1} (a), NAO (b), density (c), and southwest wind (d).

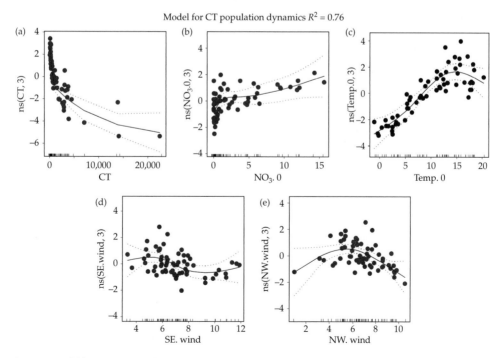

Figure 8.5 GAM Model for *C. tripos* population N_{t-1} (a), nitrates NO_3 (b), temperature (c), southeast wind (d), and northwest wind (e).

presented here suggest the importance of considering non-linear dynamics to investigate the relationships between changes in phytoplankton species abundance and a high number of covariates. At the interface of a dynamic system (Lindahl 1987; Tiselius and Kuylenstierna 1996) we can expect high phytoplankton variability, although species composition and coexistence may reflect well-defined assemblages that are taxonomically related. This is suggested by the Connell (1978) hypothesis of intermediate disturbance and the role of disturbance acting at a variety of scales. Phytoplankton species have a unique covariance relationship, resulting in a transfer function that indicates that each species can react to the same environmental fluctuations in different ways. As suggested by Harris (1986) the amplitude of this transfer function may increase at longer timescales providing an indication of the integrative properties of cellular physiology. A reduction in the vertical mixing from a few days to a period of about two weeks may reduce the vertical mixing leading to increased biomass (Harris 1983). Therefore an increase in diversity (as the one observed in our example) may be related to the variance in the physical structure.

Climate variability has been shown to be potentially linked to changes in phytoplankton biovolume in relation to primary productivity and upwelling of nutrients. This aspect, as pointed out by Lehman (2000), needs further attention since changes in phytoplankton biovolume are linked to the fate of biogenic carbon in the pelagic food web.

CHAPTER 9

Effects of climate variability on benthic communities

Jacob Hagberg, Björn G. Tunberg, Gunther Wieking, Ingrid Kröncke, and Andrea Belgrano

9.1 Introduction

Variability of population abundance is a typical feature of marine benthic communities at a variety of temporal and spatial scales (Cushing 1981; Botsford *et al.* 1982; Gray and Christie 1983; Baumgartner *et al.* 1992). Climatological factors are believed to be responsible for many aspects of temporal variability of marine communities (Aebisher *et al.* 1990; Francis and Hare 1994; Beamish *et al.* 1995; Stein and Lloret 1995; Ware 1995). The comparable role of periodic climatological variation in regulating physical and biological oceanographic processes in the North Atlantic region has been increasingly studied during recent years (see, e.g. earlier chapters in the book).

In order to elucidate the large-scale (oscillatory) climatological impacts on the benthic communities it has been necessary to analyse long-term benthic data from quantitative samples that have been collected at many sites on a regular basis over a large number of years (see, e.g. Chapter 1). Unfortunately, very few time-series are available that meet these terms. Recent shallow coastal benthic research has shown evidence of long-term periodicity in biotic communities, which appears to be associated with climatic periodicity (Kröncke *et al.* 1998; Tunberg and Nelson 1998). Tunberg and Nelson (1998) presented evidence for a periodic component in benthic community parameters and demonstrated that this periodicity is generally synchronized regardless of water depth. They also proposed that the periodic behaviour of benthic community processes is driven by a periodicity on timescales similar to the periodicity of the North Atlantic Oscillation (NAO). This has been further confirmed by studies reported by Hagberg and Tunberg (2000).

A similar oscillatory pattern (7-year periods) has also been reported for the abundance of the benthic amphipod *Pontoporeia affinis* in the Bothnian Sea of the Baltic over the period 1961–74 (Gray and Christie 1983). Benthic foraminiferans have been studied in sediment cores from Gullmarsfjorden on the Swedish west coast and from the Skagerrak/Kattegat area. The foraminiferan community show a shift in the 1930s and in the 1970s in the Skagerrak (Alve 1996) and this shift has also been found in the Gullmar fjord between 1974 and 1976 (Nordberg *et al.* 2000). Both these shifts correspond to a shift in the NAO index from a predominantly positive phase to a negative phase in the 1930s and back to a more positive phase again in the late 1980s (Hurrell and Van Loon 1997; Nordberg *et al.* 2000). An interesting feature of these changes is that the fjord community of these foraminiferans exhibited a decrease of agglutinated foraminiferans, whereas in the open Skagerrak/Kattegat region they increased instead. Nordberg *et al.* (2000) suggested that the communities are affected by a factor connected to the NAO but that this factor has different effects in inshore and offshore areas.

Climatic variability is most likely to have an effect on marine populations and it is important to understand further the synchronicity between climatic oscillations and the life cycle of benthic species. In this chapter, we are summarising several examples from different regions in order to illustrate these connections.

9.2 The North Sea

Several investigations in the Wadden Sea and coastal areas indicated that cold and mild winters affect macrofaunal communities (Ziegelmeyer 1964; Zeiss and Kröncke 1997). Highly successful recruitment after cold winters results in increased biomass (Beukema 1990, 1992) indicating the importance of cold winters for the structure of littoral benthic communities (Beukema 1990, 1992; Reise 1993; Beukema *et al.* 1996). After a period of warm winters, Beukema (1990, 1992) found an increase in species number, stable total biomass, but a decrease in individual biomass of bivalves.

Off the island of Norderney sublittoral macrofaunal communities were severely affected by cold winters, but mild meteorological conditions during winter resulted in an increase in total biomass since 1989 (Kröncke *et al.* 1998). They showed that abundance, species number, and biomass in the second quarter of the year correlated with the NAO index. The mediator between the NAO and benthos is probably the sea surface temperature (SST) in late winter and early spring. Kröncke *et al.* (1998) explained this as a result of lower mortality and higher production and reproduction in mild winters in combination with an earlier spring bloom and possible synergistic effects between climate and eutrophication. According to the species composition off Norderney, Kröncke *et al.* (2001) found a higher percentage of arctic-boreal species after the cold winter of 1978/79 until 1982, and from 1984 to 1987 (a period with three rather cold winters and a negative NAO index) together with a rather high percentage of cold-temperate species. But after 1988 (in connection with a positive NAO index) the percentages of warm-temperate species increased.

In the southern Bight of the North Sea the *Abra alba* community also showed a response to climate variability (Fromentin and Ibanez 1994). The analysis of a time series from 1977 to 1991 indicated that maxima of density always occurred during mild winters (1981–83, 1988–91) while very low densities of *A. alba* were concomitant with cold winters. That study confirmed that the *A. alba* time-series showed a periodicity of about 7.5 years corresponding to the same cycle of the air temperature pointing out the importance of the relation between the winter season and the fluctuations in the density of this

species. A similar cycle was also found by Glémarec (1993) in the interannual variability of benthic communities in Bretagne (France).

At a benthic station off Northumberland, Buchanan and Moore (1986) identified cold winters as an ephemeral destabilizing factor only. Cold winters favoured the survival in the dominant species at the expense of the lesser-ranked species due to reduced primary production. Most attempts of explaining the fluctuation in macrobenthos communities have been focused on factors that effect the food-availability (e.g. Rosenberg 1995). For example, the benthic fauna on the Northumberland coast showed a stable and repeating biennial pattern (Buchanan 1993). Food limitation was suggested to be responsible for this pattern. A high abundance in the spring resulted in a large number of competitively competent mature individuals. This resulted in few settled, less competitively, competent juveniles which lead to low abundances in the fall, and vice versa (Buchanan 1993). However, this pattern broke down in 1981 when abundances increased instead of decreasing, which was attributed to an increased food input to the benthos correlated with the intensity of inflow of Atlantic water masses into that area (Austen *et al.* 1991; Buchanan 1993). Strong connections between the inflow of North Atlantic Water to the North Sea and the NAO (Planque and Taylor 1998) suggest a possible relationship between the NAO and the Northumberland benthos.

In the frame of a long-term comparison between 1985–87 and 1996–98, Wieking and Kröncke (2001) found marked changes in macrobenthic communities of the Dogger Bank (central North Sea) as a result of the rise in the NAO. Due to an increase in bottom temperatures southern species such as the amphipod *Megaluropus agilis* (Fig. 9.1) and the ophiurid *Amphiura brachiata* increased in abundance on top and at the southern slope of the Dogger Bank and occurred even in the deeper parts in 1996–98. In contrast abundances of northern species (e.g. *Corophium crassicorne*, *Siphonocoetes kroyeranus* [Amphipoda], *Nuculoma tenuis* [Bivalvia]) decreased on top and south of the Dogger Bank. The additional increase in abundances of interface-feeding species such as the polychaet *Spiophanes bombyx* coincided with a higher primary production in the central North Sea (Reid *et al.* 1998).

Benthic communities along the northern slope of the Dogger Bank were strongly affected by increasing wind stress and stronger currents at the northern

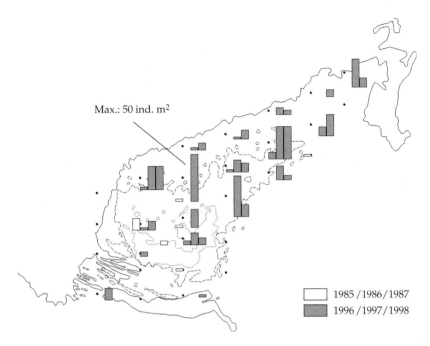

Figure 9.1 Spatial distribution and abundance of *M. agilis* on the Dogger Bank in 1985–87 (white columns) and 1996–98 (dark columns) (from Wieking and Kröncke 2001).

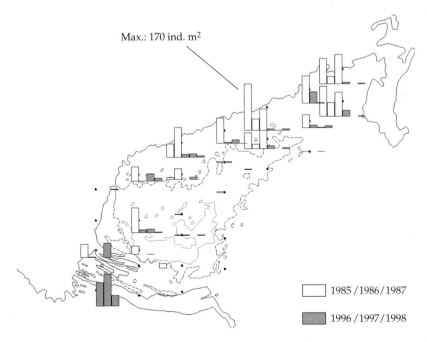

Figure 9.2 Spatial distribution and abundance of *O. borealis* on the Dogger Bank in 1985–87 (white columns) and 1996–98 (dark columns) (from Wieking and Kröncke 2001).

slope of the Dogger Bank. Factors which are also influenced by the positive NAO index during the 1990s (Siegismund and Schrum 2002; Siegismund 2001). Changes in larval supply, food availability, and sediment composition caused by resuspension of fine material lead to a decrease in species occurring on fine sand (*Ophelia borealis* [Polychaeta] Fig. 9.2) compared to the 1980s whereas abundances and total number of species preferring coarser and unstable sediment (e.g. *Echinocyamus pusillus* [Echinodermata]) increased in the 1990s. The decrease of total abundances, changes in trophic structure such as the increase in hyperbenthic predators (*Cerianthus lloydii* [Anthozoa], *Corymorpha nutans* [Hydrozoa]) and higher diversity of feeding types as well as the increase of total number of northern species were related to a stronger inflow of northern water masses and a connected decrease in food quantity and quality. These changes resulted in a pronounced separation of northern and southern macrofauna communities along the northern slope of the Dogger Bank during the positive NAO index period in the 1990s.

Pearson and Mannvik (1998) described an increase in detrital carbon supply to the benthos between 1993 and 1996 resulting in a considerable increase in macrobenthic faunal densities and species richness in the central North Sea, north of the Dogger Bank. They assumed that these changes are driven by climatic forces influencing the overlying water masses and some increase in pelagic productivity and benthic pelagic coupling. However, these observations are well in line with the NAO-driven increase in primary production in the Central North Sea.

With regard to the northern North Sea, Witbaard (1996) showed that year-to-year variation in the wind-driven component of the East Shetland Atlantic Inflow (ESAI) explains a significant part of the growth variations of the bivalve *Arctica islandica* from the Fladen Ground. Variations in the ESAI and the Dooley Current may influence the strength of the eddy system over the Fladen Ground and consequently the accumulation of material in its centre and the eddy mediated food-supply.

9.3 The Skagerrak and the Baltic Sea

9.3.1 Skagerrak

The first indications of a linkage between the benthos and the NAO in the Skagerrak, were

reported by Tunberg and Nelson (1998). They described a correlation between the macrobenthic soft bottom fauna and the NAO index, which indicated that the abundance and the biomass of the benthos generally decreased when the NAO index increased. Earlier studies reported a strong covariation of benthic abundances at different sites on the Swedish Skagerrak coast (Josefson 1987; Josefson *et al.* 1993). They suggested that land runoff was the key factor, driving this covariation. It was found that chlorophyll-*a* and dissolved inorganic nitrogen correlated positively with runoff and suggested that increased primary production, promoted by increased land runoff, increased the input of particulate organic nutrients to the benthos (Josefson *et al.* 1993). During this period the eutrophication debate was very intense and answers were sought primarily with this particular disturbance factor in mind.

Tunberg and Nelson (1998) found the same correlation between the benthos and land runoff as Josefson *et al.* (1993) found in the 1980s. They also concluded that the land runoff from the Swedish west coast correlated negatively to the NAO index ($r = -0.57$). This may therefore explain the close correlation between the NAO index and the macrobenthic abundance. However, Tunberg and Nelson (1998) also found that the temperature at 600 m depth in the Skagerrak correlated positively to the NAO index (Fig. 9.3), but at a higher significance level than the land runoff ($r = 0.75$). A possible explanation for this correlation has been given by Hagberg and Tunberg (2000) who suggested that the water exchanges of the Skagerrak deep water are driven by the NAO through the cooling of the North Sea surface water (see also Ljøen and Svansson 1972; Svansson 1975). Hagberg and Tunberg (2000) also concluded that the benthos in the Skagerrak correlated closer to the temperature at 600 m, than to runoff.

To illustrate further, some of the changes that occurred in the macrobenthos along the Swedish west coast we are using the data from station L4 at 40 m depth located at the entrance of the Gullmar fjord (58°14.68'N, 11°25.58'E). The Simpson diversity index (Simpson 1949; Magurran 1988) was calculated from the time series from 1986 to 96 and presented in relation to the NAO winter index. The results (Fig. 9.4) showed for the benthos an increase in diversity concomitant with the switch of the NAO in the late 1980s from a negative to a positive phase

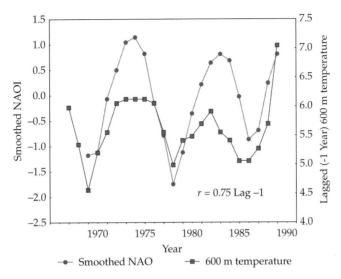

Figure 9.3 Correlation between the temperature at 600 m depth in the Skagerrak and the NAO index (from Tunberg and Nelson 1998).

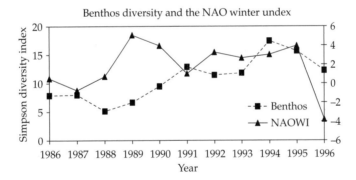

Figure 9.4 Simpson diversity index 1986–96 presented in relation to the NAOWI (the winter-NAO index).

and this is indicated from the correlation coefficients ($r = 0.12; p < 0.05$). At this station the *A. alba* showed the highest density recorded in 1989 corresponding to a positive phase of the NAO and mild winter temperature thus reflecting a species-specific response to the NAO that was different compared with the changes in the total abundance at L4. This result confirms further the hypothesis of (Fromentin and Ibanez 1994) in relation to the density fluctuations in the *A. alba* in the southern Bight of the North Sea.

Data from site L4 also show a negative and significant correlation between changes in the NAO and benthic macrofaunal abundance (Fig. 9.5). As shown in the figure, abundance was lagging changes in the NAO significantly with 2–3 years. When combining benthic data from four sites between 18 and 40 m depth in the same area, a similar pattern was observed, with a significant time lag of 2 years ($r = -0.86$) for the abundance. This indicates that the benthos generally increased ca. 2 years later than when the lowest NAO was

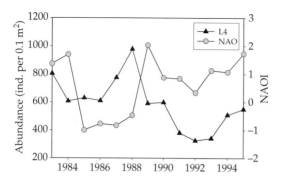

Figure 9.5 Correlation between changes in the NAOWI and benthic macrofaunal abundance at site L4 off the Swedish west coast.

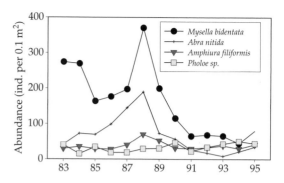

Figure 9.6 Abundance pattern of the four most abundant species at site L4 during the period 1983–95.

recorded. Figure 9.6 shows the abundance pattern of the four most abundant species at site L4 during the period 1983–95, showing the occurrence of a peak in 1988 and a second increase in 1995, that is, ca. 7 years after an indication of the potential presence of a cycle.

The role of exogenious factors (Skagerrak deep-water temperature) and endogenous factors (competition, predation, and density dependence) in the Skagerrak macrobenthic communities was further studied by Hagberg *et al.* (2003), using autoregressive models. The results indicated very weak second-order interactions, suggesting that interspecific interactions did not affect the overall community dynamics very much. Also, the strong covariation of the Skagerrak deep-water temperature and total abundance could not be seen on single species abundance. The conclusions of these analyses were that the communities were loosely coupled entities, largely driven by large scale exogenious factors and that the composition is more of a random collection of species, allthough affected by local conditions.

9.3.2 The Baltic Sea

The highly enclosed Baltic Sea is eutrophicated (e.g. Cedervall and Elmgren 1980). This was demonstrated by comparing sampling results of macrofauna from 1923 with samples taken in 1976 and 1977. Cedervall and Elmgren (1980) were able to show an increase in biomass that could not be accounted for by either sampling methods or interannual variation. Brey (1986) showed a

similar increase in the macrozoobenthos between 1961 and 1965 and a study made in 1982–83. However, Andersin *et al.* (1978) found that the abundance of the amphipod *Pontoporeia affinis* showed a cyclicity of 7–8 years with clear peaks in 1966 and 1972. Low values were recorded in 1962 and 1969. There is also some evidence of a similar oscillation of *P. affinis* in the Gulf of Finland (Segerstråle 1969). Andersin *et al.* (1978) suggested intrinsic factors to be responsible for the observed cycles but they also found connections with the primary production changes in the Baltic. More recently Laine *et al.* (1997) studied changes in the macrofauna in the Eastern Gotland Basin and in the Gulf of Finland covering the period between 1965 and 1994. In this study, the major inflows of North Sea water during 1975–76 and 1993–94 led to highest oxygen condition in the 70–100 m depth zone as well as down to 250 m. This change in the hydrographical regime was followed by a recovery of the macrozoobenthos and recolonization in 1994 by polychaetes in the deepest part of the basin (243 m). These results pointed towards the importance of relating shifts in the hydrographical regime to changes in the climate and ultimately to the observed fluctuations in the macrobenthos. Salinity changes in the Baltic Sea have been regulated by the pulses of North Sea water entering via the Danish Straits and by the freshwater runoff. Hänninen *et al.* (2000) indicated that the NAO may be the forcing factor for a chain of events that regulate the inflow of more saline water to the Baltic as well the freshwater runoff with different lags. These results open up the possibility to make better and more accurate predictions concerning oceanographic and biological interactions in the Baltic Sea.

9.4 Outlook: comparison between the North Sea, the Skagerrak and the Baltic

The major common feature in the different regions is the strong indication of a presence of a 7–8 year climate-driven cyclicity. This pattern has also been elucidated for the macrobenthos on the Swedish Skagerrak coast by Tunberg and Nelson (1998). They found that this cycle was synchronous with the NAO index but with a certain time lag. Since the NAO index also has been found to have a periodicity of 7–8 years (Rogers 1984; Tunberg and Nelson 1998), a possible explanation for the corresponding *P. affinis*

cyclicity in the Baltic could most likely be a close connection to the NAO. A possible explanation for this cyclicity (as suggested by Fromentin and Ibanez (1994)) could be the relation to the pole tide, which has a similar cycle (Gray and Christie 1983).

The variety of studies indicate that single species as well as whole communities and also functional groups are influenced by changes in the NAO, and dependent on the species composition of the different communities the impact also may differ geographically.

A modelling study by Paeth and Hense (1999) suggests that the NAO is likely to remain positive for the next four decades. We may therefore expect major changes in the benthic communities of the North Sea and the Baltic Sea.

CHAPTER 10

Climatic influences on marine fish community ecology

Kyrre Lekve and Nils Chr. Stenseth

10.1 Introduction

Biodiversity may be studied from two different perspectives: as a dependent variable responding to external factors or as a predictor variable influencing ecosystem properties such as biomass (Brown *et al.* 2001). In order to understand the influence of climatic variability on marine fish community ecology, we need to consider the responses of each species as well as understand how the interactions within the community are influenced. Our knowledge of community ecology is growing and so is our understanding of the relationship between climate and biology. However, the responses of fish communities to climatic variability have been poorly investigated. In a discussion on biodiversity and ecosystem processes, Lawton (2000) claimed that one of the current challenges in the area is 'To start to consider the role of animal and microbial diversity for ecosystem processes'.

In this chapter, the main focus will be on biodiversity as a dependent variable. We will review some proposed mechanisms for fish community responses to climatic variability. Then, we will review some case stories illustrating possible examples of these mechanisms. We will start with the simplest example, that of two interacting species (or really two complexes of species) , moving on to studies on different aspects of community responses to climatic variability.

10.2 Climate and dynamics of interacting fish stocks

Around the globe, sardine *Sardinops sagax* and anchovy *Engraulis* spp. populations fluctuate out of phase; sardine having high abundance while anchovy having low abundance and vice versa. These oscillations have been linked to global climate (Schwartzlose *et al.* 1999). Similarly, complexes of sardines (genus *Sardinops* and *Sardina*), anchovies (*Engraulis* spp.), younger stages of jack (horse) mackerels (*Thrachurus* spp.) and chub mackerels (*Scomber* spp.) have been demonstrated to vary out of phase in upwelling areas around the world (Silvert and Crawford 1988). In the northwestern Atlantic a Swedish stock ('Bohuslän') of herring *Clupea harengus* has been shown to fluctuate out of phase with other herring stocks (Norwegian spring-spawning) and Sardine *Sardina pilchardus* (Alheit and Hagen 1997; see also Chapter 6). Alheit and Hagen (1997) identified nine periods of rich herring abundance since the tenth century intervened with poor periods, linking these periods to climatic variability (ice cover around Iceland and wind regimes measured by the North Atlantic Oscillation (NAO) index (Hurrell 1995)). The understanding of interactions between two species (complexes) and how they are influenced by climate is a first step towards an improved understanding of the community response to climate change.

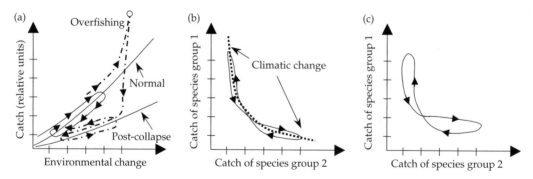

Figure 10.1 Relationship between catches and environmental changes. (a) response to environmental change of one fish stock under normal fishing pressure and after an incidence of dramatic overfishing ('post-collapse'), (b) environmentally driven replacement of one fish stock by another. The switches between the two stocks move along a gradient of environmental change (e.g. climatic change; stippled), (c) replacement of one fish stock by another driven by fishing or biotic interactions. The switch between the two stocks moves in a unique direction along the trajectory (modified from Gulland and Garcia 1984).

The response of fish species to environmental change must be distinguished from the response to fisheries and/or the effect of biological interactions (e.g. predator–prey relationships) in creating alternation between fish species. Gulland and Garcia (1984) analysed the relationship between environmental variables, fisheries, and population fluctuations (in Fig. 10.1(a) this is depicted for one species; figure 5 in Gulland and Garcia 1984). They demonstrated that the catch of a fish stock responded along a gradient of climatic change without changes in the dynamic of the fish (Fig. 10.1(a)). Strong fishing pressure may temporarily transfer the population trajectory to a different level. However, the changes in abundance move along a gradient of environmental change (Fig. 10.1(a)).

Silvert and Crawford (1988) extended the above analysis to two species and developed a model describing the 'inverse correlation between stock groupings' (S_i with i signifying the different stocks) under fishing pressure:

$$\frac{dS_i}{dt} = (\gamma G - qE_i)S_i$$

where E_i is the fishing effort directed against stock i and q is the catchability coefficient (assumed common). G is the common specific net population growth rate of the two stocks under partial competition while γ is the influence from climate. The role of the fishing effort, E_i, is to provide historical memory in the system. Thus, a long-lived predator can also play this role (e.g. birds; cormorants

Phalacrocorax spp. feeding on *Engraulis* spp. and penguins *Spheniscus* spp. feeding on *Sardinop* spp.). The crucial assumption in the model is that the fishing effort (or the predator effort) is independent for the two stocks.

Prediction from such a model serves as a useful starting point to evaluate to which extent fishing (or independent predation) or environment is the mechanism driving asynchronous fish population fluctuations.

By extending the analysis of Gulland and Garcia (1984), we may predict that if environmental change is responsible for the switching, the trajectory of abundances of the assemblages plotted against each other will be reversible (Fig. 10.1(b)). The abundance of the two stocks will move back and forth along a gradient of, for example, changes in temperature. The asynchronous fluctuations of herring and sardine since the tenth century (Alheit and Hagen 1997) may very well be illustrative of such a pattern. If, on the other hand, fishing pressure (or predation) is responsible for the swing of the pendulum, it is expected that the abundances will move in a *unique* and *irreversible* trajectory (Fig. 10.1(c); Silvert and Crawford 1988). The demarcation between pure environmental and pure processes of interaction is, of course, artificial. In reality there will be a continuum, and it can be very hard to distinguish between the different scenarios.

Several mechanisms may be responsible for the replacement of one fish stock by another. Climatic fluctuations may be one such mechanism. Locally, a change in, for example, temperature may give one species an advantage and thus shift the competitive balance between (groups of) species. For stationary fish species (e.g. labridaes) or species connected to upwelling areas we will predict to see environmental switches such as those illustrated in Fig. 10.1.

10.3 Climate and shifts in species ranges of distribution

Decadal changes in temperature have occurred repeatedly during the twentieth century (see Chapter 2). While some species seek their environmental optimum by vertical movement, the simplest response to warming or cooling of the climate is southward and northward shifts in distribution of species. Such changes have been observed during warming in 1940s in the Gulf of Main (Taylor *et al.* 1957) and subsequently cooling from 1953–67 (Colton 1972). In terms of stress these behavioural changes are of different magnitudes: thousands of kilometres of latitude are probably equivalent in terms of stress to the organisms of a few metres up or down the shore on sharp gradient between land and sea (Southward *et al.* 1995).

The most classic and thoroughly studied example of such range shifts is the so-called Russell cycle in the English Channel (Russell 1973; Cushing 1982; Southward *et al.* 1995). Following Southward *et al.* (1995) the changes that occurred during the cycle can be summarized as the following.

Since the turn of the twentieth century a climatic warming from the early 1920s, then a cooling in the early 1980s, with recent resumption of warming, has been observed in southwest Britain and the western English Channel (a change in annual mean temperature of approximately $+/-0.5\,°C$). Marked changes occurred in plankton community structure; the distribution of both plankton and intertidal organisms was affected, with latitudinal shifts of up to 190 km; there were increases or decreases of two to three orders of magnitude in abundance. Warm-water species increased in abundance and extended their range during periods of warming, while cold-water species declined or retreated. The reverse occurred during the period of cooling.

Initially a cold-water plankton community existed. Herring was the main fish species in this community with large diatoms in the spring, abundance of the large copepod *Calanus helgolandicus* in the summer and presence of intermediate trophic levels. The chaetognath *Sagitta elegans* has become

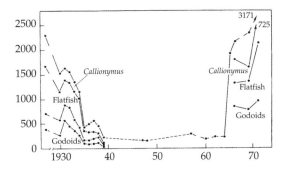

Figure 10.2 The Russel cycle. Changes in numbers of some groups of species of fish larvae. Sum of montly averages of planktonic stages of teleostean fish, excluding clupeids. For 1924–29 an average of the 6 years are used (modified from Russel 1973).

the hallmark of this community (Cushing 1982). During the warming period a warm-water community with pilchard and the chaetognath *Sagitta setosa* prevailed. Dominating plankton in the spring were small diatoms while dinoflagellates and flagellates dominated during summer. Intermediate trophic levels were reduced (Southward *et al.* 1995). During the warming period, larvae of non-clupeid fish decreased dramatically (Russell 1973; Cushing 1982, Fig. 10.2).

The Russell cycle is an example of rather predictable responses of communities to climatic change. A warming trend in climate leads to the replacement of a cold-water adapted assemblage of species with a warm-water assemblage of species. Directional shifts as a response to temperature change seem to be the main mechanism responsible for the community change observed during the Russel cycle. Although further studies are necessary to prove the link between climate and community change in the English Channel, the correlation between temperatures and community change seems compelling.

10.4 Levels of climatic effects on fish communities

The physical agents affecting fish communities may be categorized at the following two levels.

At the first level, the changes taking place during the Russell cycle may serve as an example of a *direct* response to climate change, as opposed to *indirect* responses (Southward *et al.* 1995). Changes in temperature that influence, for example, growth and recruitment (see Chapter 6), may lead to a *direct* response in fish species in the community, with the abundances of individual species changing in a predictable manner according to each species' optimum for growth and reproduction (Wootton 1990). On the other hand, climatic change may lead to, for example, changes in oceanic circulation patterns and temperature-induced changes in prey abundance. The fish communities may then display an *indirect* response to climate change by way of changes in abiotic or biotic conditions. A change in oceanic currents was launched as a causal agent for the changes in the Russell cycle, although this is not supported by later findings (Southward *et al.* 1995).

At a second level, it may be useful to distinguish between *linear* and *non-linear* responses of fish communities to (direct or indirect) climatic change. The linearity in the former case is understood as linear among species, that is, most species are affected in a similar fashion, without changing interaction effects among species. Linear responses are described in the literature (e.g. see Cushing 1982) and somewhat in Chapter 6. The changes in fish species described above for the Russell cycle are mainly of an analogous linear type: each fish species change its range of latitudinal distribution in a linear manner. The aggregated linear responses of a community may, however, be highly non-linear. If the tolerance limits of the species in the community are rather wide, the consequences of climate change need not be very dramatic. However, if changes in climate render the environment inhabitable for key species in the community, large (non-linear) community effects may be observed.

Fish community responses (linear or non-linear) to climatic change are not well understood. Before we investigate possible community responses to climate change we need to look closer into the available community theories that can be used to understand the observed phenomena.

10.5 Climatic change and theories of fish communities

Community theory has mostly been developed within a terrestrial or limnic setting, with marine benthic organisms as a notable exception (e.g. Paine 1966; Roughgarden *et al.* 1988). Terrestrial community

ecology has traditionally focused on *aggregate* variability (i.e. variables combining all species in the community, such as species richness or total abundance). Recently this focus has been successfully extended to community processes such as functional aspects of biodiversity (see, e.g. Naeem and Li 1997; Tilman *et al.* 1997). Although the dynamics within communities necessarily must be seen as resulting from the dynamics within each of the populations constituting the community (Micheli *et al.* 1999), communities has yet to be modelled explicitly at the species level for more than three species (see McCann *et al.* 1998). Initial approaches have been developed (see Hughes and Roughgarden 1998, 2000), but much progress in modelling communities can be expected in the future.

In the marine environment descriptive studies dominate. For fish communities the focus has been primarily on *compositional* variability (i.e. changes in the relative abundance of component species and species composition; Micheli *et al.* 1999). Studies have identified changes in the relative abundance of species within fish assemblages (e.g. see Murawski and Idoine 1992; Fromentin *et al.* 1997; Fogarty and Murawski 1998; Stevens *et al.* 2000). However, functional approaches have not been applied to fish communities (see Duarte 2000 for a study on seagrass; Emmerson and Raffaelli 2000; Emmerson *et al.* 2001 for studies on benthic invertebrates).

Anthropogenic impact is typically very strong in the marine environment (Botsford *et al.* 1997). All major fish stocks are exposed to heavy fisheries and the global catch has passed its peak (Safina 1995). In addition, it is very hard to sample fish communities in a reliable and consistent manner due to the lack of geographic barriers and the great amount of stochasticity in the systems (Hjort 1914).

However, important general community theory is still applicable. For several benthic systems bottom–up control has been demonstrated (Menge 1992; Raffaelli and Hawkins 1996). The bottom–up control of fish communities may very well be effectuated by drift and advection of both progeny (as buoyant egg and larvae) and by plaktonic prey. Roger Lewin termed such a bottom–up control as *supply-side ecology* (Lewin 1986). The approach of supply-side ecology is one way of understanding indirect effects of climate change.

In order to better understand such community effects, theoretical modelling may help. In the following section we develop and discuss such a model.

10.6 Modelling dynamic response to climate: a marine fish community example

Recently it has been observed that species richness remains remarkably constant over time (Brown *et al.* 2001; Parody *et al.* 2001), and furthermore that communities are not saturated (i.e. that there are still niches in the community available for colonization; Loreau 2000). Such stability may be expected from the Equilibrium Theory of Island Biogeography (MacArthur and Wilson 1967) and derived theories (Brown *et al.* 2001). However, modelling work is needed to describe the observed mechanisms. When long time series of both climatic variables and fish communities exists, an attempt can be made to model the role of climate on fish species richness (i.e. *aggregate* variability).

Along the Norwegian Skagerrak coast (Fig. 10.3) data on fish communities have been sampled in a highly standardized manner by beach seine each

fall since 1919 (see Gjøsæter *et al.* 2004a,b for descriptions of the system and sampling protocols). In order to account for non-detectability and sampling variability, we have used multiple sampling on a fjord scale to estimate the number of species, *S*. From the area covered by the sampling 12 fjords/areas that contained at least four sampling stations having run continuously from 1957 were used. By applying the DOS-version of software COMDYN (Hines *et al.* 1999; www.mbr-pwrc.usgs.gov/comdyn.html), which permits computation of the Jackknife estimator of Burnham and Overton (1979), we constructed time series of estimated species richness for each of the 12 fjords/areas based on four stations within each location (see Lekve *et al.* 2002a for details).

For coastal systems, like the one analysed in this study, patterns of currents and temperature influence growth and survival of fish (Wootton 1990; Mann and Lazier 1991). Along the Norwegian Skagerrak coast, alongshore wind stress in the

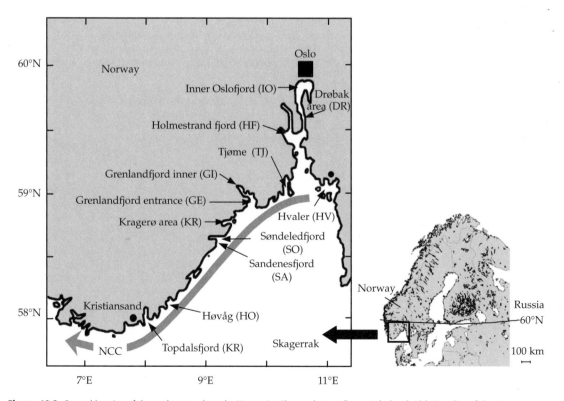

Figure 10.3 General location of the study areas along the Norwegian Skagerrak coast (lower right-hand side). Coastline of the Norwegian Skagerrak coast (main) indicating the location of the 12 fjords/areas of the analysis. The main current, the Norwegian Coastal Current (NCC) is indicated along the coast.

spring (Ottersen and Sundby 1995; Lekve *et al.* 2002b) seems to be a good proxy for the advective processes influential on recruitment (Lekve *et al.* 2002b). The influence of environmental conditions on species richness in a given year is probably not limited to that year. However, the indirect influence of environmental conditions the previous year(s) will most likely be highly diluted. A proxy measure of climate (the NAO winter index) was used to represent past and residual environmental variability (Hurrell 1996; Fromentin *et al.* 1997). The NAO winter index is strongly correlated with local environment (Fromentin *et al.* 1997; Hurrell and van Loon 1997; Ottersen *et al.* 2001), and is thus a predictor of localized climatic variability that influences the growth and survival of fish.

For our study area the effect of climate on species-richness is thus composed of wind stress and temperature in the present year and the NAO in the present and previous year as possible influential factors (Fig. 10.4). Data on the NAO winter index is available from the internet site www.cgd. ucar.edu:80/cas/climind/nao_winter.htm. Data on wind and sea-surface temperature (SST) were obtained from the Norwegian Meteorological Institute (for details, see Fromentin *et al.* 1998; Lekve *et al.* 2002b).

10.6.1 A model for climatic influence on fish communities

Fundamentally, the number of species in a fjord is determined by recruitment {i.e. immigration and settlement; the function $R(\bullet)$} and persistence {the function $P(\bullet)$} of species, represented by individuals. Based on the work by Brown *et al.* (2001) and generalizations from population dynamics models (Royama 1992), we (Stenseth *et al.* 2002; Lekve *et al.* 2003) developed a model consisting of three components. First, the resources limiting the carrying capacity for species in a community are modelled as a maximum species settlement potential. Second, there is a highly variable supply of resources (Hjort 1914), which in the ocean is strongly influenced by climate and currents (Cushing 1982; Mann and Lazier 1991), and in our case constitute the extrinsic regulation of the community. Third, the within-community competition for variable resources (Hughes 1986) constitutes the intrinsic regulation on the community corresponding to density dependence in population dynamics theory (e.g. see Turchin 1995).

We define the species richness in fjord f at time (year) t as $S_{f,t}$, and the baseline species settlement potential as $R_{0,f}$. A model in which $S_{f,t}R_{0,f}$ is defined

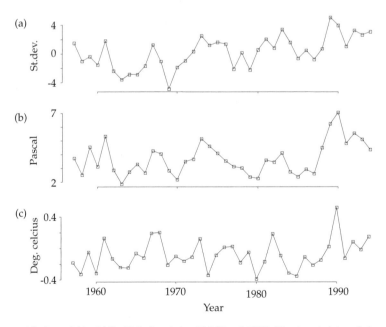

Figure 10.4 Environmental forcing variables. (a) The NAO winter Index, NAO (Hurrell 1995), (b) spring wind stress (Lekve *et al.* 2002b), and (c) spring SST (Lekve *et al.* 2002b).

such as to account for the food-web-dependent community build-up together with intrinsically and extrinsically determined components of species settlement, $R_f(\bullet)$, and species persistence, $P_f(\bullet)$, is given as

$$S_{f,t} = S_{f,t-1}\, R_{0,f} R_f(\bullet)\, P_f(\bullet). \qquad (10.1)$$

Notice the multiplicative nature of the model. The settlement rate may reasonably be seen as a multiplicative function of species number already present when considering how the succession process works: certain species are opportunists and must be present before other species can utilize the resources of the community. Thus, when there are few species, the overall effective settlement rate will be low, since a suitable combination of prey or competitors might not yet be present.

Very little can be assumed about the shape of the functions $R(\bullet)$ and $P(\bullet)$. By analogy to population dynamics (Royama 1992; Turchin 1995), we used exponential functions to describe these relationships. Furthermore, $s_{f,t} = \ln(S_{f,t} + 1)$ is included in the model on the log scale in order to accommodate for the fact that the addition of one species to a community with few species will have a more profound effect than the addition of one species to a community with many species. Combining the recruitment/immigration and persistence functions, we obtain

$$S_{f,t} = S_{f,t-1} R_{0,f} \exp\left\{-a_f s_{f,t-1} + \underline{b}_f^T \underline{w}_t\right\} \qquad (10.2)$$

where a reflects the effect of species-richness (cf. Brown *et al.* 2001) of settlement and persistence, the vector $\underline{b}_f^T$ represents the extrinsic effects on the rates of settlement and persistence, while $\underline{w}_t$ is a vector containing environmental variables such as wind, temperature, and the NAO (see Fig. 10.4).

An order-one process is indeed found appropriate for our data (see below). This suggests that delayed feedback loops may properly be seen as negligible (at the community level) in this system.

10.6.2 Determining the parameters of the model

The negative-log likelihoods of the Poisson distribution (McCullagh and Nelder 1989) for the derived model of the number of fish species were minimized by the Splus routine 'nlminb' (Venables and Ripley 1997) for each of the fjords independently. Different combinations of wind and temperature at time t and NAO at time t and $t-1$ entered the models as environmental forcing variables. Among biologically appropriate models (Lekve 2001; Lekve *et al.* 2003), the Corrected Akaike Information Criterion (AIC$_C$; Brockwell and Davies 1991) was applied for model selection (Table 10.1) displays the climatic variables that were used for the 12 fjords/areas).

This rather simple model of the community processes of fish in the coastal zone is able to capture important features of the dynamics of species-richness in the fjords/areas studied along the

Table 10.1 Environmental forcing variables entering biologically appropriate models of fish species richness fitted separately for 12 fjords/areas along the Norwegian Skagerrak coast

	Forcing at time t	Forcing at time $t - 1$
Kristiansand (Topdalsfjord)—KI	—	—
Høvåg (Steindalsfjord)—HO	Wind$_t$, Temp$_t$	NAO$_{t-1}$
Sandnesfjord, Risør—SA	—	NAO$_{t-1}$
Søndeledfjord, Risør—SO	Wind$_t$, Temp$_t$	NAO$_{t-1}$
Kragerø area—KR	Wind$_t$, Temp$_t$	NAO$_{t-1}$
Grenlandfjord Entrance—GE	Temp$_t$	NAO$_{t-1}$
Grenlandfjord Inner—GI	Wind$_t$, Temp$_t$	NAO$_{t-1}$
Tjøme—TJ	Wind$_t$, Temp$_t$	NAO$_{t-1}$
Holmestrand fjord—HF	Temp$_t$	NAO$_{t-1}$
Hvaler—HV	NAO$_t$	—
Drøbak area—DR	NAO$_t$	—
Inner Oslofjord—IO	NAO$_t$	—

'Temp' is spring SST, 'Wind' is spring wind stress (Lekve *et al.* 2002b) and 'NAO' is the North Atlantic Oscillation winter Index (Hurrell 1995).

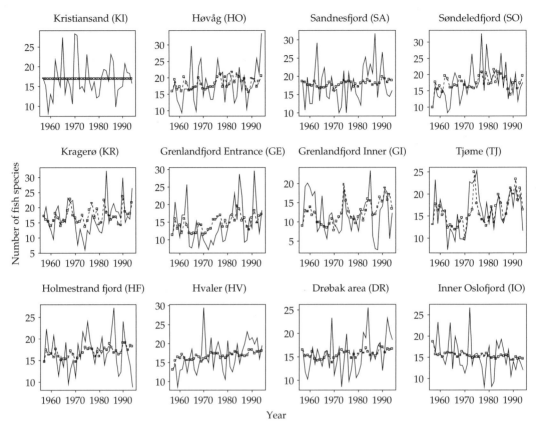

Figure 10.5 Model predictions of fish species-richness regulation of fjords along the Norwegian Skagerrak coast (cf. Fig. 10.3). Displayed are the observed number of species (e.g. estimated from repeated sampling of four stations (Lekve 2001); solid line) and the one-step prediction of the model using only the environmental data as forcing variables (see main text; stippled line with boxes).

Norwegian Skagerrak coast (Fig. 10.5). For several of the fjords, the model estimates closely follows the observed number of species (see, e.g. SO, KR, GE, GI, TJ, and HF in Fig. 10.5). In Lekve *et al.* (2003) and Stenseth *et al.* (2002) the results of fitting this model are analysed.

10.7 Equilibrium number of species and the effect of climatic/environmental perturbations

Climate represents a disturbance agent for fish communities along the Norwegian Skagerrak coast. The fact that environmental forcing variables enter the models outlined above demonstrates that environmental fluctuations indeed affect fish communities along the Norwegian Skagerrak coast.

Climatic change also has the potential to influence the fish communities in a more indirect way by modifying the habitat and thus the number and types of niches available. This can be investigated using the model outlined above.

Assuming that an equilibrium number of species exist, this can be found by setting $S_t = S_{t-1} = S^*$ and fixing the environmental variables by using the estimated coefficients, $\hat{\underline{b}}^T$ and the mean values of the variables, $\overline{w}$. Starting with Eq. 10.2, we obtain

$$S^* = S^* R_0 (S^{*(-a)} \exp(\hat{\underline{b}}^T \overline{w})), \qquad (10.3)$$

which may be rearranged as

$$S^* = (R_0 \exp(\hat{\underline{b}}^T \overline{w}))^{-a}. \qquad (10.4)$$

Using this expression for the equilibrium species number, we may ask how the number of species may change as the environment and overall climate changes. We have chosen to explore this by taking two approaches—one considering the effect on species number assuming extreme values of the climatic covariates, and another considering the climatic covariates to be interrelated (correlation coefficient between NAO and temperature of 0.76 and between NAO and wind stress of 0.43) and then assume what the effect on species number will result, given certain climatic scenarios.

Figure 10.6(a) shows the effect, for each fjord, when adopting the first approach. It can be seen that the equilibrium number of species will not change a lot—remembering that the plotted extremes indeed may be considered as profound extremes.

In Fig. 10.6(b), we have then shown the effect of two climatic-change scenarios, one for which the NAO index changes by two (up and down) and one by which the NAO index changes by four. As can

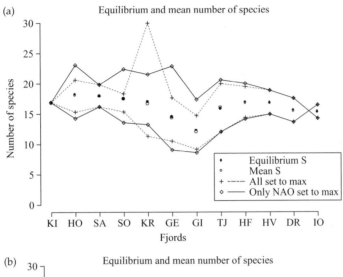

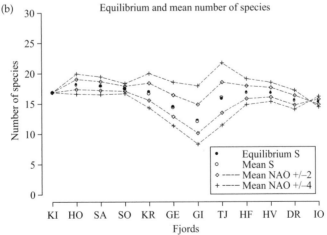

Figure 10.6 Equilibrium number of species ($S^* = (R_0 \exp(b^\mathsf{T} w))^{-a}$) and environmental perturbation. (a) Equilibrium number of species (filled circles) and mean number of species (open circles) and the perturbed equilibriums of the number of species when all the three environmental variables are set to its maximum and minimum (plusses with stippled lines) and when the NAO index is set to its max/min value while the other two variables are set to its mean values (diamonds with solid line), (b) equilibrium number of species (filled circles) and mean number of species (open circles) and the perturbed equilibriums of the number of species when the values of the environmental variables are all expressed in terms of NAO equivalents. Diamonds and stippled lines signify ±2 NAO index units while plusses signify ±4 NAO index units.

be seen, the effect on the species number is relatively slight—the reason being that the various climatic factors counterbalance each other, a result, which indeed seems very reasonable.

10.8 Climate and fish communities: further questions

The effects of climate change on marine fish communities are not well understood. For single species and sometimes two species (or complexes of species) much modelling has been undertaken (e.g. see, May 1984). There are, however, major challenges in understanding the response of community dynamics to changes in climate.

10.8.1 Community dynamic rates

The results presented in this chapter deals primarily with aggregated variables. Much less is known about the processes creating the observed patterns. Climatic variability will expectedly influence important rates determining community dynamics, that is, recruitment, immigration, and extinction. First, climatic change will lead to changes in growth and survival. Such changes will be species-specific, and affect species at the edge of their range of distribution more than species in the centre of distribution. The aggregated effect at the community level may thus be strongly non-linear.

Climatic variability may also affect rates of immigration and extinction, possibly in a non-symmetric manner (Kirchner 2002). In particular, climatic change may lead to responses in circulation patterns that may change systems of drift of egg and larvae from spawning grounds to nursery area (e.g. the eggs and larvae of the Arcto-Norwegian cod follow currents 600–1200 km from spawning grounds along the Norwegian coast to the nursery areas in the Barent Sea; Ottersen and Stenseth 2001; see also Chapter 6). Such physical changes have had profound effects in the past (the rise and fall of cod stocks along the western Greenland coast; Cushing 1982) and possibly in the future. Bottom–up controlled communities will be strongly affected by changes in the availability of recruits (Underwood and Fairweather 1989). Thus, the rate of colonization and consequently the rate of immigration will change. The rate of (local) extinction will be affected in a different manner. Indirectly, abundance of food (i.e. plankton) may be influenced by changes in

circulatory patterns. If the food vanishes, mortality will increase and thus the extinction rate may increase. Climate may also influence rates of extinction in fish communities directly as temperature change. Some species will then experience that the environmental conditions are beyond their tolerance limits, and the species may become locally extinct.

10.8.2 Hierarchy of division rules

In the North Atlantic, the cod (*Gadus morhua*) is an omnipresent top predator. The omnipresence of cod is an indication that the division rules of communities in the North Atlantic are deterministic and hierarchical. Consequently, these communities are characterized by top–down control. During the previous century cod stocks in the North Atlantic have experienced pronounced changes. Some of these are probably caused by overfishing (Myers *et al.* 1997), possibly in combination with changing environmental conditions (Rose *et al.* 2000). Other phenomena have been mainly attributed to climatic factors; the gadoid outburst in the northeast Atlantic has been attributed to temperature fluctuations, while the rise and fall of the West Greenland cod stock has been related to changes in the Irminger Current (Cushing 1982). Thorough knowledge of the variation in these cod stock has been obtained. However, the *indirect* community effects of the varying abundance of cod has not been studied with equal intensity. We expect that quite profound changes in the communities took place as a consequence of changes in cod abundance. The effect of changes in cod abundance on fish communities should be further investigated in order to better understand the secondary effects of climate change on fish communities (but see Stevens *et al.* 2000).

10.8.3 Community effects of changes in large-scale circulation patterns

Another possible source of response to climatic factors in fish communities is changes in circulation patterns. The Gulf Stream has been shown to display latitudinal displacement as a response to climatic change (Hurrell 1995; Taylor and Stepens 1998; Chapter 2). As the Gulf Stream furthermore is a determinant of plankton abundance, which is the major food source of fish, there is probably a robust link between climate and spatial distribution of fish communities (by way of currents and plankton).

10.8.4 Biodiversity functioning

In terrestrial studies, the main focus of many ecologists has recently been biodiversity functioning (Naeem and Li 1997; Tilman *et al*. 1997; Lawton 2000; Pfisterer and Schmid 2002). The biodiversity functioning research approach has scarcely been applied to marine biology and no study on marine fish biodiversity functioning exists (see Duarte 2000 for a study on seagrass; Emmerson and Raffaelli 2000; Emmerson *et al*. 2001 for investigations on benthic invertebrates). The extension of the functional approach to marine fish will be a valuable contribution to the understanding of marine community ecology, and by incorporating the effects of climate variability a further contribution can certainly be made.

PART V

Climate impacts on North Atlantic marine ecology—views from outside

The overall focus of the book is the marine systems of the North Atlantic. Many of the features of these North Atlantic systems are also generally seen in other marine systems as well as in terrestrial and freshwater systems. The purpose of this section of the book is to make these generalities more explicit, and by doing so hoping to stimulate a cross-fertilization between these often semi-isolated fields of science. After the earlier chapters of the book were ready, we invited six authors—or teams of authors—working outside the North Atlantic marine systems to provide their views on how the reported findings and summaries related to their field of study. With perspectives ranging from marine biology in the central and northeast Pacific, through limnology in central and Northern Europe to terrestrial biology, respectively, in North and South America, we are as editors convinced that new light is cast in both directions from and to the North Atlantic marine field of studies.

From New Caledonia in the middle of the Pacific, Lehodey is perfectly located to provide in Chapter 11 an overview of two of the main inter-annual to inter-decadal large-scale climate fluctuations in that ocean, the El Niño-Southern Oscillation (ENSO) and Pacific Decadal Oscillation (PDO). This is followed by a discussion of their effects on marine populations spanning from phytoplankton to small (anchovies, sardines) and large (tuna) pelagic fish. Finally, linking to the models described in Chapter 3, Lehodey introduces us to a spatial environmental population dynamics model (SEPODYM). Results from applying this coupled physical–biological interaction model to predict Skipjack and Albacore tuna responses to climate fluctuations are provided.

In Chapter 12, Bailey and colleagues focus on ecological effects of ENSO, PDO as well as the Arctic Oscillation, but for a different region and different species. Walleye Pollock is a gadoid fish with great abundance in the Bering Sea and Gulf of Alaska. Its role in the ecosystem parallels that of cod in some parts of the North Atlantic, but being even more dominant. Bailey and co-workers briefly review process studies on Pollack. They point to the concept of 'Niño North', extension of an El Niño signal to mid- and higher latitudes, and describe possible effects on ecology in the Gulf of Alaska. Bailey and co-workers emphasize the complexity of climate–marine fisheries dynamics, not only in the Northeast Pacific, but with explicit links to Chapters 3 and 6 also in the North Atlantic. They point to how established climate–fish links often deteriorate with time and underline the role of regime shifts and switches of climatic controlling factors.

In Chapter 13, the limnologist Straile points to horizontal transport being an important factor for many of the North Atlantic populations and communities discussed earlier in the book, both phytoplankton (see Chapters 4 and 8), zooplankton (Chapter 5), and fish (Chapters 6 and 10). Since freshwater systems typically are a lot smaller than marine, advection through large currents systems is a far less important issue than in the oceans. Thus, freshwater systems allow for studying the local ecological impact of large-scale climate patterns such as the North Atlantic Oscillation (NAO) without the confounding effects of horizontal transport.

By isolating local physical and biological responses from those advected, new insight applicable also to marine ecosystems can be gained. Straile documents effects of large-scale climate variability, as represented by the NAO, across a wide range of freshwater systems spanning great geographical distances, small and large, shallow and deep, oligotrophic and hypertrophic lakes with or without winter ice cover. Fluctuations originating with the NAO affect hydrology as well as phytoplankton and zooplankton biology. Effects include timing of phytoplankton blooms, balance between abundance of diatoms and other phytoplankton species, and population dynamics of *Daphnia*.

Also with a freshwater perspective, but focusing on different issues, Hessen asks and endeavours to answer three central questions in Chapter 14. First, to which extent do climatic effects on freshwaters per se impact marine ecosystems? Second, can the enclosed and comparatively simple ecosystems of lakes be used to gain insights in marine effects? And third, are there any pronounced and general differences in expected responses between limnetic and marine ecosystems? Hessen shows that climatic fluctuations indeed may influence marine ecosystems indirectly by way of freshwater systems. Processes related to weathering rate, temperature changes, and precipitation are key determinants of the fluxes of dissolved organic carbon and key nutrients to coastal marine regions, and thus play an important role in regulating productivity in the ecosystems of such areas. This is not least the case in the North Atlantic Ocean, which is fed by a number of large rivers.

Like Straile, Hessen sees the possibility of extrapolating information on climate effects from freshwater to marine systems. However, he points to a number of differences that question the relevance of freshwater findings for marine systems. Pronounced seasonality, strong thermal stratification during summer, absence of horizontal mixing, and totally different zooplankton communities dictating precaution with regard to a direct comparison between lakes and marine ecosystems.

From a terrestrial point of view Post in Chapter 15 accentuates the importance of time lags in the response of both terrestrial and marine populations to a climate signal. He demonstrates that in marine systems lagged responses span many trophic levels, and refers to examples including benthos, zooplankton fish, and seabirds. He emphasises that unlike terrestrial systems, where responses to a climate signal like the NAO typically is direct at lower levels and lagged higher up, in marine systems trophic position is not a good indicator of responses being direct or lagged. Post describes three different mechanisms through which time lags in the response of populations to climatic variability might arise: influence of atmospheric processes, life history, and trophic interactions. Furthermore, a formal, time-series analysis-based approach to discerning the mechanisms or processes underlying lagged population response to climate is launched in Chapter 15.

Lima and Jaksic take a terrestrial and South-American perspective in Chapter 16. While linking to NAO effects in the North Atlantic sector, they provide examples of how ENSO affects plants, invertebrates, birds, and mammals. They build upon their own results and those of colleagues working on the coast of Chile, but they point to general principles valid far beyond. The implications of the way climatic effects emerge in population fluctuations being dependent on their density-dependent structure is far-reaching. Indeed, this fundamental aspect of population dynamic theory has deep implications for analysing and interpreting the effects of climate on numerical fluctuations of natural populations, terrestrial or marine.

We hope that through the six brief commentary chapters included in this section, we are able to demonstrate the generality of the findings reported in the main sections of the book.

CHAPTER 11

Climate and fisheries: an insight from the Central Pacific Ocean

Patrick Lehodey

When reading the previous chapters, it is fascinating to see how it would be easy to replace the name Atlantic with Pacific without changing the meaning of the observations and analyses that are presented. Of course, there are differences in the atmospheric and oceanographic signals observed in both oceans, but the mechanisms and consequences in the marine ecosystems appear very similar, and even sometimes connected.

As in the tropical Pacific, trade winds in the Atlantic Ocean blow from east to west, creating an increasing Sea Surface Temperature (SST) gradient in the same direction, associated to a thermocline deepening westward and an upwelling regime in the east. Also, in both oceans, heavy precipitation is located to the west, over Indonesia and Amazonia, respectively. However, because the size of the Atlantic basin is about one half that of the Pacific, the amplitude of oscillation and the contrast between east and west sides of the Atlantic equatorial belt remains relatively limited (Jin 1996). This is likely the reason that explains the lack of a strong interannual tropical climate signal in the Atlantic, similar to the El Niño-Southern Oscillation (ENSO) developing in the Pacific. Nevertheless, it is possible to identify some ENSO-like warm events, as the strong warming anomaly that occurred in the eastern Atlantic in 1984. As in the Pacific during El Niño, the warm anomaly was accompanied by a deepening of the thermocline and a decrease in intensity of the upwelling in the east. A similar event was also observed along the Angolan–Namibian coast in 1995 (Gammelsrød et al. 1998). There is also an evident similarity in the interdecadal signals in Atlantic and Pacific Oceans, that is, NAO and Pacific Decadal Oscillation (PDO), respectively. As there are likely teleconnection between interannual and decadal signals in each basin and between them.

Interestingly, the term PDO was proposed from analyses based on salmon fisheries data (Francis and Hare 1994; Mantua et al. 1997) showing a decadal-scale regime shift in the north-east Pacific in the mid-1970s. Following this regime shift, most Alaska salmon populations increased while the populations of the west coast of North Pacific decreased. These findings followed an increasing number of observation illustrating persistent widespread changes in the North Pacific basin in the 1970s with strong linkages between the atmosphere, oceanic mixing, and productivity across the trophic levels (Venrick et al. 1987; Brodeur and Ware 1992; Polovina et al. 1994). A chain of mechanisms very similar to what has been described for the Atlantic in the previous chapters. After a brief description of the climate oscillations in the Pacific Ocean and their biological consequences, this chapter provides another illustration of the connection between climate and fisheries using the example of tuna species in the Pacific Ocean. A modelling approach is also presented and used to explore the underlying mechanisms linking changes in climate and tuna populations.

11.1 Major climate oscillations in the Pacific Ocean

In the last two decades, the interannual ENSO has monopolized the attention and rapidly became the most famous climatic event in the world (Philander 1990). However, with the NAO in the Atlantic, another challenger in the media is raising more and more consideration in the Pacific: that is the PDO. The term decadal gives a simple order of magnitude, since the variability, as for the other well-known climate oscillations, is spread over a range of frequencies.

11.1.1 ENSO

Though ENSO effects are felt globally, the major signal occurs in the equatorial Pacific. ENSO is a low-frequency oscillation between a warm (El Niño) and cold (La Niña) state, that evolves under the influence of the dynamic interaction between atmosphere and ocean, with an irregular frequency between 2 and 7 years. During the last 20 years, powerful El-Niño events occurred in

1982–83, 1986–87, 1991–92, and 1997–98 and La Niña events in 1988–89, 1996–97, and 1998–2000.

Briefly, the system takes its energy from the contrasted situation between east and west of the equatorial Pacific (Fig. 11.1). The western equatorial Pacific is called the warm pool because the warm waters in the surface layer have a temperature above 28 °C year-round. Contiguous to the warm pool, the equatorial upwelling in the central and eastern Pacific is generated by the trade winds that result in a divergence (the cold tongue) and a vertical circulation bringing towards the surface relatively cold and nutrient-enriched deep water. This pattern induces a shallow thermocline (~50 m) in the eastern Pacific that deepens progressively westward (~150 m in the warm pool). The equatorial divergence occurs within a mean westward zonal flow forced by the trade winds, the South Equatorial Current (SEC). During their transport from east to west in the SEC, the surface waters are warmed up and finally accumulated in the warm pool in balance with the trade winds forcing. The warm waters in the west induce an atmospheric convection connected to the colder eastern Pacific Ocean through

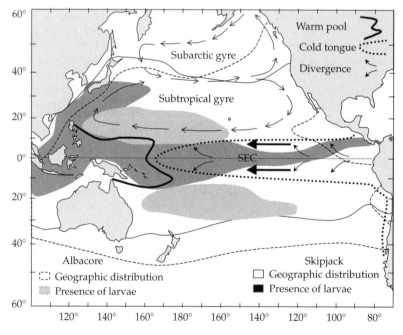

Figure 11.1 Schematic map of the Pacific Ocean showing the main physical features detailed in the text. Characteristics of the skipjack and albacore populations have been superimposed.

the atmospheric Walker circulation. At the sea surface, the temperature gradient from east to west drives the trade winds, closing the loop of the coupled ocean–atmosphere system.

A La Niña situation is an intensification of the mean state described above. Stronger trade winds increase the intensity of the SEC and push the warm pool to the extreme west of the equatorial Pacific. Upwelling intensity in the east also increases, rising the thermocline closer to the surface while it deepens in the warm pool. Conversely, during El Niño events, the trade winds relax and allow the warm waters of the warm pool to spread far to the east in the central Pacific. In general, the most powerful events reach the west coast of California and Peru in the Christmas season and create catastrophic conditions, with devastating storms and floods. The upwelling decreases in intensity and the thermocline deepens in the central and eastern Pacific while it rises abnormally in the western Pacific. These zonal (east–west) displacements of the warm pool are accompanied by changes in the Walker circulation that are reflected by the Southern Oscillation Index (SOI), calculated from the difference in sea-level pressure (SLP) between Tahiti and Darwin (Fig. 11.2(a)). A strong negative index indicates an El Niño while a positive index reveals a La Niña event.

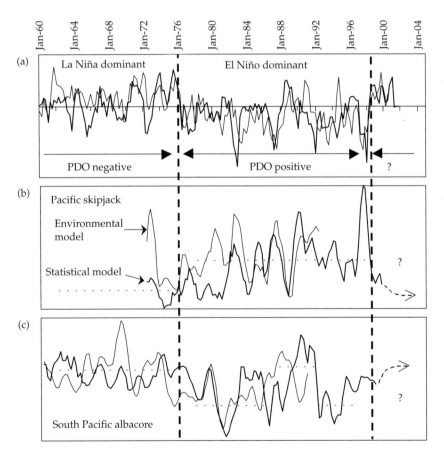

Figure 11.2 Climate oscillations and tuna recruitment in the Pacific. (a) Fluctuation of the SOI (thick curve) and PDO (thin curve). Recruitment series of skipjack from the statistical model MULTIFAN-CL (Hampton and Fournier 2001) and the spatial environmental model SEPODYM (Lehodey 2001) for skipjack (b) and south Pacific albacore (c). There is a direct and opposite correlation between the interannual ENSO and decadal PDO signals and the recruitment of these species that suggests a possible new regime for the next coming years. The albacore series has been back-shifted by 2 years as the recruitment for this species is assumed to occur at age 2 years (it occurs after only one quarter for skipjack). The question mark (?) indicates unknown regime-type.

A special feature of the warm pool is the presence of haline stratification called the barrier layer (Lukas and Lindström 1991) that appears when the isohaline layer is shallower than the isothermal layer. The barrier layer lies between the bottom of the density-mixed layer and the top of the thermocline. A major consequence of this stratification is the inhibition of the downward penetration of energy generated by the westerly wind bursts occurring in the western Pacific. This results in a trapping of westerly wind burst momentum in the surface layer (Vialard and Delecluse 1998), giving rise to strong eastward equatorial currents. This surface layer zonal advection affects the extent of the warm pool and appears to be a major process in the development of El Niño events (McPhaden and Picaut 1990; Delcroix et al. 1992; Picaut et al. 2001). The intermittent eastward surface flow in the western Pacific generated by these wind bursts encounters the westward advection of the SEC and induces a convergence zone on the eastern edge of the warm pool. In an average situation, this convergence zone oscillates around 180° longitude, but spectacular zonal displacements occur in correlation with the ENSO signal (Picaut et al. 2001).

11.1.2 PDO

As for NAO and ENSO, the PDO presents characteristic patterns associated with atmospheric pressure, wind, temperature, and precipitation (Latif and Barnett 1994; Mantua et al. 1997; Zhang et al. 1997). Similarly, warm and cold phases can be defined in association with positive and negative values of climate indices, based on surface temperature and SLP (Trenberth 1990; Trenberth and Hurrell 1994; Mantua et al. 1997; Zhang et al. 1997). Positive PDO indices values (warm phases) correspond to warm SST anomalies along the Pacific Coast, cool SST anomalies in the central North Pacific, and below average SLP over western North America and the subtropical Pacific. The reversed anomalies occur for negative values of PDO indices (cold phases). These opposite phases have been associated with periods of decline (warm phase) or intensification (cold phase) in the 'Aleutian Low' pressure cell. The North American climate anomalies associated with PDO warm and cool extremes are broadly similar to those connected with El Niño and La Niña events. That is why the PDO is often described as a long-lived El Niño-like pattern of Pacific climate variability

(Zhang et al. 1997). During warm phase of PDO, winter and spring seasons in North America show temperatures above average in the northwest and below average in the southeast, while rainfall are above average in the southern US and northern Mexico, and below average precipitation in the interior northwest Pacific and Great Lakes regions (Mantua and Hare 2002).

In the past century, just two PDO cycles of 20–30 years have been identified. Cool phase prevailed from 1890 to 1924 and from 1947 to 76, and warm phase from 1925 to 46 and from 1977 through the end of the century. Within the 20–30-year regimes of the PDO, short periods of reversals in the indices occurred, as the most recent in 1989–91 (Mantua et al. 1997; Minobe 1997). Since 1998, recent changes in Pacific climate suggest a possible reversal to cool PDO conditions, coinciding with the end of the 1997–98 El Niño event that was, with the 1982–83 event, the most powerful of the past century. However, it is difficult to predict if this reversal is a short inversion period or a new decadal regime shift. One mechanism suggested to explaining the short reversal periods would be the existence of two superposed decadal-scale climate oscillations, one at a period of about 50–70 years, the other at a period of 15–25 years (Minobe 1999). While major regime shifts would be due to reversion of both oscillations, the short regime shifts would correspond to reversion of only the 15–25-year oscillation (Minobe 1999; Hare and Mantua 2000). However, these results rely on limited observations.

The lack of understanding is a barrier to the long-term predictability of climate anomalies linked to the PDO and to their impacts on the ecosystems. However, the question is an exciting challenge for the scientific community as it has been shown that including PDO climate information in ENSO statistical predictive models may improve climate forecasts. In particular, El Niño and La Niña typical patterns would appear only when ENSO and PDO signals are in phase, that is when El Niño events occur during warm phase of PDO, or conversely La Niña events during cool phase of PDO (Gershunov and Barnett 1998; Gershunov et al. 1999; McCabe and Dettinger 1999). A number of theories exist to explain Pacific decadal climate variations. They consider several types of feedbacks in the physical ocean–atmosphere system (see the review provided by Miller and Schneider (2000)). Interestingly, some of the proposed mechanisms include more or less strong interaction with the tropics.

11.1.3 Climate connections

As suggested above by the increased predictive capacity when combining both ENSO and PDO signals, connections should occur between these different climate oscillations. A correlation between both ENSO and PDO signals is evident for the last 50 years (Fig. 11.2(a)). Until 1976, cold La Niña events were dominating in the tropical Pacific while the situation reversed from 1977 to day with less and weaker La Niña events and strong more frequent El Niño events. These observations would suggest that the decadal variability is running independently from the ENSO cycle but would influence the development mode of these inter-annual fluctuations. However, an alternative has been proposed that explain the decadal changes as the low-frequency residual (called red-noise) of random or ENSO fluctuations in ocean–atmosphere interactions (Kirtman and Schopf 1998; Rudnick and Davis 2003). For example, model simulations have shown that the tropical Pacific Ocean alone can develop decadal climate oscillations of the equatorial thermocline (Jin 2001).

Whether ENSO fluctuations are influenced by or the cause of the PDO is still unclear. Nevertheless, there are evidences of connections between tropics and extra-tropics. An 'atmospheric bridge' has been proposed by Lau and Nath (1996) and Alexander *et al.* (2001). But it is the oceanic circulation that provides the best evidence of such a connection. Using observations over the past 50 years, Mc Phaden and Zhang (2002) have shown that the wind-driven meridional overturning circulation between the tropical and subtropical oceans has been slowing down since the 1970s, causing a decrease in upwelling of ~25% and a rise in SSTs of ~0.8 °C in the equatorial belt. It is worth noting here that this slowdown of the meridional overturning circulation concerned both northern and southern hemispheres. Thus, decadal variability in the Pacific Ocean is not limited to the northern hemisphere.

Climate connections also seem to occur between oceanic basins. Though mechanisms remain unclear, teleconnection with El Niño events in the Pacific appear responsible for warming of tropical Atlantic and Indian oceans. In the Atlantic, the major region affected by the Pacific ENSO is the north–west tropical Atlantic, where about 50–80% of the anomalous SST variability may be explained by the Pacific ENSO variability (Enfield and Mayer 1997). The scenario that emerges from these observations may be briefly summarized as follows. A warm episode in the eastern Pacific would set favourable conditions for a 'cold episode' or rather, a reinforced seasonal signal in the tropical Atlantic by increasing the easterlies in its western region. The development of the last powerful El Niño in 1997–98 agree with this scenario, as the equatorial upwelling in the eastern tropical Atlantic began earlier in the year, that is in May when normally it starts in the boreal summer (Signorini *et al.* 1999). The warm event would follow in the Atlantic as the wind stress is relaxed a few months later—Enfield and Mayer (1997) indicate a delay of 4–5 months after the mature phase of Pacific El Niño events—therefore corresponding to the end of the El Niño in the Pacific. Such climate teleconnection seem the most plausible mechanism to explain apparent worldwide synchrony observed in fluctuations of small pelagics populations associated with upwelling ecosystems (Schwartzlose *et al.* 1999).

11.2 Biological changes associated to climate oscillations

11.2.1 Plankton

In the tropical Pacific, the biological consequence of the equatorial upwelling is a large zonal band with high primary production (the cold tongue), which contrasts with the generally low primary productive waters of the western Pacific (the warm pool) and the north and south tropical gyre (Fig. 11.1). The equatorial divergence and the westward zonal flux of the SEC create a spatial shift in the planktonic communities both on the meridional (poleward) and zonal (westward) axes. However, this productivity in the tropical Pacific is strongly affected by ENSO variability. In a La Niña or mean situation, a rich chlorophyll cold tongue is observed, extending as far west as 160°E, while the warm pool from Indonesia to 160°E is typically in a low primary productivity situation. During the development of El Niño events, as trade winds and equatorial divergence decrease in intensity, there is a lower primary productivity in the cold tongue. Therefore, the cold tongue retreats east of the date line in an eastward movement accompanying the warm waters extend to the central Pacific and the displacement of the atmospheric convective zone associated with these warm waters. During the same period, an increase in primary production is observed in the western equatorial Pacific, in response to stronger wind stresses and to the eastward shift or vanishing of the barrier layer (Vialard and Delecluse 1998).

In the North Pacific, interannual and interdecadal changes in the intensity of winter winds modify the circulation of the subarctic gyre in the northeast Pacific and Ekman pumping at the centre of the gyre. Brodeur and Ware (1992) have proposed that by changing the intensity of upwelling of nutrient into the euphotic zone and the advection patterns, these physical mechanisms are responsible of large fluctuations of the summer biomass of zooplankton. During decadal regimes with high winter wind conditions, there is a ring-like structure of high zooplankton biomass around the gyre, while in reverse conditions the biomass is lower and distributed throughout the region.

Though the PDO effects are most visible in the North Pacific, there are some evidences that PDO also affects biological events in the tropics. However, long time series of observation in biomass of phytoplankton and zooplankton are missing in these regions to investigate the decadal variation. An interesting result comes from hindcast simulations by coupled physical–biogeochemical models similar to some of those listed by Werner and coworkers in Chapter 3. They predict that in addition to ENSO variability, decadal variations also occur in the equatorial region for primary and secondary production. For instance, a recent simulation (Chai *et al.* 2003) predicted that nitrate concentration in the Niño 3 box area (5°N–5°S, 150°W–90°W) is about 2 mmol/m³ higher before 1975–76 (1965–76) than the period after (1977–92). This would be primarily due to the wind change in this region, which is a decrease in the trade wind after 1975–76 in the tropics.

Therefore, these observations with those presented for the Atlantic in the previous chapters suggest that the primary forcing resulting from climate change is a change in the field of atmospheric pressure and the associated surface wind stress. This change affects the depth of the surface layer, the horizontal and vertical flow and mixing within it, as well as spatial patterns in sea temperature. This leads to variations in primary production that are reflected in the secondary production.

South American coast, where a strong coastal upwelling normally sustains an exceptional biological productivity. Besides, the most productive fishery in the world is the coastal Peruvian anchovy fishery with an annual catch that reached a peak of 10 million metric tons at the end of the 1960s. This huge biomass of anchovy feeds many predators including fish, guano producing birds, and mammals. Therefore, exploitation of anchovies and guano (used in fertilizers) are major industries in Peru. The impact of the 1972 El Niño event was all the more catastrophic for the economy of this country. As the warm waters spread along the South American coast, the upwelling stopped and the Peruvian anchovy stock collapsed. Anchovies died or migrated in cooler waters, as did their predators, including the guano birds. The annual anchovy catch drop to 4 million metric tons and remained below this level during several years, likely because of a combination of low recruitment and overfishing given the lower carrying capacity of the stock. In contrast, with the spreading of warm waters tropical species like tuna and billfish made unusual incursion in the Peruvian waters and were caught in large numbers by local fishermen.

Though not comparable in terms of biomass, similar productive upwelling regions occur on the east side of Pacific and Atlantic basins along North America (California Current), and North and South Africa (Canary and Benguela Currents), and on the west side in the north-western transition zones of the Kuroshio (Pacific) and the Gulf Stream (Atlantic). All these regions support large stocks of small pelagics including sardines, anchovies, herrings, saury, and mackerels. Examples of changes and fluctuations in stocks of herrings and sardines in the North Atlantic are presented in Chapter 6. In the Pacific, in addition to the interannual variability related to ENSO and illustrated above by the Peruvian anchovy, similar decadal fluctuations have been demonstrated in stocks of small pelagics (Schwartzlose *et al.* 1999) with periods of collapse and recovering and out-of-phase patterns between species (e.g. sardine versus anchovy). These fluctuations provide additional evidences of climate dependence on these populations.

11.2.2 Small pelagics

El Niño events affect traditional fisheries exploiting small pelagics (anchovies, sardines) along the

11.2.3 Large pelagics

Important interannual fluctuations have been observed in Pacific tuna fisheries, in apparent correlation with the ENSO events as indicated by the SOI (Lehodey *et al.* 2003). Causes for these fluctuations include most of the processes through which climate may influence the biology and ecology of fish, detailed in Chapters 1 and 6. Though climate and environmental variability likely affect the individual growth and natural mortality of tuna, such relationships are difficult to demonstrate. However, other processes may explain fluctuations of catch rates: migration, catchability, and spawning and recruitment are more obvious.

Spatial distributions of purse seine catch (skipjack and yellowfin tuna) have been used to demonstrate a spatial shift in the catch that follows the eastward extension of the warm pool during an El Niño event. Tagging data have shown that this shift corresponded effectively to real displacements of fish (Lehodey *et al.* 1997). These movements occur in relation with the displacement of the convergence zone on the eastern edge of the warm pool. Model simulations have been used to investigate the mechanisms leading to such displacements (Lehodey *et al.* 1998; Lehodey 2001). The large fluctuations in the catch rates of skipjack tuna would be driven by changes in stock size, and by horizontal spatial extension (during El Niño) or contraction (during La Niña) of the habitat. Spatial extension of the skipjack habitat during El Niño events would have a negative effect on the catchability in the west, but would increase it in the warm pool–cold tongue convergence zone during the eastward displacement associated to the development of El Niño events. A higher availability and concentration of forage in this oceanic convergence zone is proposed to explain the movement of skipjack and the high level of catch observed during this phase.

As skipjack inhabit the epipelagic layer, changes in catchability due to changes in the vertical structure seem negligible for this species that is caught by surface gears (purse seine, pole-and-line). This is not the case for yellowfin and bigeye, as the adults are also caught by deep fishing gears (longline). In the western Pacific, the vertical change in the thermal structure during El Niño (La Niña) events result in the rising (deepening) and vertical extension (contraction) of their temperature habitats, but with a deeper habitat for bigeye. This change would increase (decrease) purse seine and pole-and-line catch rates of yellowfin and longline catch rates of bigeye. In the eastern Pacific, El Niño events would have a negative impact on yellowfin

longline catch, while the negative effect is also well-known on the purse seine fishery. Indeed, the large decline in the catch during the 1982–83 El Niño is one of the main factors responsible for the displacement of the US fleet towards western and central Pacific Ocean.

As already highlighted, decadal fluctuations in the abundance of salmon in the North Pacific have been associated with the PDO. Following the regime shift in 1976, catches of almost all the major Alaska salmon stocks showed a nearly synchronous dramatic increase in a large part of their fishing grounds. Conversely, in the same time many salmon stocks of the west coast of North America declined. However, salmon is not the only concerned species by the PDO. Similar decadal fluctuations have been observed in Hawaiian lobster, seabirds, and marine mammals (Polovina *et al.* 1994), several groundfish species of the northeast Pacific and jellyfish in the Bering Sea (Hare and Mantua 2000), and the North Pacific albacore (Polovina *et al.* 1995; Au and Cayan 1999).

11.3 Modelling the pelagic tuna ecosystem

To explore the underlying mechanisms by which the climate and environmental variability affects the pelagic ecosystem and tuna populations, a spatial environmental population dynamics model (SEPODYM) has been developed (Bertignac *et al.* 1998; Lehodey *et al.* 1998; Lehodey 2001; Lehodey *et al.* 2003). The model is a coupled physical–biological interaction model at the scale of ocean basin. It combines a forage (prey) production model describing the transfer of energy of stored biomass through the trophic levels of the food web with an age-structured population model that is appropriate for simulating space and time distributions of successive cohorts of targeted (predator) species and their fisheries. In comparison to the different modelling approaches described in Chapter 3, this model would be an intermediate approach, with greater complexity at both top and bottom levels (i.e. high predators and physical–biogeochemical dynamics, respectively). Intermediate level (forage organisms) for which there is little information is described in the simplest form.

11.3.1 SEPODYM

The forage component is considered as a single population, using the same mathematics than in a single species population model with constant and continuous mortality and recruitment. However, the spatial dynamic is considered and described with an advection–diffusion equation using horizontal currents for the advective terms. The concept of this tuna forage model may be simplified and discretized as follow: in the ocean, at any time, anywhere, there is a mixing of all kinds of eggs, cells, etc. that are the germs of the future organisms of the pelagic food web. In some places and time, the input of nutrient in the euphotic zone allows the almost immediate development of phytoplankton (new primary production) that is the input (corresponding to spawning) in the forage population model. This new production allows the development of a new 'cohort' of organisms, that is, true zooplankton like copepods as well as all larvae of fish and other larger organisms (meroplankton). The organisms having the longer life span and larger growth potential (larvae and juvenile of fish, squids, shrimps, etc) feed on the expense of the organisms with short life span and lower growth potential (phytoplankton, zooplankton). As the water masses are maturing, they are advected with these organisms (but a part of them can also diffuse due to the diffusion of water and their own random movements) and the currents create fronts of convergence where forage is aggregated. Of course this dynamic occurs as a continuous process in time and space.

Tuna movement and recruitment in the model are constrained by environmental variables combined in two habitat indices. The adult habitat index used to constrain the movement simply combines the spatial distribution of forage with temperature and oxygen functions defined for each tuna species. The spawning habitat index is used to constrain the recruitment to environmental conditions. The basic assumption to define this index is that the spawning area is limited by the presence of mature tuna and of SST above a limit value. This latter condition is supported by the high correlation found between SST and occurrence of reproductively active tropical tuna (Schaefer 1998). However, other environmental effects are included. With temperature and physical constraints like the advection creating favourable zones of retention for larvae, and that are already considered in the model, the food availability and the predation are likely the other major factors that affect larval survival and pelagic fish recruitment. Food of larvae is directly dependent on primary production, while their predators are organisms included in the tuna forage. Therefore, these two opposite effects have been integrated in the spawning habitat index by using the ratio primary production over forage (P/F).

The tuna populations are age-structured to account for growth and gear selectivity. Growth and mortality-at-age parameters are provided from independent studies. The total level of recruitment (or spawning) is adjusted, so that the stock biomass estimates are roughly equal to those obtained independently by statistical population dynamics models. Finally, the different fleets exploiting the tuna are described, their catch depending from gear-related selectivity functions and catchability coefficients. Fishing efforts of each fleet vary by month and in space, with a $1°$ square resolution. Results of the simulation are compared to observed fishing data by fleets, such as total monthly catch, spatial distribution of catch, and distribution of length frequencies.

The input data set (physical and primary production data) of the most recent simulations are provided by a coupled 3Dphysical-NPZD (nutrients-phyto-zooplankton-detritus) model (Chai *et al.* 2002) with ten components and two nutrients (nitrate and silicate). The physical ocean general circulation model (OGCM) that drives the NPZD model is derived from the modular ocean model (MOM) developed at the Geophysical Fluid Dynamics Laboratory/National Oceanic and Atmospheric Administration (GFDL/NOAA) (Princeton, USA). There are 40 layers with 10 m resolution within the euphotic zone (120 m). The surface forcing uses comprehensive ocean–atmosphere data set (COADS) monthly averaged fields from 1950 to 1993, with a restoring surface salinity process back to climatological value. The model simulates the equatorial region very well with elevated high biomass in the cold tongue region, as well as the subarctic area. In the equatorial upwelling region, the total phytoplankton biomass for the top 120 m decreases during all five El Niño events, not just near the surface but throughout the water column. The strongest reduction of the phytoplankton biomass occurred during the 1982–83 El Niño, and followed by 1972–73 El Niño. Besides the strong interannual variability, two different regimes are predicted before and after 1976, with higher nitrate concentration in the equatorial upwelling before 1976 compared to the 1980s. Currents and primary production are averaged over the euphotic layer to be used with SEPODYM.

11.3.2 Predicted skipjack and albacore climate-related fluctuations

While skipjack is a typical tropical tuna, spending its entire life in tropical waters, or tropical water transported into temperate latitudes by currents (e.g. the Kuroshio current), albacore tuna is subtropical with adults widely distributed in tropical waters as well as in subtropical and temperate waters. These two species have many contrasted or even opposite characteristics. For stock assessment studies, albacore is considered with two independent north and south Pacific stocks, and skipjack with two more or less independent east and west stocks. However, the maximum biomass of this latter species is in the western Pacific Ocean, in relation with the warm waters of the Pacific warm pool. Skipjack tuna have a rapid growth, early age at first maturity, year-round spawning, high natural mortality rate and short lifespans (corresponding to 4 years), and high production to biomass ($P{:}B$) ratio. Albacore is characterized by slower growth, first maturity at 2 or 3 years of age, apparent seasonal spawning, relatively low natural mortality and long lifespans (~15 years), and low $P{:}B$ ratio. However, both species have high fecundity, and spawn in warm waters over wide areas.

Recruitment estimates of skipjack and south pacific albacore tuna (Fig. 11.2) have been estimated recently with the statistical population model MULTIFAN-CL (Fournier *et al.* 1998; Hampton and Fournier 2001). This model uses all the information available from the fisheries and their monitoring (i.e. catch and effort statistics, length frequencies, and tagging data) and the robust statistical likelihood function to produce in one single set, the best estimates of all the parameters of the population and fisheries. While MULTIFAN-CL produces recruitment estimates from robust statistical methods, SEPODYM recruitments are predicted only from environmental constrains. Therefore, comparisons between these two independent estimates are very useful to investigate hypotheses on the mechanisms of recruitment.

These two series show interesting opposite trends with apparently two different low and high regimes that could be associated to the two different climate regimes of the PDO, and characterized by a high frequency of either La Niña, or El Niño events. Therefore, the effects of ENSO on the recruitment of skipjack and albacore appear opposite, El Niño events apparently being favourable to skipjack recruitment while south Pacific albacore recruitment would benefit from La Niña events. Simulations with the model SEPODYM predict the same patterns (Fig. 11.2), and recruitment estimates of both statistical and environmental models are converging. However, in the case of SEPODYM, mechanisms that produce this variability can be identified.

For skipjack, the model predicts the main spawning ground in the western Pacific between Indonesia and Papua New Guinea, then a decreasing gradient from west to east. However, large interannual variations occur (Lehodey *et al.* 2003), the recruitment increasing during El Niño events (1972, 1982–83, 1987, 1990) and decreasing during La Niña events (1988–89). The main reasons would be the extension of the spawning ground with the warm waters during the development of El Niño, in association with a higher primary production in the western Pacific that increases the $P{:}F$ ratio of the spawning habitat index. An exceptional catch record of skipjack at the end of 1998 seems to support this mechanism. The catch was concentrated in a relatively small area centred on the equator at 165 °E and contained an unusual large number of juveniles (20–35 cm in length). Four to eight months before, a huge phytoplankton bloom was observed with SeaWiFS satellite images at the same place (Murtuggude *et al.* 1999). Interestingly, the size of the juveniles corresponded to an age of 4–8 months, and scientific observers onboard purse seiners indicated the presence in this zone of large schools of oceanic anchovies (*Encrasicholinus punctifer*), a key prey species of surface tuna in the western and central Pacific, mature at age 4–6 months.

While the main skipjack spawning grounds appear associated to the warm pool, those of albacore are roughly extending through the tropical Pacific on each side of the equator (Fig. 11.1), and consequently under the influence of the productivity of the equatorial cold tongue. This would explain that the albacore recruitment is higher (lower) during La Niña (El Niño) events when the primary production and the $P{:}F$ ratio of the spawning habitat index are high (low). The prediction also shows a clear regime shift in the abundance of juvenile due to the decadal change in primary production that is reproduced by the NPZD model. As they are propagated in the population structure, the high frequency (seasonal and interannual) fluctuations are smoothed. Finally, adult biomass of albacore presents only large decadal variation. It is worth to note that the short inversion period of the PDO in 1989–91 is also apparent in the albacore recruitment series. These preliminary results are extremely encouraging, as the recruitment mechanisms are the key process for fish stocks prediction.

11.4 Conclusion

From this review on climate variability and its impacts in the Pacific and from the results presented for the Atlantic, the need of integrating climate considerations in the future management of exploited fish stocks appears as an evidence. However, the evidence of multiple correlations and the development of theories and conceptual models are not yet sufficient to predict the trends of fish populations. Numerical models are needed to test hypotheses and mechanisms linking climate physical forcing and biological processes. The task is particularly complex because of the spatial and temporal dynamic nature of these phenomena, the occurrence of multiple interactions and feedbacks, and the succession of scales. Understanding how physical processes are transferred and transformed across scales is likely one of the critical question to consider in the development of these models. Progress and findings of the past decade are encouraging and suggest that simple mechanisms may occur but would be masked by the dynamics at several temporal and spatial scales.

Complexity of marine fisheries dynamics and climate interactions in the northeast Pacific Ocean

Kevin M. Bailey, Anne B. Hollowed, and Warren S. Wooster

12.1 Introduction

Much of the progress made in understanding climate–fisheries interactions can be traced to the long time series of data collected on commercial fisheries in the northeast Atlantic Ocean. Some European catch records date back to the sixteenth century or before (Southward *et al.* 1988). Backed by such a wealth of data, Ottersen and coworkers in Chapter 6 have posed the general question 'how does the physical environment affect fish and shellfish?' They conclude that 'the interaction between ocean and atmosphere may form dynamical systems, exhibiting complex patterns of variation, which may profoundly influence ecological processes in a number of manners'. While that answer may seem general, from our point-of-view it is accurate; the impact of climate on fisheries is highly variable, indirect, and complex. Several dominant themes related to climate change and fisheries interweave in chapters of this volume (Chapters 2, 3 and 6), including: complex interactions, formation of patterns over large-scales, high variability over small-scales, and the desirability of forecasting tools. These themes are important not only to North Atlantic fisheries, but are focal issues in many of the world's seas, including the northeast Pacific Ocean as we describe here.

In the northeast Pacific Ocean, the time series of fisheries data are not very long; in a few cases they may date back a hundred years, and for most species accurate statistics have been collected for decades rather than centuries. In the absence of long time series of catches, other methods have been employed to study fisheries–climate interactions. For example, process studies of ecological oceanography offer a different approach. Process studies are mechanistic and reductionist by necessity, and as described below, understanding climate–fisheries interactions from process studies is confounded by the complexity of extrapolating from small to large time and space scales. Strong climate disturbances, like El Niño, present opportunities for natural, but uncontrolled, experiments. In large part, the interest in climate–fisheries interactions in the northeast Pacific Ocean has been driven by the dramatic effects of El Niño events, especially those involving fisheries collapses. More recently, there have been gains in understanding of potential interactions between low frequency climate effects and fisheries. Large-scale responses in the dynamics of fish populations to changes in the Pacific Decadal Oscillation (PDO) and Arctic Oscillation (AO), the Pacific counterparts of the North Atlantic Oscillation (NAO), have been hypothesized, largely based on observations of a single cycle. Finally, sometimes catch statistics can be supplemented by archaeological and geological records. Process studies, natural experiments, and climatological studies of fisheries are diverse, but complementary approaches, often offering a variety of perspectives of climate–fisheries issues on different scales.

12.2 A case study of processes: walleye pollock in the Gulf of Alaska

The impact of ocean conditions on the recruitment process of the central Gulf of Alaska walleye pollock population has been studied since 1985. This population of pollock, generally about 1 million tons, aggregates to spawn during the first week in April over a 40-km by 90-km area just off Cape Kekurnoi in the Shelik of Strait. Such a predictable concentration of fish has provided a natural laboratory to study fisheries oceanography in one of the storm centres of the Western Hemisphere. The preliminary guiding hypothesis of these studies was that recruitment to the fishery was determined by wind effects on the drift of larvae (Hinckley *et al*. 1991). Initial studies showed that feeding interactions of early stage larvae (Canino *et al*. 1991; Theilacker *et al*. 1996), storms (Bailey and Macklin 1994), and predation on eggs and larvae (Brodeur and Bailey 1995) had significant effects on early mortality. Later, with accumulation of more years of data on the abundance of different life history stages and by analysis of life tables, it appeared that recruitment was not solely determined in the egg and early larval stage, but that in some years devastatingly high mortality occurred during the juvenile stages, meaning that the 'critical period' biologists have long searched for may vary from year-to-year (Bailey *et al*. 1996).

The 'variable life stage control' hypothesis took another complexion as more data accumulated. In the late 1970s the North Pacific Ocean experienced a 'regime shift' in environmental conditions that was accompanied by a major change in community structure (Anderson and Piatt 1999). Small pelagic fishes and shrimps decreased spectacularly in abundance and there was a long-term build-up of groundfishes, especially piscivorous flatfishes. While the mechanisms behind the shift in community structure are poorly known, it apparently had an effect on the process of recruitment in walleye pollock. During the early phase of the new regime period, larval mortality was coupled to eventual recruitment and mediated by environmental conditions, but during the later phase of the regime larval mortality and recruitment became uncoupled, apparently due to increasing mortality of juvenile pollock with the increasing number of flatfish predators (Bailey 2000). Therefore, it appears that 'control' of recruitment depends on the status of a continually changing ecosystem. This serves as an example of the scenario described by Ottersen and coworkers in Chapter 1, where 'one variable may be a dominant forcing mechanism for a time only to be replaced later by another variable...'.

A recent and evolving perspective is that recruitment of pollock is a complex process, governed by the rules of complexity theory, but bounded by low frequency changes in the ecosystem (Bailey 2002). Larval mortality is highly variable and subject to many interacting high frequency factors, with feedback and non-linearity. Larvae show sophisticated behaviours involving choice and decisions when confronted with multiple and perhaps conflicting stimuli (Olla *et al*. 1996). For example, they avoid turbulence by descending (Davis 2001), taking them out of the photic zone and into colder water where growth is less optimal and prey are less abundant (Kendall *et al*. 1994). On the other hand, although juveniles also show complicated behaviours in response to the environment (e.g. Sogard and Olla 1993), they are less impacted by small-scale physics, and juvenile mortality seems to be more stable and predictable, occurring largely as a result of predation and cannibalism. The build-up of predators in the community represents a low frequency, slowly changing pattern with lagged effects. Patterns in recruitment have been well-described by preliminary models incorporating stochastic mortality weighted by environmental conditions during the larval period and by deterministic factors and constraints during the juvenile period.

12.3 A natural experiment: El Niño and fisheries

The effects of a strong environmental perturbation like El Niño on fisheries and community structure are well-documented in the tropical and subtropical waters off the west coast of the United States (e.g. Lea and Rosenblatt 2000). How El Niño has impacted subarctic fish stocks is less well-understood. Associated changes in environmental conditions may be due to local effects, remote forcing, or atmospheric teleconnections.

El Niño is the warm phase of the El Niño-Southern Oscillation (ENSO). Extension of this signal to mid- and higher latitudes has been called Niño North (Hollowed *et al.* 2001). These warm events along the west coast of North America are characterized by poleward propagation of ocean long waves associated with deeper thermoclines, higher salinity and temperatures, increased sea-level, stronger poleward flow, and relaxed upwelling. Atmospheric teleconnections also impact the northern Gulf of Alaska, including intensification of the Aleutian low-pressure system, changes in the wind field, increased storminess and relaxed coastal upwelling. Individual El Niño-related events are characterized by their remarkable variability and strength in the subtropics and the occasional northward propagation of these effects. Strong tropical events are not always evident in subarctic waters. For example, the 1972–73 event was even anomalously cold in the northern Gulf of Alaska. On the other hand, moderate or even weak tropical events, such as that of 1976–77, can have relatively strong effects in that northern region.

Even moderate El Niño events clearly influence the distribution of fishes (Pearcy and Schoener 1987). These effects are most pronounced in the subtropics and diminish polewards. Furthermore, small pelagic species are expected to be affected more than groundfish species (Bailey *et al.* 1995).

The effect of El Niño on recruitment is not as clear. Bailey and Incze (1985) proposed that El Niño events initiated strong year-classes at the northern end of the species range through the beneficial effects of warming, and poor year-classes at the southern end. In a more recent evaluation, this picture has become more complicated (Bailey *et al.* 1995) and individual El Niño events have inconsistent effects on any one species. The confounding effect of many interactions with opposite effects appears to introduce complexity to the process. For example, warming is generally beneficial, but is accompanied by reduced nutrient input, which decreases prey production. Some generalizations may be possible; for example, it appears that pelagic species are more adversely affected than groundfishes, and this is demonstrated by catastrophic mortalities of piscivorous seabirds that feed on pelagic fishes. As well, offshore spawning species with inshore nurseries may benefit from enhanced onshore transport during warm events.

However, this may affect different populations dissimilarly. In the Gulf of Alaska, enhanced transport onshore and warming during Niño North seems to be associated with exceptionally strong year-classes of halibut, at the northern end of their range. On the other hand, arrowtooth flounder experience enhanced onshore transport as well, but are at the warm end of their range in the Gulf of Alaska and may not have benefited from warming. Arrowtooth flounder do not have strong year classes associated with El Niño (Bailey and Picquelle 2002). Temporal patterns of production of Pacific cod, walleye pollock, and Pacific hake tend to follow a pattern that is somewhat consistent with El Niño North. Warm years typically persist after tropical El Niño events that have reached northward. Patterns of production of these stocks show that strong year-classes occur more frequently in such years (Hollowed *et al.* 2001).

12.4 Regime shifts and marine fisheries

Our case study and the contributions by Werner and coworkers in Chapter 3 and Ottersen and coworkers in Chapter 6 illustrate the complexity of processes underlying survival of individual larvae, patches of larvae, or cohorts from a specific population. Historically, fisheries scientists have utilized statistical inference to identify potential relationships between climate and fish production, and many times find interesting relationships. However, strong correlations between recruitment and environmental variables often deteriorate with time suggesting that the relationships might have been spurious (Drinkwater and Myers 1987), or perhaps indicating a shift in controlling factors (Bailey 2000). The results of single species statistical inference models can be strengthened, if similar responses are observed in several species. Comparative studies provide some evidence of the linkages between large-scale atmospheric forcing, ocean processes, and fish production. Time-series comparisons of recruitment patterns of several of the most abundant marine fish stocks have revealed that they exhibit temporal patterns of production that appear to follow spatial and temporal patterns of large-scale climate forcing (Francis *et al.* 1998; Hare and Mantua 2000; Hollowed *et al.* 2001).

Besides ENSO, two primary modes of atmospheric forcing have been identified in the North Pacific: PDO and the AO. These modes of forcing, which reflect the large-scale atmospheric circulation of the region, impact ocean conditions in the north-east Pacific and Bering Sea.

The PDO is a decadal-scale oscillation in the North Pacific sea surface temperature (SST) with alternating positive and negative phases that have lasted 20–30 years during the twentieth century. Effects of the PDO are felt most strongly in the extratropical North Pacific, in contrast to the inter-annual ENSO signal, which is best developed in the equatorial region.

Indices of PDO and ENSO are commonly computed for boreal winter when the signals are strongest. Comparison of the time history of the two indices (Mantua *et al.* 1997) shows the difference in their dominant frequencies and emphasizes the inter-decadal, regime-like nature of the PDO. During the warm phase of the PDO, throughout much of the 1980s and 1990s, SSTs of the West Coast of North America have been warmer than usual and the thermocline deeper, conditions which appear to have reversed in the early 1990s.

The AO is a measure of the spin up (or spin down) of the polar vortex, which is indexed by the amplitude of the first principle component of sea-level pressure in the region during winter months (Thompson and Wallace 1998; see also Chapter 2). Temporal patterns of the AO revealed a potential shift from a negative to a positive phase in 1989 (Overland *et al.* 1999). The influence of the AO on ocean conditions is most notable in the Bering Sea. There, wind forcing is influenced by the impact of the PDO, El Niño, and the AO on the distribution and intensity of winter storm tracks as measured by indices of the Aleutian Low (Bond and Harrison 2000; Hare and Mantua 2000). Shifts in the intensity and distribution of storm tracks also influence the timing and distribution of sea ice in the Bering Sea. The distribution of sea ice and timing of its retreat has been shown to have a strong influence on the timing and intensity of spring blooms over the shelf and the location of a cold pool of water over the shelf (Wyllie-Echeverria and Wooster 1998).

Temporal patterns of fish production based on smoothed indices of recruitment of some species in the Gulf of Alaska and California Current ecosystems tend to be coincident with the two major atmospheric forcing processes influencing the region: El Niño (as discussed above) and the PDO (Hollowed *et al.* 2001). In these two systems, some stocks appear to follow temporal patterns of the PDO: Pacific sardine, arrowtooth flounder, Pacific

halibut, and Pacific salmon (Hare and Mantua 2000; Hollowed *et al.* 2001). Further evidence of the importance of large-scale forcing on fish production is noted in the marked synchrony of occurrence of strong year-classes of marine species in the North Pacific and Bering Sea (Hollowed *et al.* 1987; Beamish 1994). These findings suggest that fish (at least several commercial species) have adapted to take advantage of a common suite of conditions controlled by a number of climate–ocean processes.

12.5 Paleo-oceanography studies

The presence of fish remains in sediments and archaeological sites has been a valuable source of information when examining climate–fisheries–human interactions in the Pacific Ocean (Finney *et al.* 2000; Jackson *et al.* 2001). In particular, fish remains in varved sediments of anoxic basins have shown high resolution recording of marine fish abundances that can be compared to other indices of ancient climate. These studies have revealed historical colonization events and cycling of populations dating back thousands of years.

Colonization by marine fishes of Saanich Inlet on the south-eastern side of Vancouver Island has been traced back to events subsequent to deglaciation about 12,000 years ago (Tunnicliffe *et al.* 2001). After deglaciation and formation of an open marine bay, the area was colonized by hake and herring. Interestingly, it took the northern species of cod and pollock another 6000 years for their remains to deposit in the sediments. What explains the earlier arrival of the southern species hake? Pollock and cod are believed to spawn in fixed geographic locations, while hake spawning varies in location depending on environmental conditions. Perhaps this difference in spawning strategy could allow hake to opportunistically invade new habitats during a warming period, while pollock and cod may have gradually colonized coastal embayments in a stepping-stone manner. Alternatively, regional disruption of atmospheric circulation during the glaciation that caused a collapse of the California Current and a sudden warming of the ocean off California (Herbert *et al.* 2001) may have forced hake northward, and a more gradual warming off Oregon and Washington allowed them to invade new habitats.

A record of coastal fishes dating back around 1700 years is available off the coast of southern

California (Soutar and Isaacs 1974; Baumgartner *et al.* 1992). These data have shown that anchovy and sardine populations fluctuate with a high frequency period of about 60 years and with a low frequency period of 500–700 years. Comparison of these series to paleo-oceanographic climate records are being conducted, although contemporary harvest records indicate that sardine harvests and temperature vary in phase (Omori and Kawasaki 1995).

12.6 Forecasting

Can we predict the effects of climate changes on fisheries? Werner and coworkers in Chapter 3 state that 'understanding and predicting the response of marine ecosystems to global change is essential'. On the other hand, climate and climate–ocean effects may themselves be hard to predict from shifts in the NAO, as Hurrell and Dickson (Chapter 2) report: 'much of the atmospheric circulation variability in the form of the NAO arises from processes internal to the atmosphere, in which scales of motion interact with one another to produce random (and thus unpredictable) variations'. Although, the NAO may explain a lot of the variability in climate over the North Atlantic, the chaotic nature of atmospheric circulation causes substantial departures from the caricatured effects. Therefore, a first question might better be phrased: can we predict the impact of climate changes on the ocean, with what detail, and with how much lead time? In the case of El Niño, its development in the equatorial region can be followed from satellites and the TOGA mooring array leading to predictions of the event in the eastern tropical Pacific months in advance. The strength of the event and precise timing are still difficult forecasts. Whether these effects extend to higher latitudes may be a consequence of interaction between the El Niño and PDO signals. Predicting effects of large-scale climate shifts on the ocean and then on the biology within oceans will be even more difficult due to the compounding of uncertainty both on large and local scales.

Process studies examine interactions of climate and biology on relatively small scales, themselves imbedded within ever-increasing larger-scale processes. The interactions of many such small-scale processes are complex. Complex processes like a population trajectory through time may be unpredictable (a possible exception is where initial conditions generate mortality that is catastrophic, so the population is unable to recover). However, coalescence of small-scale processes may lead to patterns at higher levels of ecosystem structure. Holistic effects of low frequency or large-scale changes, such as regime shift impacts on the flow of energy in the food web, species succession, and shifts in species ranges may be predictable. It is clear that large-scale climate cycles have a major effect on fish populations and productivity. And, while climate impacts may not accurately forecast a population's trajectory through time, environmental regimes can set up boundary conditions that may influence the probability of an outcome. General patterns in recruitment, such as warming that favours strong recruitment of herring stocks at the northern end of the species range (Ottersen *et al.* Chapter 6), may also lead to predictions. However, these patterns may also fail, evidenced by disruptions in the cycle of Bohuslän herring. Likewise, with the latest warm regime in the California Current, recovery of the Pacific sardine lagged behind regime-shift related warming by 15 years or more, and population levels have not attained those observed in the 1930s. In the case of Pacific sardine and other heavily harvested populations, perhaps some population segments became extinct during a population retraction resulting in loss of local adaptation, or changes in trophic interactions have hindered recoveries. It may be reasonable to make short-term forecasts based on mechanisms or existing conditions to the next level or scale up, but forecasts become more imprecise at farther points, as expected in a complex system. For example, recruitment can be forecast from the number of pre-recruit juveniles but is poorly predicted from initial conditions during larval life (Helle *et al.* 2000; Bailey 2002). Better predictions in the future from a set of environmental conditions may arise from new tools, hybrid modelling techniques mixing stochastic and determinstic models, better understanding, or more general and larger targets for the forecaster's arrow.

CHAPTER 13

A fresh (water) perspective on the impacts of the NAO on North Atlantic ecology

Dietmar Straile

13.1 Introduction

North Atlantic climate variability does not only have an influence on North Atlantic ecosystems as described in the previous chapters and by Drinkwater et al. (2003), but influences terrestrial and freshwater ecosystems across the Northern Hemisphere (Ottersen et al. 2001; Mysterud et al. 2003; Stenseth et al. 2002; Straile et al. 2003b). Its impact on freshwater ecosystems has been documented on an area ranging from Lake Mendota in the United States to Lake Baikal in Russia, from Lake Kallavesi in Finland to the Caspian Sea (Straile et al. 2003b). In an increasing number of studies, the North Atlantic Oscillation (NAO) is shown to influence the physics, chemistry, as well as the biology of rivers and lakes, with biological impacts encompassing various trophic levels. A review on NAO impacts was recently provided by Straile et al. (2003b) and will not be repeated here. Instead, this chapter is an essay on selected aspects of NAO impacts on freshwater ecosystems, which might shed some light or pose new questions on the NAO impacts on the North Atlantic.

There is one aspect of lake ecosystems, which makes them easier to study compared to marine ecosystems: their size. Because of their smaller size, the physical environment of most freshwater plankton ecosystems may be considered to some extant as simplified because large current systems do not exist as in the ocean and hence cannot influence

the abundance of plankton organisms at distinct sampling sites. Horizontal transport is a major topic in this book's chapters of climate impacts on phytoplankton (Chapter 4), zooplankton (Chapter 5), as well as fish populations (Chapter 6) in the North Atlantic. The impact of horizontal transport on plankton abundance is of such pronounced importance that the term 'translation' has been coined to account for this effect (Chapter 5). Ecological consequences of changes in horizontal transport are manyfold. Direct and indirect effects may be distinguished: the direct effect of changes in horizontal transport is that a distinct species or community will be transported to the site A instead of site B. This, however, might imply important secondary and indirect effects for the transported species/community but also for the local communities at both sites A and B. Site A might be less or more suitable for the growth and reproduction of the transported species than site B because of local differences in temperature and/or food and/or predator abundance (spatial match or mismatch; Cushing 1990). Likewise, local communities at sites A and B will probably experience a different future as the transported species or community is now more or less abundant, which might change food availability and/or predation pressure. Additionally, local and transport responses might act simultaneously. For example, NAO related changes in growth and survival of cod larvae in the Barents Sea are thought to result from changes in temperature and prey availability, the latter resulting from increased inflow of *Calanus finmarchicus* into the Barent Sea and high *C. finmarchicus* growth and survival in the Barent Sea due to higher local temperature (Ottersen and Stenseth 2001). Given, this complexity, it is not surprising that distinguishing between the relative influence of physical processes such as climate-induced translations versus biological processes is considered to be a fundamental problem in biological oceanography (Chapter 5). Obviously, freshwater ecosystems offer the possibility to study the local impact of climatic variations such as the NAO on, for example, the population dynamics of phytoplankton, zooplankton, and fish species without the confounding effects of horizontal transport. Hence, a look across the salinity border might provide insights into the role and importance of local physical and biological responses of aquatic ecosystems to the NAO and climate variability in aquatic ecosystems.

13.2 The impact of NAO on phytoplankton

The relative importance of local processes versus transport processes in determining local plankton abundances will depend on the intrinsic growth rates of species, which in turn will be largely determined by their body size. Due to their small size, phytoplankton will respond fast to local changes in climatic conditions, which then might be as important as transport responses. Clearly, NAO related changes of transport processes are also of importance for phytoplankton. For example, increased oceanic inflow into the North Sea resulted into increased numbers and abundances of oceanic species such as the diatom *Thalassiothrix longissima* (Edwards *et al.* 2001). However, high phytoplankton growth rates should allow phytoplankton populations to show a fast numerical response to local conditions. Marine phytoplankton populations have been shown to respond to the NAO in terms of seasonal phenology, growth rates, species composition, and biomass as described by Smayda and coworkers in Chapter 4. Because of the multitude of possible mechanisms by which the NAO might influence phytoplankton populations it is, however, extremely difficult to understand the reasons for NAO related changes within a specific local phytoplankton assemblage.

In lakes, NAO related changes of light availability seem to be important to understand the impact of the NAO on phytoplankton. Light availability in many temperate lakes during winter/early spring is controlled by ice cover. In some of these lakes, ice-break-up in late winter/early spring is tightly related to the NAO (Livingstone 1999; Weyhenmeyer *et al.* 1999; Straile *et al.* 2003b). Warm winters in Central and Northern Europe in the early 1990s resulted into a pronounced reduction of ice cover duration compared with, for example, the 1980s (Adrian *et al.* 1999; Weyhenmeyer *et al.* 1999; Weyhenmeyer 2001). As ice cover strongly reduces light availability for phytoplankton, ice-break-up is usually associated with the start of the phytoplankton bloom. Earlier ice-break-up hence was shown to result into a considerable advancement of the timing of the phytoplankton peak in, for example, Lake Erken, Sweden and Müggelsee, Germany. In Lake Erken, the timing of the spring peak advanced by approximately one month during the last decades and the advancement was significantly related to the NAO (Weyhenmeyer *et al.* 1999). The timing of the phytoplankton maximum in Müggelsee was also related to the NAO and the average date of the phytoplankton peak advanced by approximately one month from the 6th of May during the years 1979–87 to the 7th of April during 1988–98 (Gerten and Adrian 2000). In addition to the influence of the NAO and ice-break off on bloom timing, the timing of ice-break off was also related to phytoplankton species composition. In both Müggelsee and Lake Erken, high NAO years favoured the growth of diatoms relative to other phytoplankton species as diatom blooms developed only under ice-free conditions when turbulent conditions prevented diatoms from sinking off the euphotic zone (Adrian *et al.* 1999; Weyhenmeyer *et al.* 1999). In contrast to lakes with winterly ice-cover, no significant relationship between bloom initiation nor phytoplankton peak abundance and the NAO was observed in a deep lake, Lake Constance, Germany, although water temperature in this lake was closely associated with the NAO (Straile 2000). In Lake Constance, a complete ice cover is a rare event (it only occurred once in this century in 1963, being a year with a very low NAO index value) and the start of the phytoplankton bloom is controlled by the absence of water mixis below 20 m (Gaedke *et al.* 1998). Mixis events below this critical depth seem to occur in Lake Constance independently from the NAO. In fact, the vertical distribution of nutrients and oxygen during winter and early springs suggests that winter mixis is complete within the upper 100 m of the water column independently from winter conditions, while after warmer winters (high NAO years) mixis below 100 m was less likely than and after cold winters (Straile *et al.* 2003a). Phytoplankton blooms in Lake Constance were also observed at cold-water temperature during short periods with absence of mixis (Gaedke *et al.* 1998), suggesting that direct effects of water temperature on phytoplankton growth rates might be only of secondary importance.

Phytoplankton might be indirectly influenced by climatic variability due to its influence on herbivore grazing rates. For example, Smayda and coworkers suggest in Chapter 4 that increased grazing rates due to higher abundances of cladocerans and *Acartia hudsonica* in high NAO years resulted into decreased abundance of the diatom *Detonula confervacea*. NAO associated grazing impacts on phytoplankton seem to be common in Central European Lakes (see below). In many temperate lakes, temperature controlled mass development of herbivores results into a so-called clear-water phase due to the elimination of algae from the water column (Lampert *et al.* 1986). The timing of the clear-water phase increased during the last three decades on average by approximately two weeks in several European lakes and the interannual variability is tightly associated with the NAO (Straile 2002). Seasonal advancement of the clear-water phase was less than that of ice break-up and in magnitude similar to phenological responses in terrestrial ecosystems (Chmielewski and Rötzer 2002) (Fig. 13.1). The

advancement of the timing of the clear-water phase demonstrates that impacts of the NAO need not be confined to winter, when the NAO has its strongest teleconnections to European climate variability. Clear-water timing, which occurs usually in late spring/early summer, is still rather closely related to the winter state of the NAO. The high heat-carrying capacity of water and the growth and grazing of herbivore populations carry on the meteorological signal towards summer.

13.3 The impact of NAO on zooplankton

The chapter of Pershing and coworkers on NAO impact on North Atlantic zooplankton (4) demonstrates how important horizontal transport mechanisms are for the understanding of the abundance of copepod species. Interannual variability of *C. finmarchicus* in the northwest Atlantic as well as *C. finmarchicus* and *Calanus helgolandicus* in the northeast Atlantic cannot be understood without considering horizontal transport. This chapter further summarizes work, which shows that a thorough understanding of the seasonal dynamics and the life history of species is necessary to understand the association between climate and species dynamics.

The impact of the NAO on copepod populations in lakes is less evident as compared to marine copepods, but also as compared to other freshwater zooplankton taxa. The winter abundance of *Eudiaptomus gracilis* in Esthwaite water, United Kingdom, has been shown to be significantly related to the NAO (George and Hewitt 1999). Additionally, Adrian (1997) has shown that winter conditions in Heiligensee in Berlin, Germany, influence population dynamics of copepods species. Given the strong influence of the NAO on winter temperature in this region (Straile and Adrian 2000; Gerten and Adrian 2001) a relationship of copepod population dynamics with the NAO in Heiligensee is likely. Interestingly, in both cases increased food abundance was considered to be the cause for the increased abundance of copepods: In Heiligensee, increased food abundance, that is, phytoflagellates and rotifers, during warm winters probably increased the survival of nauplii and copepodid stages of the cyclopoid copepod *Cyclops vicinus* (for resulting indirect effects see Adrian 1997). Copepod species in Lake Constance show no strong relationship with the NAO (Straile, unpublished data). In this deep lake, water temperatures but not phytoplankton abundance during winter and early

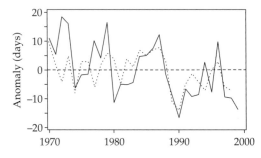

Figure 13.1 Anomalies of average European tree leafing dates (dotted line) from Chmielewski and Rötzer (2002) and average Central European clear water timing (thick line). Clear-water anomalies are adjusted for the effects of altitude, latitude, and lake depth (Straile 2002). Over the whole study period, the average date of leafing was the 23rd of April, and the average clear-water date was on the 9th of June. Both time series are significantly related to the NAO winter index and also significantly related to each other (all $p < 0.05$, after linear detrending: all $p < 0.05$).

spring was closely associated with the NAO (Straile 2000; Straile *et al.* 2003a). However, in contrast to copepods, spring population dynamics of cladocerans, for example, daphnids, were strongly associated with the NAO in Lake Constance. As food abundance is high and predation pressure is low during the period of *Daphnia* spring increase, interannual variability in population growth was most probably a direct consequence of water temperature variability (Straile 2000). Abundance of daphnids increased during spring within less than two months approximately 1000-fold. As the speed of population growth during this highly dynamic period seems to be controlled largely by temperature, even small interannual differences in water temperature may have huge effects on abundance. For example, average daphnid biomass during May varied by a factor of 200 between a rather low NAO year (1987) and a high NAO year (1990) in Lake Constance (Straile and Geller 1998). Within a subsequent study, it was shown that *Daphnia* spring dynamics in the small hypertrophic Müggelsee was also related to the NAO (Straile and Adrian 2000). Despite differing in several limnological characteristics and being 700 km apart from Lake Constance, interannual variability in *Daphnia* spring population growth in Müggelsee was closely related to *Daphnia* interannual variability in Lake Constance. Also in Müggelsee, the response of growth rates to changes in water temperatures was considered to link *Daphnia* population dynamics to the NAO (Straile and Adrian 2000).

Indirect evidence suggests an influence of the NAO on *Daphnia* population dynamics during spring in many Central European lakes. The timing of mass occurrences of daphnids can be inferred from its grazing impact on phytoplankton. Massive *Daphnia* grazing in late spring results in many temperate lakes into the so-called clear-water phase, a short and distinctive period of high water transparency (Lampert *et al.* 1986). Due to the impact of the NAO on *Daphnia* population growth, the timing of this 'foot print' of *Daphnia* grazing is also tightly related to the NAO in both Lake Constance and in Müggelsee (Straile 2000; Straile and Adrian 2000) but also in several other Central European lakes of variable morphology and trophic status (Scheffer *et al.* 2001; Straile 2002). Hence, the NAO seemed to have a striking synchronizing effect on the population dynamics of cladoceran herbivores including the predator–prey interaction with its food resources in Central European lakes during the last decades.

The analysis of freshwater zooplankton long-term data sets suggests that species with high intrinsic growth rates as parthenogenetically reproducing cladocerans (and rotifers; Gerten and Adrian 2000) seem to be more likely to respond fast to local changes in environmental conditions than species with complex life cycles and a comparatively long developmental time, for example, copepods. Hence, local responses of zooplankton populations in the North Atlantic might be also more important for taxa, which have higher maximum population growth rates as copepods. For example, in coastal regions the microphagous cladoceran *Penilia avirostris* might be able to respond fast to changed environmental conditions. Likewise, growth rates of tunicates are among the highest reported for metazoans (Madin and Deibel 1998). As a consequence of its high growth rates, tunicates have been shown to build up mass occurrences leading to clear-water phases (Bathmann 1988) comparable to the impacts of mass developments of daphnids. Obviously such mass occurrences will have important consequences for carbon fluxes and export production in the ocean (Bathmann 1988; Andersen 1998). More information on the population dynamics of these species seem to be needed in order to obtain a more complete picture on the impacts of the NAO on the functioning of North Atlantic ecosystems especially regarding the fate of biogenic material (Chapter 3). A second potentially very important missing link in the understanding of NAO effects or more generally climate change effects on carbon flows and export production in oceans, but also in lakes, is the impact of the NAO/climate change on protozoans. It is now generally agreed that protozoans are important in both lakes and the oceans as grazers of primary production (Müller *et al.* 1991; Gaedke and Straile 1994; Neuer and Cowles 1994), as remineralizers of nutrients (Caron and Goldman 1990; Ferrier-Pagès and Rassoulzadegan 1994) and in the transfer of carbon and nutrients to higher trophic levels (Carrick *et al.* 1991; Gaedke *et al.* 2002). Unfortunately, until now it is not clear whether and to what extent protozoan population dynamics and the protozoan share of total nutrient remineralization and secondary production are influenced by climate variability. Models of the response of carbon fluxes in the North Atlantic to climate variability will definitely have to consider zooplankton species in addition to copepods.

13.4 Conclusions

Overall, the work on NAO impacts in freshwater ecosystems underlines the importance of a thorough understanding of the seasonal dynamics and life history of species for a better understanding and prediction of climate variability on aquatic ecosystems (Chapter 5). Due to differences in seasonal occurrence and in life history characteristics species will respond differently to the NAO and may also be affected by the NAO via different mechanisms. In this respect, it is tempting to note that responses of freshwater copepods to the NAO have been suggested to be due to changes in food concentrations, whereas the responses of freshwater cladocerans seemed to be due to altered water temperatures. Effects of the NAO on a distinct population can be strong and may be passed on to other taxa, for example, due to grazing or predation. Such food web interactions may prolong the influence of the NAO beyond the time period of teleconnections between the North Atlantic and local climate. Furthermore, evidence from European lakes suggests that such indirect effects may be as strong that they can be observed synchronously across a large area.

A freshwater perspective on climate variability and its effect on marine ecosystems

Dag O. Hessen

There are several points of departure when assessing the effects of climatic variations from a limnological perspective. First, to which extent do climatic effects on freshwaters per se impact marine ecosystems? Second, can the enclosed and comparatively simple ecosystems of lakes be used to gain insights in marine effects? And third, are there any pronounced and general differences in expected responses between limnetic and marine ecosystems? In this commentary essay, I will touch upon these subjects where appropriate for the various chapters, focusing the riverine efflux to marine areas.

14.1 Freshwater impact on marine ecosystems: effects of global warming

Terrestrial, freshwater, and marine ecosystems may respond quite different to the same climatic signals, but there is also a causal linkage between these three systems; all terrestrial areas represent catchments, and climatic effects on catchment properties will affect rivers and lakes and subsequently the marine recipients. Such effects may be quite substantial in coastal areas or marine habitats impacted by riverine runoff. For productivity in coastal areas, influx of key elements like nitrogen (N), phosphorus (P), silica (Si), and carbon (C) from rivers are highly important, and the flux of these elements depend both on catchments properties, various human activities and climate (Smayda 1990; Meybeck 1993; Howarth et al. 1996; Hessen 1999). The riverine input of freshwater, elements and

substances may be pronounced not only in coastal areas, but also for semiclosed marine waters like the Baltic Sea (cf. Granéli et al. 1990; Pettersson et al. 1997) and major parts of the Arctic seas (Opsahl et al. 1999). In fact, the Arctic shelf seas constitute about 25% of the global shelf area, and receives freshwater inputs from some of the worlds largest rivers like Yenisey (ca. 600 km^3 $year^{-1}$) and Ob (ca. 400 km^3 $year^{-1}$) that provides major inputs of nutrients and dissolved organic carbon (DOC) to the Arctic seas. Both isotopic signatures and vertical light attenuation profiles (e.g. Aas et al. 2002) clearly demonstrate that this terrestrially derived influx significantly impact high Arctic marine ecosystems. Under a future climatic scenario with permafrost thawing (Majorowicz 1996) and increased precipitation (Hanssen-Bauer and Førland 1998), this riverine export of nutrients and DOC could increase strongly.

Nitrogen export from most catchments has increased strongly during the past decades, mainly resulting from increased inputs of fertilizers and increased atmospheric deposition of oxidized and reduced N, the first originating mainly from combustion processes, the second mainly from volatilization of manure (Vitousek et al. 1997) While 'pristine' watersheds export some 100–200 $kgN\,km^{-2}\,year^{-1}$, agricultural and urbanized watersheds may export more than 10,000 $kgN\,km^{-2}\,year^{-1}$ (Howarth et al. 1996; Hessen 1999). For the North Atlantic shelf as a whole, Nixon et al. (1996) estimated that N-fluxes from major rivers and estuaries exceeded atmospheric deposition by a factor of 3.5–4.7, but for

Skagerrak and the North Sea as a whole, the huge input from northward coastal currents and vertical mixing by far exceeds the riverine and atmospheric inputs on an annual basis. Atmospheric deposition of N seem to correlate well with catchment export (Howarth *et al.* 1996), and for many catchments receiving in excess of 2000 $kgN \, km^{-2} \, year^{-1}$ from wet and dry deposition, there are signs of a progressing 'N-saturation' yielding increased area-specific runoff rates.

The flux of N is tightly coupled to the pools and the turnover of organic carbon (C) in many temperate watersheds, and a major fraction of total N export is in the form of organic N, associated with organic C (mainly dissolved humic matter). Northern ecosystems typically have large stores of C and N in the soil. This decreases southward, to the opposite extreme in tropical soils, which are almost devoid of such organic stores. Thus, southern and northern ecosystems will behave essentially different with regard to C and N cycling. In northern ecosystems, the largest pool of organic material in the water column is dissolved organic material (DOM) of humic origin, and dissolved fulvic acid is the largest fraction of the humic matter (40–60%). Pettersson *et al.* (1997) estimated humic compounds to account for approximately 80% of the organic matter in 24 Swedish and Finnish rivers entering the Gulf of Bothnia. In such northern, boreal catchments, dynamics of humus DOM is instrumental also to catchment export of organic N and to some extent also inorganic N. Potentially limiting metals such as Fe may also to a large extent be associated with fluxes of organic C (Pettersson *et al.* 1997). The high amounts of DOC in humus DOM may also be a major determinant to heterotrophic production in estuarine ecosystems. Some general patterns of DOC concentration across climatic zones have been reported (Meybeck 1982), although there is a strong within-zone variation due to different altitude, soil, and vegetation cover and hydrology. For rivers, Meybeck (1982) concluded that concentrations of DOC were greatest in taiga rivers followed by rivers in the wet tropics, temperate, and tundra zones, respectively. Boreal catchments will show strong gradients in their export of DOC. Heathland and alpine areas may have DOC concentrations less than $1 \, mgC \, l^{-1}$, while catchments dominated by wetlands, bogs, and coniferous forests may yield more than $50 \, mgC \, l^{-1}$.

Thus, the flux and fate of organic C (ranging from 1 to 8 tonnes $km^{-2} \, year^{-1}$ in most temperate catchments) may be a main determinant of annual fluxes of N, though the role of organic N for marine productivity is not finally settled. Also DOC provides a major source of iron to marine areas, since iron is commonly adsorbed (or complexed) with humus molecules. Similarly, export of P will show pronounced variability among catchments, depending on population and land use, and range from 'pristine' minima of less than $4 \, kgP \, km^{-2} \, year^{-1}$ to maxima of more than $1000 \, kgP \, km^{-2} \, year^{-1}$. While fluxes of both N and P increase with catchment disturbancy, there is a strong tendency towards higher N:P-ratios in more disturbed watersheds, and N:P ratios may show extreme variability in agriculturally dominated watersheds. Although, Si fluxes are mainly governed by natural processes (weathering rate), a strong site specific variability of total Si export is also recorded, typically ranging from less than 500 to more than $-5000 \, kg \, SiO_2 \, km^{-2} \, year^{-1}$ (Hessen 1999).

Land use, soil and vegetation, and hydrology are the major factors controlling export or retention of the various elements in watersheds. These are important processes, since a considerable share of marine productivity in coastal areas are based on these riverine inputs. Anthropogenic effects such as fertilizer inputs and atmospheric deposition are superimposed on these catchment properties. Also short- or long-term climatic perturbations may add to this altered flux of N from watersheds (Meybeck 1993). For areas in the North Atlantic region, climate dynamics linked with the North Atlantic Oscillation (NAO) index is particularly relevant. A high NAO index coincides with heavy rainfalls and mild winters (cf. Hurrell and Dickson in Chapter 2), and global circulation models do in general predict a considerably higher temperature increase over the northern parts of the North Atlantic compared with the global average (IPCC 2001) and increased precipitation over most of Europe. Patterns of rainfall may show substantial regional variation, however, and could increase by 20% of wet parts like western Norway (www.nilu.no/regclim/Engelsk/default.htm), while decrease in dry areas with high potential evaporation. Increased winter temperatures may shift annual runoff patterns, and could also cause increased export of organic N and C to marine areas. For example, for Norwegian watersheds an increased winter temperature associated with high precipitation and a high NAO index gave a shift in the annual flux of N to coastal areas from spring to late winter (Fig. 14.1). Prevailing mild winters in northern ecosystems could thus reduce the snow-melt peak of riverine transport of N in spring. Since runoff or organic N (but not nitrate) is closely linked with runoff of DOC (Fig. 14.2), such annual shifts in runoff patterns will also affect the load of DOC to coastal areas. Correspondingly, also runoff

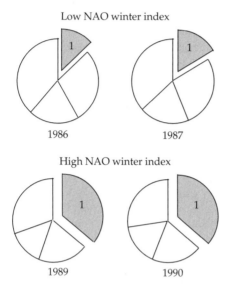

Low NAO winter index

1986   1987

High NAO winter index

1989   1990

Figure 14.1 Quartenary runoff of total nitrogen from river Bjerkreimselva in typical years with low and high NAO winter index (from Hessen 1999). Shaded area represents January–March. These patterns are typical for southeastern watersheds of Norway. Data on NAO index obtained from www.cgd.ucar.edu/~jhurrell/nao.html.

of other key plant nutrients like P and Si will reflect these changes, meaning that the coastal spring bloom may occur earlier.

More systematic shifts in climate may also have strong effect on nitrogen budgets of catchments by increased soil mineralization (Hessen *et al.* 1997; Hessen 1999). Typical northern, boreal soils contains 2000–7000 tonnes $N\,km^{-2}$, while soils further south commonly have only 1000–4000 tonnes $N\,km^{-2}$. This would imply a potential for increased runoff due to elevated mineralization rates. By use of models for soil mineralization of N, using sulphur and N deposition scenarios, and various degrees of N saturation and climatic variables (precipitation and temperature) as dynamic variables, a marked increase in mineralization and watershed export of inorganic N was predicted for a Norwegian watershed (Hessen and Wright 1993).

Climate also has effects not only on seasonal runoff patterns, but also on total export, concentrations, and probably quality of dissolved organic matter. Long-term monitoring at the Experimental Lakes Area, northwestern Ontario, provided data on reference lakes over a period of 20 years, coincident with increased drought and a warming of two degrees (Schindler *et al.* 1996). Following these climatic changes, total runoff of DOC from catchments showed a pronounced decrease. The opposite seem to occur for northern areas where increased precipitation and warming yields increased export of DOC to marine ecosystems (Forsberg 1992; Freeman *et al.* 2001).

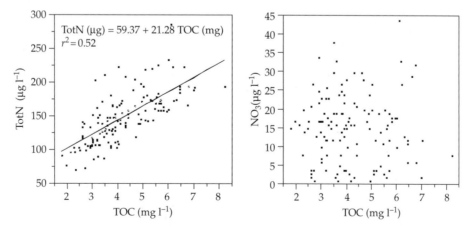

Figure 14.2 Relationship between runoff concentrations of total nitrogen and total organic carbon (TOC) in rivers (left panel). Total N is associated with DOM, hence there is no correlation between runoff of inorganic N and TOC (right panel). (From Hessen 1999)

For similar reasons as for C, the export of P, N, and Si to coastal areas could also decrease with warming, increased evaporation, and reduced precipitation, while it will increase under scenarios of increased precipitation. This will be most pronounced for watersheds with lakes or reservoirs, where changes in retention times will affect the export of all major elements to marine areas. These effects will be most pronounced for changes in water bodies with short renewal times (<1 year), since the retention equations have the form of asymptotic saturation curves approaching maximum retention for renewal times more than 1–3 years.

These changes in biogeochemical cycling may affect not only supply rates but indeed elemental supply ratios (e.g. N:P) that may affect autotroph community composition. Finally, changes in transport of DOC may also affect fluxes of humus associated iron that may be potentially limiting to marine photosynthesis (Watson *et al.* 2000) and DOC play a dual role in supporting organic C for marine bacteria and provide an efficient screen for short waved radiation. In fact, the comparatively high attenuation coefficients for UV-radiation in northern marine areas may be accredited riverine fluxes of DOC, not the least from the large Russian rivers (Opsahl *et al.* 1999; Aas *et al.* 2002).

The sum of these fluxes and climate effects will surely be relevant for costal productivity, and increased episodes of flooding over central Europe have provided episodes of substantial export of N and P to the North Sea and Skagerrak, and may be one of the causes for the blooms of toxic algae that have occurred in these regions (cf. Maestrini and Graneli 1991, see also Chapter 4). Hence, the conceptual food web models (cf. Chapter 3) should consider these riverine inputs, especially since the North Atlantic Ocean (NA) is fed by a number of major rivers. Watershed input of N to the NA shelf area is estimated to some 600 Gmol year^{-1}, which by far exceeds estimated inputs from atmospheric deposition (Galloway *et al.* 1996; Howarth *et al.* 1996). For the open ocean, N-fixation becomes the totally dominant source of 'new' N. This in contrast with P, where watershed export (in total near 50 Gmol year^{-1}) is totally dominant both for the shelf and open ocean. While the controversy of P versus N limitation is not settled, P could be seen as the ultimate limiting nutrient for marine production (Falkowski 1997) and where the weathering rate and flux of P depend both on climate and vegetation cover (cf. Lenton 2001). The strength of the biological pump as depicted by Werner and coworkers in Chapter 3 can respond both to changes in external forcing, for example, influx of N, P, Si, and Fe, and also changes in circulation, which might occur as a result of climate warming. Nutrient supply is thought to have mediated variations in the sequestration of CO_2 in the oceans that have accompanied the glacial–interglacial cycles, which have occurred over the past two million years (Falkowski *et al.* 2000; Ganeshram *et al.* 2000). Also the input of DOC could profoundly affect bacterial activity and heterotrophic processes in nearshore areas.

14.2 Climate effects in freshwater systems: lessons to learn for the marine environment?

There are a number of corresponding properties between marine and freshwater systems, that is, physico-chemical factors, nutrient fluxes, and food-web dynamics. On the other hand, there are also major differences between lakes and oceans, the most striking being that while lakes actually are closed or well defined systems (yet with influx and efflux of water and nutrients), oceans are open systems without real boundaries for species migrations and currents. Also the effect of climatic perturbations may have widely different effects on freshwater systems in coastal versus inland areas as well as between lowland and alpine areas (depending on changes in snow cover). While coastal, lowland freshwaters typically have extended ice-free season and reduced effects of snow-melt, alpine lakes may paradoxically experience reduced growth season due to strongly increased snow-fall, increased snow-pack depth, and thus delayed snow-melt. In fact, a high NAO index correlates well with reduced growth season and reproductive failure in alpine populations of Brown trout for the very same reason (Borgstrøm 2001).

Most lowland lakes will, however, experience increased growth season and increased surface temperatures. A strong correlation between the NAO winter index and average lake temperature has also been predicted (Straile and Adrian 2000; Scheffer *et al.* 2001). This will most probably speed up biological processes, yet the changes in system productivity and standing stocks will strongly depend on changes in nutrient regimes (N and P) and changes in thermal profiles and mixing depths.

The predicted net grazing effects was assumed to be low under a warming scenario, but could be severely reduced with lowered temperatures (Scheffer *et al.* 2001). They concluded, however, that the 'clearwater phase', that is, the period with maximum grazing pressure and low algal biomass would occur earlier and could be extended under a warming scenario, as clearly demonstrated also by Straile *et al.* (2003). Since, growth rate of zooplankton depend strongly on temperature (at any given food concentration), increased temperature will directly translate into increased population growth in freshwater zooplankton which can then be predicted quite accurately from established equations relating growth rate to temperature. These effects are commonly described by simple power functions like the pooled data on egg developmental time (D, in days) for crustaceans on the form $LnD = a + b \, Ln \, T + c \, (Ln \, T)^2$, where the constants Ln a, b, and c are 3.79, -0.15, and -0.26 for pooled data on freshwater crustaceans (e.g. Bottrell *et al.* 1976). Thus, even moderate shifts in the lower range of temperatures may have strong impacts on zooplankton population dynamics and thus food web interactions. For species with extended life cycles, like many copepod species, shifts in life cycle timing may be expected.

Since, growth rates of autotrophs will also be affected, the net outcome of increased temperature is not obvious. North-American studies do in fact suggest increased algal biomass in spite of reduced nutrient inputs due to a long-term trend of warming (Findlay *et al.* 2001).

Increased temperature and reduced precipitation yield reduced inputs on nutrients and a pronounced reduction in DOC with profound effects on lake water transparency (Schindler *et al.* 1996). Increased surface temperature, early thermocline development and hence reduced mixing will also strongly affect the biota. In particular the preferable habitat of several fish species may be reduced by decreased oxygen concentration and suboptimal epilimnetic temperatures, squeezing the spatial niche from above and below (Stefan *et al.* 1995; DeStasio *et al.* 1996). Since growth rates of most freshwater organisms increase up to temperatures of 20–25°C, growth rate of organisms in the epilimnion will increase unless food availability

decreases. Above 25°C net growth will normally decrease due to a relative increase in respiration rates (cf. Lampert 1977). With increased thermal stratification, heath diffusion to hypolimneon could decrease, however, hence the net outcome for a lake ecosystem per se is not easily predicted. Sweeping statements on climate effects in different types of lakes are premature even within the temperate region, since shallow and deep lakes or lowland and alpine lakes will behave differentially. For reasons given above, also increased or decreased precipitation would be a major determinant of climate-induced effects on lake biota.

The relevance of these freshwater findings and predictions for marine systems is not straightforward, although the NAO seem to play an important role for both. The pronounced seasonality, snow and ice-cover and the strong thermal stratification that occur in most temperate lakes during summer calls for precaution with regard to a direct comparison between lakes and marine ecosystems. Also the direct predictions of temperature-dependent processes is less straightforward in the marine systems. The other major difference is the absence of horizontal mixing in lakes, with exception of very large lakes or systems with short renewal times. Finally, it should be kept in mind that zooplankton communities differ strongly between freshwater and marine systems. The key players in freshwater systems are commonly cladocerans and notably daphnids that reproduce asexually and may have very rapid responses on shifts in nutrient and temperature regimes. This in contrast with the more diverse marine systems where pelagic metazoan grazer communities are dominated by copepods. Hence, the freshwater aspects that perhaps are most relevant to marine ecosystems are the terrestrial/freshwater processes related to weathering rate, changed temperature and precipitation, which are key determinant of the fluxes of DOC and key nutrient elements to coastal marine ecosystems. Over long timescales these processes are believed to regulate both marine productivity and hence global climate (Lenton and Watson 2000; Lenton 2001) and this calls for a joint effort from both terrestrial, freshwater, and marine disciplines to predict climate effects on marine ecosystems.

CHAPTER 15

Time lags in terrestrial and marine environments

Eric Post

Temporal lags in the response of populations to climatic variation associated with the North Atlantic Oscillation (NAO) are widespread in both terrestrial and marine environments (Forchhammer 2001; Ottersen *et al.* 2001). In marine systems, lagged responses span many trophic levels, and have been observed in benthic macrofauna off the western coast of Sweden (Chapter 10), zooplankton in the North Sea (Chapter 5), fish in the North Atlantic (Chapter 6) and seabirds off the coast of Scotland (Thompson and Ollason 2001; see also Chapter 7). Notably, however, in marine systems trophic position is not a perfect indicator of whether populations exhibit temporally direct or lagged responses to the NAO; zooplankton of the genus *Calanus*, for example, display both (Chapter 5). In contrast, responses to the NAO in terrestrial systems appear most often to be direct at low trophic levels (i.e. among plants), and lagged at higher trophic levels (i.e. among herbivores and predators) (Post *et al.* 1997; Post and Stenseth 1999; Post *et al.* 1999a), though on Isle Royale, USA, direct responses to the NAO have been documented at all three trophic levels (Post and Forchhammer 2001).

The existence of both immediate and lagged responses to climate presents conceptual and analytical challenges to the study of ecological consequences of large-scale climatic variability, as well as to our ability to forecast population responses to future climatic change. In this chapter, I describe three fundamentally different mechanisms through which time lags in the response of populations to climatic variability might arise, as well as conceptual

and analytical approaches that might be used to distinguish them.

15.1 Influence of atmospheric processes on time lags

Lagged population responses to large-scale climatic variability may arise when the proximal abiotic factor influencing population dynamics is itself correlated with regional atmospheric processes at some time in the past. Abundance of *Calanus* sp. in the Gulf of Maine, for example, is influenced by sea temperature, which correlates with the NAO index in the previous year (Chapter 5). Though such lags are probably more likely to arise in marine systems, where water temperature is an important proximal driver of biological dynamics and is likely to lag behind changes in air temperature, similar processes may occur in terrestrial systems where precipitation is important. In sub-Saharan Africa, for example, latitudinal variation in the 200 mm rainfall isocline responds to the NAO with a time lag of one year, as does vegetation productivity throughout the Sahel (Oba *et al.* 2001).

15.2 Influence of life history on time lags

Life history traits, such as the timing of reproduction, number of progeny produced, and birth mass, show sensitivity to the NAO and other climate indices in many terrestrial species (Forchhammer *et al.* 1998; Forchhammer 2001; Ottersen *et al.* 2001).

Because such life history traits can influence population dynamics years later (Fig. 15.1(a)), when cohorts reach reproductive maturity, direct effects of climate on development and reproduction can produce lagged population responses to climate (Post and Stenseth 1999; Post *et al.* 1999b).

This process can be visualized mathematically through numerical expression of the relationships depicted graphically in Fig. 15.1(a). Doing so leads to the following model of population dynamics as a function of the influence of climate (U) on some, in this case unspecified, life history trait that influences density (N) (see Post *et al.* 2001 for details):

$$N_t = (a_0 + a_2b_0 - a_0b_1) + (a_1 + b_1)N_{t-1} + (a_2b_2c_1$$
$$- a_1b_1)N_{t-2} + a_2b_2c_2U_{t-1}. \quad (15.1)$$

Here, we can observe that the delayed influence of climate (U_{t-1}) on density results from the concerted influences of climate on resources (coefficient b_2 in Fig. 15.1(a)), of resources on the life history trait (coefficient a_2), and of the life history trait on density (coefficient c_2). In deriving Eq. (15.1), I have assumed that the life history trait of interest influences density in the following year; longer delays are possible (e.g. Thomspon and Ollason 2001), and

will result in longer time lags for the influence of climate on density.

Life history-induced time lags can be dramatic in long-lived species with delayed reproductive maturity. In the northern fulmar (*Fulmarus glacialis*), for example, individuals do not breed until they are several years old, and influences of the NAO on cohort development and recruitment success in populations in Scotland result in lagged population responses of up to 5 years (Thompson and Ollason 2001). A similar cohort-induced, delayed population response to the NAO occurs in Soay sheep (*Ovis aries*) in the United Kingdom (Forchhammer *et al.* 2001), though at a shorter time lag.

15.3 Influence of trophic interactions on time lags

Density interactions between species at adjacent trophic levels may also produce time lags in the response of a population to climatic variation (Post and Forchhammer 2001). If, for example, the population dynamics of a species at one trophic level are directly influenced by climate, then the species at

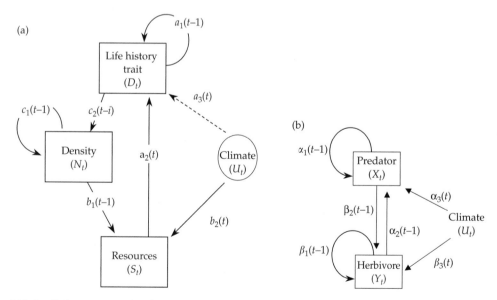

Figure 15.1 Graphical, process-oriented models depicting two mechanisms through which time lags in population response to climate change may arise: (a) an influence of climate on some life history trait that influences density at a number of years (*i*) later; and (b) an influence of climate at adjacent trophic levels when density at both levels is dependent on density at the adjacent level. See Post *et al.* (2001) for a full exposition of the model in (a), and Post and Forchhammer (2001) for an exposition of the model in (b).

the adjacent trophic level may show a lagged response to climate (Fig. 15.1(b)).

These relationships can also be visualized mathematically through numerical expression of the process expressed graphically in Fig. 15.1(b). This gives us the following model of population dynamics at one trophic level (in this case the predator, X) as a function of the influence of climate (U) on dynamics at its own level and the adjacent trophic level (in this case, the herbivore, Y; see Post and Forchhammer 2001 for details):

$$X_t = (a_0 + a_2b_0 - a_0b_1) + (a_1 + b_1)X_{t-1} + (a_2b_2c_1 - a_1b_1)X_{t-2} + a_3U_t + (a_2b_3 - a_3b_1)U_{t-1}. \quad (15.2)$$

In this case, the delayed influence of climate on predator dynamics (U_{t-1}) results from trophic interactions with the herbivore and climatic effects on the herbivore (coefficients α_2 and β_3 from Fig. 15.1(b) as well as from an interaction between the influence of climate on the predator and its effect on herbivore density (coefficients α_3 and β_2).

Such interactions are apparent in community-level dynamics on Isle Royale, where a lagged response of balsam fir (*Abies balsamaea*) to the winter state of the NAO was revealed to be a direct effect of the NAO on moose (*Alces alces*) abundance that cascaded down to fir growth 1 year later (Post *et al.* 1999a). Similar effects are likely to be important in the dynamics of fish and their predators in marine environments, where the NAO influences zooplankton abundance (Fromentin and Planque 1996). Indeed, the existence of extremely lagged responses to the NAO in marine systems (e.g. Thompson and Ollason 2001) may reflect, in part, the ubiquity of high numbers (i.e. >3) of trophic levels in marine systems.

15.4 Can we discern the mechanisms behind lagged responses to climate?

The first approach in identifying causal mechanisms underlying time lags in ecological response to climate is purely conceptual. We may start by asking questions such as the following. Is there likely to be (and does data analysis reveal) a lagged relationship between large-scale atmospheric dynamics and the proximal abiotic factor influencing biological dynamics in the focal system? Does the focal species undergo complex life history transitions that may give rise to climatically induced developmental delays or cohort effects on population dynamics? Are the dynamics of the focal species influenced by interactions with adjacent trophic levels that may also be influenced directly by climate?

A more formal approach to discerning the mechanisms or processes underlying lagged population response to climate involves time-series analysis, which may be informative in the absence of data at multiple trophic levels or on multiple life history stages (Forchhammer 2001). In this approach, the scenario in which it will be most difficult to distinguish between the different mechanisms generating time lags is when population response to climate occurs at a lag of one year. Nonetheless, manipulation of the relationships depicted in process-oriented time-series models such as those in Fig. 15.1 may prove useful as a first step. We might, for example, attempt to distinguish between life history-induced and trophic-induced lagged responses to climate by manipulating Eqs. (15.1) and (15.2). Starting with Fig. 15.1(a) and Eq. (15.1), we can see that setting the climate terms involved in the life history response to climate (i.e. coefficients b_2 and, in another possible scenario, a_3) equal to zero, the lagged climate term in Eq. (15.1) drops out. Switching to Fig. 15.1(b) and Eq. (15.2), we see, in contrast, that when the climate term influencing herbivore dynamics (i.e. coefficient b_3 or β_3) is set to zero, we may still arrive at a lagged effect of climate on predator dynamics that derives from the delay induced by the trophic interaction between predator and herbivore (i.e. a non-zero term for a_3b_1 in Eq. 15.2; see also Post and Forchhammer 2001).

Although time lags in the response of populations to climate present conceptual and analytical challenges, their existence in a multitude of systems can be exploited to one obvious advantage: prediction. Hence, an empirically derived basis for improving our conceptual and analytical understanding of lagged responses to climate should prove valuable in the pursuit of scientifically robust predictions of population and community response to future climate changes.

CHAPTER 16

The impacts of ENSO on terrestrial ecosystems: a comparison with NAO

Mauricio Lima and Fabian M. Jaksic

16.1 Introduction

Natural ecological systems (individuals, populations, communities, and ecosystems) experience climatic fluctuations through the effects of local weather, temperature, winds, snow, currents, and rainfall. As it has been noted, these local meteorological and oceanographic variables are governed by large-scale climatic phenomena, so that climate impacts on ecological systems are complex and we are only at the beginning of an understanding. Recently, there has been an increasing interest for studying the connections between climatic fluctuations and the dynamics of ecological systems (as discussed in Chapter 1). Climate has important effects on individuals, by directly influencing growth, birth, and mortality, but also indirectly by influencing the interactions among individuals (competition, predation, parasitism, and diseases). These more subtle effects of climate may have profound consequences in population dynamics because of changes in intrinsic properties of population systems. On the other hand, because of the non-stationary nature of climatic variation in the long-term, climatic forces have been influential in determining the evolutionary history of expansion and contraction of populations and also the patterns of biota diversification.

In this chapter, we aim, through our terrestrial point of view, to cast new light upon the preceding main chapters. While our starting point is the eastern side of the tropical Pacific, where El Niño-Southern Oscillation (ENSO) plays a highly significant role, we link closely to the North Atlantic Oscillation (NAO) impacts in the North Atlantic sector. Like the NAO, the impacts of ENSO on terrestrial ecosystems have come under close scrutiny only very recently (Jaksic 1998, 2001a; Holmgren et al. 2001), but the data being synthesized points to a pervading, if elusive, influence on ecological patterns and processes in terrestrial biomes. Here, we draw upon our experience along the western fringe of South America, especially on our own data from Chilean ecosystems.

16.2 Plants

That high rainfall associated with ENSO incursions somehow affects the vegetation, was first noted by Hamman (1985) in the Galapagos Islands (Ecuador), by Soto (1985) in Chile, and by Arntz and Fahrbach (1991) in Peru. Dillon and Rundel (1990), Nicholls (1991) and Armesto et al. (1993) were the first botanists that associated the massive germination and flowering of desert plants in Chile and Australia with periods of high precipitation. At least in Chile, this phenomenon is sufficiently spectacular to be called 'blooming desert.'

At least two ENSO intrusions have been relatively well documented from the botanical viewpoint, those of 1991–92 and of 1997–98. Gutiérrez *et al.* (1993) started analysing vegetation characteristics during 1989 and showed that shrub species richness and shrub cover remained about constant from until 1994, but that herb cover changed substantially from pre-El Niño (1989–90), to El Niño (1991–92), and to post-El Niño (1993–94). Herb species richness and particularly herb cover increased markedly during El Niño in comparison to pre- and post-periods. The same pattern was observed in the seed bank (Gutiérrez and Meserve 2003). Perennial seeds in the soil peaked immediately on arrival of El Niño, whereas ephemeral seeds peaked the following year. This study of Gutiérrez *et al.* (1993) indicates that two consecutive rainy years resulted in important increases in primary production, specifically that of herbs and seeds.

Gutiérrez *et al.* (2000) analysed the response of the vegetation in two contrasting solar exposures (north-facing versus south-facing slopes) of northern Chile, to the dry year of 1996 and to the wet one of 1997 associated to the 1997–98 El Niño event. In comparison to the dry year, shrub cover increased only slightly on both slopes (2–13%), but there were marked increase in the cover of annual herbs (200–400%), and more so in the case of seed bank density (400–500%). It should be noted that shrub species richness did not change between slopes or between years, whereas there were clear increases in herb species richness in both slopes (60–190%). In general, increases were more extreme in the north-facing slope. One must remember that in the Southern Hemisphere, the north-facing slope is that which faces to the Equator, not to the Pole. This study by Gutiérrez *et al.* (2000) shows that El Niño effects differ over small spatial scales, contributing to increasing the degree of patchiness in primary production.

From the two case studies above it may be concluded that the response of the vegetation in northern Chile is broadly similar, even when faced with El Niño events that occurred in different years (1991 and 1997). All results obtained thus far indicate that a single wet year suffices to trigger important increases in primary production, particularly among herbs and seeds. The output

in primary productivity may be immediate or lagged by one year, depending on the life forms involved.

In comparison to NAO, the ENSO database is lacking in information on plant phenology, and thus no parallels may be drawn. However, there is hope that as research in San Carlos de Apoquindo is sustained over the long term (Jaksic 2001b), this type of data may eventually be obtained. No information on tree rings is available from Chile either, but there is a good paleohistorical record of palinological profiles (Villagrán 1993). The issue of desertification is important along the coastlines of Peru and Chile, but here ENSO dampens its impact by setting off the blooming desert mentioned above. As for agriculture, no one has attempted an analysis of crop production in relationship with ENSO incursions. The blessing of rainfall in desert areas of Chile is counteracted by pest outbreaks that include both small mammals and insects (Fuentes and Campusano 1985). And, based on a short-time series, the quality of wines seems to improve when it is based on vintages post-ENSO. For instance, fewer medals were awarded to Chilean red wines of the vintages of Autumn 1993–1995 (corresponding to grapes that grew during the preceding year, when Southern Oscillation Index (SOI) was below −106), in comparison to those of the 1996 and 1997 vintages (corresponding to years when SOI ranged from −30 to +62, respectively). It is yet too early to determine, but the vintage of March 1999 (1998 SOI = −14) seems better than that of March 1998, right after the ENSO event of 1997 (SOI = −132).

16.3 Invertebrates

Jaksic and Lazo (1999) failed to note clear increases in large, terrestrial arthropod abundance from 1993 to 1997 in northern Chile, even when an ENSO event started in 1997. However, clearly, insect outbreaks are associated to ENSO events in arid and semiarid Chile (Fuentes and Campusano 1985). Nevertheless, there is no information on the phenology of any invertebrate species, not even of the conspicuous butterflies.

16.4 Amphibians

Reports of 'sapadas' (frog irruptions) are part of Chilean folk knowledge, especially in the northern part of the country, but no reliable records exist on this phenomenon for testing whether it is related to ENSO.

16.5 Birds

Chapter 7 covers seabirds in the North Atlantic region. Here we present additional and comparable information relating to effects of ENSO on seabirds along the South American coastline. Among the most easily spotted effects of El Niño intrusions are the massive migrations and subsequent dieoffs of seabirds (Schreiber and Schreiber 1984, 1989; Duffy and Merlen 1986; Ainley *et al.* 1987, 1995; Duffy 1990; Arntz and Fahrbach 1991, 1996; Wilson 1991; Massey *et al.* 1992). This happens because when fishes migrate to high seas, seabirds cannot keep up with them and begin to starve and soon fail to reproduce (Tovar *et al.* 1987; Guerra *et al.* 1988). In addition, predation on them increases (Spear 1993). Even terrestrial birds that inhabit oceanic islands suffer great population losses, although for different reasons (Gibbs and Grant 1987; Grant and Grant 1987, 1993; Hall *et al.* 1988; Miskelly 1990; Lindsey *et al.* 1997). In the latter case, when El Niño arrives, excessive precipitation reduces reproductive success, and the subsequent drought years that follow (the reverse phenomenon of La Niña), further decimate bird populations because of the depletion of food resources.

It is worth noting that until now all studies have referred to El Niño effects on seabird colonies or on landbirds on small oceanic islands (e.g. the Galapagos). Continental landbirds have not been studied under this perspective, except for a paper by Jaksic and Lazo (1999). These authors established a baseline spanning the dry years of 1993 through 1996 and the wet year of 1997 (associated to El Niño 1997–98). Birds had apparently reached their peak diversity and density following the wet years associated to El Niño 1991–92, and were at their lowest through 1993–96. After the rainy 1997, summer census results demonstrated an increase by 27% in diversity and by 123% in density, in comparison to the previous summer. Interestingly, the increase in bird density was not due solely to the increase in species richness: on a per-species basis, density was 75% higher in summer 1997 than during the previous summer. Jaksic and Lazo (1999) explained these increases as a consequence of augmented primary (vegetational resources) and perhaps of secondary productivity (total arthropod abundance). This study shows that increased productivity climbs up to the trophic level of birds within six months after El Niño driven rains.

As for vegetation, Chile is lacking information on bird phenology, and thus no parallels may be drawn with NAO (but see Chapter 7). In addition, at least in Chile, there is total lack of information on phenotypic plasticity, population ecology, and on interspecific interaction of birds along the succession of ENSO events. However, such information is available from the Galapagos Islands off Ecuador (Gibbs and Grant 1987; Grant and Grant 1987, 1993; Hall *et al.* 1988).

16.6 Mammals: marine mammals and rodents

Unlike the NAO-affected region, where most studies have been conducted on ungulates, none has been carried out on ungulates in the ENSO-affected region of western South America. Conversely, quite a bit is known on pinnipeds and rodents.

Like seabirds, pinnipeds such as sea lions and elephant seals also suffer from El Niño intrusions chiefly because their staple food fish and squid migrate off the coast and down into deeper waters. This results in an increase in the duration of pinniped 'forays to the sea', which in turn translates into more energy spent searching for food than what is obtained when food is actually found and processed. This negative balance shows off in loss of body fat and muscle, in higher frequency of diseased individuals, and in mass mortality of pinnipeds (Arntz and Fahrbach 1991, 1996; Harvell *et al.* 1999). For sure, the first to be affected are puppies and juveniles, unable of feeding by themselves and with lower capabilities for swimming than have adults. Massive mortality of pinniped puppies has been reported during El Niño events along the Peruvian and Chilean coasts (Limberger 1990; Arntz and Fahrbach 1991, 1996; Majluf 1991) and in the Galapagos Islands (Trillmich and Limberger 1985; Trillmich and Dellinger 1991). Indeed, the arrival of Peruvian sea lions at northern Chilean waters has been proposed as an early warning that El Niño is coming down to Chile (Torres 1985).

Equally dramatic are the rodent irruptions associated to El Niño. Suspicions that rodent outbreaks (or irruptions, or 'ratadas') are somehow linked to unusually high precipitation have existed since Pearson (1975) documented the outbreak of 1972–73 along the Peruvian coastal desert. Coincidentally, Fulk (1975) and Péfaur *et al.* (1979)

reported a rodent outbreak during 1972–73 in several localities of northern Chile. A problem with these pioneering studies was that nothing was known about the baseline or 'normal' population abundances of the rodents involved. An exception is the work of Meserve and Le Boulengé (1987), who followed for two years the irruption of 1972–73 first reported by Fulk (1975) in Fray Jorge National Park, northern Chile. Another exception is the work of Jiménez et al. (1992), who followed the irruption of 1987–88 in Aucó, northern Chile. All of these authors concurred that rodent outbreaks were triggered by unusually high precipitation. Using information on rodent irruptions published in northern Chilean newspapers, Fuentes and Campusano (1985) detected a significant association between years of high precipitation and rodent outbreaks in northern Chile. The same result was obtained by Lima et al. (1999a), using more sophisticated statistical tools.

A major problem with the studies reported above is that all took place after El Niño events. Only two recent investigations have spanned the entire period before-during-after El Niño. Meserve et al. (1995) reported population trends of small mammals in Fray Jorge National Park (northern Chile) from early-1989 to mid-1993. An El Niño event started in mid-1991 and ended in mid-1992. These authors showed that seven small mammal species started increasing during 1991, some of them immediately after winter 1991, while others lagged one year behind. In general, small mammals increased their population levels from 12 to 15/ha in pre-El Niño (1989–90) to 26–117 during El Niño, and decreased gradually in the next two years post-El Niño (from 111 to 43/ha in 1993–94). The other study, by Jaksic et al. (1993, 1996), spanned a more prolonged period at Aucó, 100 km south of Fray Jorge National Park. These authors were able to detect concurrent increases of mice with precipitation associated to two El Niño events: 1987 and 1991, to which Jaksic (2001a) added data for El Niño 1997. This time series reveals that rodent abundance follows with a lag of six months the precipitation pattern. And, it also reveals that similar rainfall levels do not result in similar rodent abundance. That is, rodent density may not be predicted solely from rainfall.

These results are very suggestive of a consistent connection between El Niño intrusions, precipitation increases, and rodent outbreaks. Indeed, Lima and Jaksic (1998, 1999a,b) have detected the rainfall 'signature' in time-series models developed for irruptive rodents in Chile, and Lima et al. (1999b) developed a population dynamic model that links El Niño driven rainfall and rodent outbreaks. It should be noted that this model involves not only rainfall, but also density-dependent population regulation.

16.7 Mammalian carnivores and others

Jaksic et al. (1997) analysed the effects El Niño 1991–92 in Fray Jorge National Park, Chile. They gathered 30 months of data before El Niño, and 36 months during and after the event (total = 5.5 years). The rainy winter of 1991 brought an immediate doubling of the seed bank (see Section 16.2). Consumers of seeds and herbs (chiefly rodents) increased their population levels 2 × immediately during 1991 and 10 × during 1992 and 1993 in comparison to 1990 (see section on mammals, above). Carnivorous predators (foxes, hawks, and owls) displayed a 1-year delayed response to small mammal abundance: from seven individuals/750 ha during 1991 to 17/750 ha during 1992, an increase of 143%.

Jaksic et al. (1997) also documented an anticlockwise trajectory of predator abundance versus small mammal abundance. That is, the population response of vertebrate predators followed with delay the changes in abundance of their main prey. This suggests that the dynamics of the predator–prey system depends on the abundance of the rodent prey, which depends on the abundance of the herbs and seeds that they eat, which in turn is determined by the rainfall pattern, which ultimately is associated to El Niño. These authors speculated that the effects of El Niño propagate from bottom trophic levels (prey) to those at the top (predators). Based on this scenario, Jaksic et al. (1997) predicted that as primary productivity varied with precipitation, the same should happen with secondary (small mammal abundance) and tertiary productivity (that of vertebrate predators). This prediction remains to be tested and is at variance with the statement of Post et al. (1999) that 'there can be an effect of the NAO on the behaviour of a top predator (in this case wolves), that may cascade down to affect the secondary (the moose) and primary producers (fir trees)'.

Unfortunately, no data are available on carnivore demography, as there are in the NAO-affected region, and no data are expected to be collected in the near future in the ENSO-affected region of Chile. Thus, no comparisons are possible at present.

16.8 Emerging ecological insights

We agree with Mysterud *et al.* (2003) that 'Compared with the effects of El Niño events on terrestrial ecosystems . . . , the effect of the NAO seems less dramatic . . . ' This may be so because ENSO effects have been chiefly studied in small oceanic islands (strongly influenced by the surrounding ocean) and on western-fringe deserts of South America, where precipitation is a major determinant of primary productivity and up. As primary productivity varies with precipitation, the same happens to secondary productivity (rodent abundance) and to tertiary productivity (predator abundance). This is what is known as a bottom-up model of ecosystem control (Jaksic *et al.* 1997; Polis *et al.* 1997; Meserve *et al.* 1999). As shown above, above ground vegetation flushes only among herbs but not shrubs (Gutiérrez *et al.* 1993, 2000). The seed bank is quickly replenished of ephemeral seeds, but perennial seeds recover one year later (Jaksic *et al.* 1997; Gutiérrez *et al.* 2000). Small rodents irrupt within months of El Niño driven rains, but larger ones take a full year to increase, and then not very dramatically (Meserve *et al.* 1995; Jaksic *et al.* 1997). Predators lag one year behind their mammal prey, and they increase by a combination of *in situ* reproduction and immigration, with smaller predators responding more quickly to prey levels (Silva *et al.* 1995; Jaksic *et al.* 1996, 1997). The feedback loop between the plant compartments (vegetation plus seed bank) and their herbivores seems weak: the evidence shows that primary productivity is the driving force, and is little affected by herbivory in comparison to rainfall (Meserve *et al.* 1996, 1999). After the rodent outbreak, the decline appears to be a complex non-linear combination of extrinsic and intrinsic factors, including rainfall, delayed density-dependence and predation pressure (Lima *et al.* 1999b).

The role of climatic fluctuations on population dynamics has been one of the most debated problems in population ecology (Andrewartha and Birch 1954). Today, ecologists agree that density dependence and climatic factors operate simultaneously in natural populations (Turchin 1995; Berryman 1999; Stenseth 1999) and that population regulation has to be understood within the context of both exogenous

forces (both stochastic and climatic) and intrinsic regulation (see Royama 1992; Turchin 1995). In fact, one interesting aspect of population dynamics not widely appreciated by ecologists, is that the way climatic effects observed in population fluctuations depends on their density-dependent structure (Royama 1992; see also Chapter 1). This fundamental aspect of population dynamic theory has deep implications for analysing and interpreting the effects of climate on numerical fluctuations of natural populations. One of these implications is that the way populations respond to climate is in part determined by the food web structure (Berryman 1999). Therefore, predicting the responses of natural systems to global change can be very difficult, if we do not have a detailed knowledge of the system. In addition, because climate effects may change the individual interactions within and between populations, the challenge is to understand the complex interactions between density-dependence, trophic structure and climatic fluctuations.

For example, the case of Soay sheep on the St Kilda archipelago provides an excellent example of how climate—through NAO and winter severity—influences intraspecific interactions and thereafter the density-dependent structure of population dynamics (Coulson *et al.* 2001). Density-dependence and climate interact in such a way that population crashes due to bad weather occur more frequently at high density than at low density. This result shows that climate may represent more than a simple forcing term; it can be part of the dynamic system properties. In the same vein, NAO appears to modulate the strength of intra-specific and interspecific competition between two sympatric flycatchers (Sætre *et al.* 1999; see also Stenseth *et al.* 2002). The result of this study strongly suggests that NAO fluctuations prevent the competitive exclusion of pied flycatcher by the collared flycatcher because the different effects of intra- and inter-specific interactions and the differential responses to climate of both sympatric species.

On the other hand, the role of food web structure and how climatic effects emerge at different trophic levels has been shown in two studies described in Mysterud *et al.* (2003). For example, the effects of NAO on the population dynamics of red deer in Norway depict a clear case of how climate effects can be direct and indirect by propagating through the trophic structure (Forchhammer *et al.* 1998). The + or − delayed effects of NAO on red deer populations are the consequence of indirect food-web responses, which has previously been hypothesized for trophic models (Chapter 15 focuses on time lags/delayed effects). The findings summarized in Forchhammer *et al.* (1998) are a clear example of how climatic forcings may have lagged and indirect effects in population dynamics. Another interesting example of how food web and climate may interact to produce complex responses is described by Post *et al.* (1999) with regard to moose–wolf interactions in Isle Royale, United States. Pack size of wolves increases with snow depth (and NAO) producing a cascade of effects through moose mortality and population density, but also affecting the growth of their main food plant, the balsam fir (*Abies balsamea*). As a consequence, climate is able to change drastically the food web structure by modulating hunting behaviour of the main predator. A somewhat similar finding was reported by Stenseth *et al.* (1999), analysing the effects of climate variation across northern Canada on the density-dependent structure of the lynx–hare population dynamics.

In a similar vein, we found that small mammal response to increases in rainfall levels (mediated by El Niño) in a semiarid region of western South America was dependent on the food web architecture in which each species was embedded (Lima *et al.* 2002a,b). Although, rainfall influences population dynamics of small mammals, the differences observed in the magnitude of population fluctuations between them were caused by a different feedback structure. While population dynamics of the olivaceous field mouse (*Akodon olivaceus*) and the mouse-opossum (*Thylamys elegans*) appear to be regulated by first-order feedback and limited from below, the population dynamics of the most irruptive species, leaf-eared mouse (*Phyllotis darwini*) was influenced by direct and delayed density-dependence caused by predation by barn owls (*Tyto alba*), in addition to rainfall influences. However, the interaction between these two populations (barn owls and leaf-eared mice) is not simple. A first-order feedback loop dominates the regulatory structure of the mouse, but during high rainfall years, a second-order feedback may dominate population dynamics. In particular, during high rainfall years, when competition for food is relaxed, the mutual feedback loop with barn owls dominates the dynamics. Interestingly, because of the large variability observed in composition and abundance of the predatory guild in semiarid Chile (Jaksic *et al.* 1996, 1997) the magnitude of the rodent responses to rainy years may depend on the

predator guild composition, which also seems to be determined by climate (Lima *et al.* 2002b). The generally hypothesized causal chain from ENSO to increased rainfall to higher primary productivity and finally to higher small mammal density (Jaksic 2001a), may be an appropriate description of population variations in non-outbreak species, regulated mainly by intra-specific competition and limited by their food base. By contrast, the dramatic population fluctuations exhibited by outbreaking species may be due to climatic effects coupled with a more complex food web architecture.

Another interesting issue is the existence of strong non-linear effects of climate on population dynamics. Recent studies have shown the existence of such non-linear effects (Sæther *et al.* 2000; Mysterud *et al.* 2001). Non-linear climatic effects will have profound implications for understanding population responses to global change. First of all, a strong non-linear relationship may produce large population responses to small climatic variation, thus facilitating large expansion or contraction (local extinction) of natural populations as a result of rather small climatic changes. On the other hand, the recent finding of non-monotonic relationships between NAO and population dynamics (Mysterud *et al.* 2001) adds more complexity to our effort at understanding population responses. Finally, non-linear and non-monotonic population responses to local weather variation (Murúa *et al.* 2003) will imply an additional challenge for understanding nature.

16.9 Future challenges

El Niño constitutes an excellent natural experiment that empirically demonstrates the teleconnections occurring on planet Earth (see Chapter 2), and at the same time illustrates how atmospheric–oceanographic phenomena may affect the terrestrial biota. The alternate phases of ENSO, namely El Niño/La Niña, should be explored in greater detail to make reasonable extrapolations about where global climate change may lead us. Global climate change was at first thought of as simply a steady increase in the temperature of the planet, but El Niño shows that warm and cool phases alternate on such large geographical scales as the entire Pacific Basin. That 33% of 12 strong El Niño events occurred in the last 20% of the twentieth century (1982–83, 1986–87, 1991–92 and 1997–98, see Jaksic 2001a), suggests that we should be more concerned about the occurrence of increasingly more extreme events of El Niño, rather than about a gradual warming of the climate. The NAO also constitutes a natural experiment that is only recently being milked, and may represent a sort of replicate for ENSO in addressing issues of global change.

One important message emerging from these studies is that the idea of climate being a linear exogenous forcing term that drives the dynamics of ecological systems is too simple for representing the complexity of nature (see also Chapter 1). In fact, the use of climate variables as proxies of population responses may lead to spurious predictions. As was stated by Royama (1992), the response of natural populations to climatic fluctuations is dependent on the density-dependent structure, which is the consequence of the food web architecture where the population is embedded (Berryman 1999). Therefore, in studies relating climate variation to population dynamics, we need a good comprehension of natural population dynamic processes. In accomplishing such a task it is necessary to put together good and systematic field data, statistical tools, and mathematical models based on clearly-stated dynamic principles. The very existence of complex interactions between climate and density-dependence and the non-linearity of climatic effects, represents a challenge for population ecologists, climatologists, statisticians, and mathematicians. However, it also represents new opportunities for improving our understanding of nature. This book and the chapters that we commented upon here represent an important step in that direction.

PART VI

Conclusion—what are the ecological effects of climate variation in the North Atlantic?

In the preceding chapters a total of no less than 40 authors have synthesized what currently is known about how climate affects the ecological systems of the North Atlantic—as well as (in Part V) placed this insight within a broader perspective. In this final chapter, we will draw the summarized insight together—and look towards the challenges ahead of us with respect to achieving a further improved understanding of how climate affects the ecological patterns and dynamics in marine systems. This we do by first giving a list consisting of one key point from each of the main Chapters (1–10), second, by a more general discussion.

Although we have focused throughout on the marine or marine-based systems of the North Atlantic, many of the findings reported apply to other regions and systems of the world—a point emphasized in Part V of the book.

Suggestions and challenges— what we must remember

1. Enhance the capability within statistical modelling of ecological responses to climate variability and change by bringing specialists on time-series analysis into the field.

2. The importance of atmosphere–ocean interaction for marine ecology must be emphasised. Of particular interest in the North Atlantic region is the long, irregular amplification of the North Atlantic Oscillation (NAO) towards its positive phase over recent decades. This climatic event has produced a wide range of effects on North Atlantic ecosystems, and a continuation or strengthening of this pattern may have far-reaching effects.

3. Marine ecosystems are diverse and the choice of modelling approach must depend on the questions asked. The modelling of lower trophic levels is, to a large degree a question of understanding the processes that control levels and fluxes of carbon. On the other hand, for the modelling of fluctuations in fish communities representing biomass alone is not enough, also representations of population structure is necessary. Further use of data assimilation, the integration of models with data to improve the estimation of a system's state, is recommended within both approaches.

4. Change in phytoplankton community structure might be direct by climatic forcing (bottom–up control) or more complex by the integrated effects in the food web (partly top–down). The effects in the North Atlantic sector appear to be region-specific. Application of new exploratory statistical analyses methods are required to be able to detect changes in the community structure for time/space structured data.

5. Climate variability is suggested to influence zooplankton populations in two main ways, 'indirect effects', and 'translations'. Indirect effects cover the influence of climate variability through its effect on ecosystem properties such as the timing or magnitude of the spring bloom. Translations are defined as the effects climate variability can have on the distribution of zooplankton through physical processes as ocean circulation patterns.

6. Observations of changes to the fish stocks due to climate variability in the past provide a basis for predicting some general future responses. However, qualitative predictions of the consequences of climate change on fish resources will require good regional atmospheric and oceanic models of the response of the ocean to climate change. Furthermore, improved knowledge of the life histories of the species involved and further understanding of the role environment, species interactions, and fishing play in determining the variability of growth, reproduction, distribution, and abundance is also needed.

7. Thanks to their position as top predators, the response of seabirds to climate change is a good index of its effect on the whole food web. In order to improve our understanding of what might happen under various scenarios of global change, the study of seabird populations might be of great value.

8. Climate variability has been shown to be potentially linked to changes in phytoplankton biovolume in relation to primary productivity and the upwelling of nutrients. This aspect needs further attention, since changes in phytoplankton biovolume are linked to the fate of biogenic carbon in the pelagic food web.

9. Studies of benthic communities in the North Sea, Skagerrak, and Baltic indicate that single species as well as whole communities and also functional groups are impacted by changes in the NAO. However, the impact may differ geographically and according to the species composition of the different communities.

10. Climate change will expectedly influence important rates determining community dynamics, that is, recruitment, immigration, and extinction. There are, however, major challenges in understanding these impacts. For example, although the variation in North Atlantic cod stocks throughout the last decades is reasonably well known, the possibly major *indirect* community effects of the varying abundance of cod have not been studied with equal intensity. The effect of changes in cod abundance on fish communities should be further investigated in order to better understand the secondary effects of climate change on fish communities.

How does climate really affect marine ecology?

The examples provided throughout the book highlight both the recent progress and the difficulties in trying to understand ecological processes and patterns as a result of climate variations. As pointed out in several earlier chapters, climate may affect individuals directly or indirectly through physiology (metabolic and reproductive processes), and populations directly (typically operating through the individual level) or indirectly through the food-web links within the ecosystem (including prey, predators, and competitors). Thus, climate variability might directly affect a particular organism, or ripple through trophic levels from primary production, through zooplankton and forage fish to predatory fish, seabirds, and marine mammals. Along the way, growth, life history traits, and population dynamics may be influenced at all levels.

Delayed effects of climate are common and important. Individuals born in a specific year may be larger or smaller than the average depending on the climatic conditions in the year of birth. Such cohort effects may also arise because of higher rates of survival among the young in some years. Also, a cohort with large individuals as young may tend to be abundant, as they grow older due to size-dependent mortality.

Climate may impose different effects also between age-classes, with the consequences for population development depending on the age-structure of the population. That older individuals produce more and healthier offspring than younger, has been described for some fish populations. Under such conditions, the impact of climate variability is likely to be more pronounced during periods when the population is dominated by a few cohorts of younger fish, than if composed by a larger variety of age groups, also including older individuals. The former is indeed the expected consequence of modern fisheries directed towards older and larger fish. Furthermore, whenever younger age classes are most strongly affected, the ecological effects of climate are more difficult to detect because of cohort effects.

The chapter on marine birds (Chapter 7) highlights another general feature, namely that different stages of any organism's life may be subject to variability and changes in highly different locations— and as a result be exposed to greatly different climate patterns. In the case of many seabirds, parent individuals need to forage in wide regions for food to their progeny—that is, being central place foragers. Similarly, the environment of copepods (Chapter 5) is completely different for mature individuals over-wintering in the deep as compared to when drifting with the currents in the upper layers of the sea. Likewise, for some fish populations spawning may take place in geographically and thus climatically different regions from feeding migration (Chapter 6).

Much of the discussion throughout the previous chapters refers to the use of indices of large-scale climate patterns (like the NAO; Chapter 2) for the study of climate effects on ecological systems. The use of such indices has proven to be rather useful, not least because climate does not affect populations through a single weather variable, but rather through a blend of weather features. Indices of large-scale climate fluctuations may act as proxies for the *overall* climate condition—representing a 'package of weather'(see Stenseth *et al.* 2003). By definition, such climate indices ought to reduce complex space and time variability into simple measures, thus helping ecologists appreciate the global nature of climate systems and providing statistically tractable climate factors. Furthermore, large-scale climate patterns describe changes in several physical characteristics of a particular ecosystem, and might thus capture more of the overall physical variability than an individual local weather index. Also, large-scale climate patterns provide a conceptual framework and a broader understanding of observed changes in the local physical environment.

The effects of large-scale climatic fluctuations on ecology have typically been assumed to be linear. However, as pointed out in Chapter 1, there are (at least) two different ways in which the effect of a large-scale climate signal on ecological systems may be expected to be non-linear. First, the large-scale phenomena may not be linearly related to local climatic variables, and individuals are indeed expected to respond to the climate they experience at a local scale. Second, the ecological response to changes in local climate may not always be linear. However, since the issue of non-linear responses may be related to scale, transforming the predictor and/or the response variable may linearize the relationships. Fortunately, non-linearities may often be assessed without any restrictive assumptions regarding the functional relationship between the predictor variables and the response variable by means of statistical methods such as generalized additive models (GAM).

In earlier chapters (Chapters 3, 5, 6 and 9), we have shown that the effects of climate may ultimately depend on other ecological factors (i.e. interactions). Intrinsic density-dependence and extrinsic processes due to a stochastic environment typically interact in rather complex ways within populations and ecosystems. Indeed, failure to include other important factors may lead to spurious relationships between climate and the trait studied.

A weakness of many climate–ecology studies is that although the observed patterns may be clear, the understanding of underlying mechanisms is often weak. There is therefore a pronounced need for proper experiments exploring the underlying causal mechanisms. The combined analysis of ecological and climatic time-series will certainly continue to teach us much about how climate variability affects ecological patterns—particularly as more advanced statistical approaches are being applied. However, such pattern-oriented studies should be linked with more process-focused studies—many of which are experimental in their design. Disentangling the ecological consequences of climatic variation is thus not a simple task: most of all we need to understand the density-dependent and density-independent structure, which may interact with the climatic signal. To study such interactions further, we need detailed knowledge about the interface of climate and ecological systems.

Indeed, much remains to be done before we with some confidence can deliver long-range predictions of the responses of North-Atlantic individuals, populations or communities to climate change. However, a lot of basic knowledge both about likely directions of climate change and ecological responses has been established. In this enterprise, we need to appreciate—and better understand—the effect of the expected higher frequency of more extreme climate events, as a consequence of increased variance and not just change in average levels (Chapter 1). Such increased environmental variability may in general have a negative effect on population growth—effects which might be difficult to predict, particularly under the influence of non-linear dynamic relationships between climate and population processes.

In spite of such challenges, we are in a fairly good position today regarding our understanding of climate–ecology interaction in the North Atlantic region. With the information summarized in the earlier chapters of this book and regional marine climate scenarios, we are ready to apply the many established climate–ecology links towards understanding how the populations and communities of the North Atlantic might function also in the future. Indeed, we hope this book—combining more climate-focused studies with more traditional ecology-focused studies may pave way to a further integration with the result of an improved understanding of the patterns and the underlying processes. In fact, as this book demonstrates, we understand quite a bit about the patterns—but need to understand more of the underlying processes generating these patterns.

Afterword

Roger Harris

The topic addressed in this volume, the ecological effects of climatic variations in the North Atlantic, is a scientifically challenging one. The Atlantic is a large ocean basin with extensive and varied shelf systems and the interactions within its marine ecosystems are complex and in some aspects still poorly understood. To identify patterns and underlying processes requires working across disciplines, at all trophic levels in the foodweb, and the collation and integration of varied and dispersed data sets. The challenge has been well met by both authors and editors.

Over the past two centuries atmospheric CO_2 levels have increased steadily as a result of human activity and it is now recognised that the Earth System is being significantly perturbed by anthropogenic activity; we have entered the Anthropocene era (Finnigan 2003). The conventional concept of 'the greenhouse effect' predicts steadily rising global temperatures as CO_2 increases. There is now ample evidence of ecological effects of recent climatic change (Walther et al. 2002) and there are early indications in the ecology of the North Atlantic of the probable consequences of such warming. For example many fish species have extended their range northward along the European Atlantic coast (Quero et al. 1998) and the work of Beaugrand et al. (2002) suggests that increase in regional sea temperature has triggered a major re-organisation in calanoid copepod species composition and diversity over the whole ocean basin. Warm-water species have extended their distribution northwards by more than 10° of latitude, while cold-water species have decreased in number and extension. However, in addition to such effects of gradual warming, increased variability and abrupt changes are equally possible and one of the most significant of such changes certainly for the North Atlantic and for the whole Earth System would be a rapid change in the thermohaline circulation (Manabe and Stouffer 1995; Rahmstorf

2000). Likely causes would be a cessation of ice formation in the northern North Atlantic or an increase in freshwater input to the ocean from melting ice on land. Recent studies have indicated that the waters of the N Atlantic are already freshening (Curry et al. 2003). The consequences of such abrupt climatic change would have dramatic ecological effects in the region not the least for human society.

As has been brought out throughout this book the North Atlantic exhibits particularly strong climate variability and is the focus of one of the best studied climate oscillation systems, the North Atlantic Oscillation (Hurrell et al. 2003). A recurrent theme of many of the chapters is the emphasis on the NAO and many ecological effects are identified that correlate with this dominant component of interdecadal climate variability in the ocean basin. Persuasive argument is made that North Atlantic ecosystems respond to such variability and better understanding of climate variability mediated through effects on oceanography can explain a significant proportion of the variability in the marine ecosystems of the region. The apparent shift of the NAO to a more persistent positive phase over recent decades is demonstrated. It is recognised that many of the contemporary ecological effects observed are a likely consequence of this change in the NAO and that persistence of this trend would have significant long term impacts on the ecology of the North Atlantic.

While some of the chapters in this book review the subject in relation to conventional trophic levels (phytoplankton: 4; zooplankton: 5; fish: 6; seabirds: 7) the problem of understanding ecological effects of climate variability really requires a highly multidisciplinary and integrative approach. All components of the system are linked and interrelated. A recent example is the demonstration by Beaugrand et al. (2003) of a suggested effect of plankton on cod recruitment in the North Sea. The mean size of zooplankton prey, seasonal timing and abundance all affect survival through a match/mismatch mechanism. Hence temperature variability affects larval cod survival and suggests that warming since the mid-1980s has modified the plankton ecosystem reducing the survival of young cod. Key elements of a multidisciplinary and integrative approach to study effects of climatic variability on ecology should combine use of retrospective and time-series datasets, new technology for sampling and observation, studies of processes and mechanisms, and use of models to provide integration, synthesis and a predictive capability. Many of these elements are reflected throughout the volume.

Marine ecosystems vary over large spatial and temporal scales. These scales of variability are much greater than the duration of single or even multi-year collaborative studies. Hence retrospective analyses of historical information are essential to place the results from present studies into climatic and global contexts (Finney *et al.* 2002). Exploitation of such past data sets is a significant foundation of many of the chapters in this book. Retrospective data analysis is often the only method available to identify characteristic temporal scales of ecosystem variability and their rates of change. These may be due to either natural or anthropogenic forcing or both combined depending on how far back the data series extend (Holm 2002). Disentangling human effects, for example resulting from fishing (Jackson *et al.* 2002), from natural climatic variability can be difficult.

The Atlantic Ocean has some of the longest established time-series. It was the focus of the first pioneering plankton expedition of Hensen in 1889, which led to the development of the concept of the spring phytoplankton bloom (Mills 1989). The plankton populations have been surveyed over the past 50 years by the unique basis-wide survey of the Continuous Plankton Recorder (Reid *et al.* 2003). Records of the fluctuations in marine living resources at higher trophic levels go back many centuries (for example, Alheit and Hagen 1997; Southward *et al.* 1988).

The pioneers of fisheries science carried out their first work along the European coasts and along the eastern seaboard of North America. The early Norwegian fisheries biologist Johan Hjort realised that most North Atlantic fish populations consist of several year classes, and that abundance was generally determined during the larval phase (Hjort 1914). Multiple forcing may make it difficult to establish clear links between changes in the physical environment and the response of fish populations. However, some links between climate and fish stocks have been well established subsequently (Cushing and Dickson 1976; Cushing 1982; Drinkwater *et al.* 2003; Chapters 6 and 10). Rich time-series and retrospective data sources make the North Atlantic a particularly fruitful ocean basin in which to explore such relations between climate variability and ecological effects. Parallel analysis of both climate and ecological time series has been a productive approach to investigating the linkages between climate variability and biological populations. However, understanding of underlying mechanisms remains limited and stronger focus on processes and mechanisms will be required in future research programmes.

The majority of the data sets described in this book have been obtained with conventional sampling systems such as nets, bottles and trawls deployed on research or fisheries survey cruises. They represent hard won information, but information that in its very nature is limited both temporally and spatially when compared to the problem of climatic variability effects over the whole ocean basin. Sampling and observational systems are advancing rapidly (Wiebe and Benfield 2003). The challenge for the future is to make full use of remote sensing with its basin-scale synoptic capability, new in situ optical (Dickey 1991) and acoustic techniques (Holliday and Pieper 1995) for real-time assessment of biological components of the system (Benfield *et al.* 1998) as well as long-term autonomous observation devices and platforms (Dickey *et al.* 1998). The revolution in new technology will lead to previously unimagined near-real time observation and data assimilation capabilities over the coming decades. One important approach will be to monitor 'hot spots' for signs of change. There are initiatives to set up integrated Earth Observing Systems employing an integrated network of satellites, aircraft and in situ data collection. The North Atlantic community is well placed to play its part in such an effort and to exploit the new technologies becoming available.

A strong case is made in this book for the effects of climate on a wide range of ecological processes involving phyto and zooplankton, fish and seabirds. Spatial and temporal distribution patterns are strongly influenced. Such ecological responses can be both direct and indirect, they can be lagged temporally, they can be either linear or non-linear and they can result from 'match-mismatch'. Despite the patterns observed, which clearly show linkage between climate variability and ecology, the mechanisms involved remain poorly understood. Process focussed studies are required.

At the small- and mesoscale, process studies are critical for the understanding of the functioning and structure of marine ecosystems. Small-scale processes such as turbulence (Rothschild and Osborn 1988) may be influenced by climate variability. However, large-scale phenomena, such as interactions between distribution, migratory behaviour and the onset of stratification and timing of phytoplankton blooms (Townsend *et al.* 1994) are probably most important. There are indications of a possible link between the NAO and the composition of the spring phytoplankton bloom (Irigoien *et al.* 2000). There are also intriguing correlations between the NAO and aspects of the distribution and ecology of the congeneric pair of *Calanus* species, *C.finmarchicus* and *C.helgolandicus* (Fromentin and Planque 1996; also Pershing *et al.*, Chapter 5).

As is pointed out in several chapters in this book, climate may have either direct or indirect effects on the ecology of individual species. These effects may be via physiology at the individual and population level (e.g. metabolic and reproductive processes) or may operate indirectly through linkages within the ecosystem, for example predator-prey interactions or competition. Individuals may be directly affected by climate variability or the effect may be through the trophic levels from phytoplankton, through zooplankton and fish to higher predators such as seabirds and marine mammals. Individual species characteristics such as growth rates, life history traits, and population dynamics may be affected at all these trophic levels in the food-web. The target species approach has been particularly pioneered in the North Atlantic. Good examples are the co-ordinated studies on the key copepod species, *Calanus finmarchicus* (Tande and Miller 2000), and the pan-Atlantic studies of the cod, *Gadus morhua* (Dickson *et al.* 1994) carried out under the auspices of the Global Ocean Ecosystem Dynamics project (GLOBEC 1997). The US GLOBEC George's Bank programme has adopted the target species approach particularly effectively, focussing on *Calanus*, *Pseudocalanus* and the larvae of two fish species, cod and haddock (Wiebe *et al.* 2001).

Looking forward, the goal must be to understand and ultimately to predict how populations respond to natural and anthropogenicaly induced changes in climate variability. Co-ordinated interdisciplinary process and modelling studies will be required to reach this goal (e.g. McGillicuddy *et al.* 1998). Development of appropriate models that help to elucidate ecosystem dynamics and responses on a range of times scales, including major climatic fluctuations will be critical (Werner *et al.*, Chapter 3). Predictive capabilities will only be realised if observational programmes and models are designed and developed together with the explicit aim of predicting marine ecosystem responses over a range of time scales. Special regional aspects will influence choice of key variables, for example the relevant target species. A future requirement is to develop better basin-scale models of the North Atlantic that reproduce the major features of the system. These can then be used to locate points where small climate changes may have large and perhaps irreversible ecological effects. The complexity of ecosystems can cause subtle and chaotic responses to changes in external forcing. Although ecosystems may not normally behave chaotically, sensitivity to external influences associated with nonlinearity can lead to amplification of climatic signals. Taylor *et al.* (2002) used an observed association between plankton populations around the UK and the position of the Gulf Stream as a means of demonstrating how a detailed marine ecosystem model can extract a weak signal spread across different meteorological variables. Biological systems may therefore respond to climatic signals other than those that dominate the driving variables.

Ecosystem responses to climate may be nonlinear (Scheffer *et al.* 2001) and particular emphasis in the Pacific Ocean has been on the concept of regime shifts (Hare and Mantua 2000; Mantua 2004; Wooster and Zhang 2004). Recently there have been suggestions from the Atlantic that there may be similar abrupt shifts in ecosystems of the North Sea and adjacent North Atlantic (Reid *et al.* 2001; Beaugrand 2004). Again progress has been made in describing and documenting such changes but underlying processes and mechanisms are less well understood (Collie *et al.* 2004).

The North Atlantic is part of a wider picture of global climatic variability and resultant ecological effects. In the latter chapters in the volume this context is provided with, for example, contributions from the Pacific where ENSO shows many parallels with the phenomena described in the North Atlantic (Chapter 11) and fisheries dynamics

(Chapter 12) can also be compared and contrasted. The inclusion of freshwater (Chapters 13 and 14) and terrestrial (Chapters 15 and 16) perspectives extends the context beyond that merely of the marine environment reminding the reader that the oceans are an intimate part of the wider Earth System. To take such a holistic approach will be increasingly important in the future as we enter the era of Earth System Science (Steffen *et al.* 2003).

Adopting innovative approaches will be required to advance our understanding of the ecological effects of present and future climatic variations in the North Atlantic. This book provides an important platform for future work. To study the ecological effects of climate variations in the entire North Atlantic basin has required, and will increasingly demand in the future, the integration and synthesis of a large number of national and multinational studies performed by countries particularly of the ICES area. Progress has already been made through the Trans-Atlantic Studies of *Calanus finmarchicus* (TASC) a project funded by the European Union (Tande and Miller 2000) and building on the ICOS project, which in turn was stimulated by the hypothesis of Backhaus *et al.* (1994). Such initiatives, attempting to build basin-scale understanding of the ecology of key species continue to bear fruit (Heath *et al.* 2004). Future programmes need to be able to consider the basin as a whole and research funding mechanisms need to be developed, which will allow concerted and integrated trans-Atlantic collaboration between the marine science communities of North America and Europe. This is a key practical requirement if the challenge of understanding the ecological effects of climatic variations in the North Atlantic is to be effectively addressed over the coming decades.

References

Alheit, J. and Hagen, E. (1997). Long-term climate forcing of European herring and sardine populations. *Fisheries Oceanography*, **6**, 130–9.

Backhaus, J. O., Harms, I. H., Krause, M. and Heath, M. R. (1994). An hypothesis concerning the space-time succession of *Calanus finmarchicus* in the northern North Sea. *ICES Journal of Marine Science*, **51**, 169–80.

Beaugrand, G. (2004). The North Sea regime shift: evidence, causes, mechanisms and consequences. *Fisheries Oceanography, in press.*

Beaugrand, G., Brander, K. M., Lindley, J. A., Souissi, S. and Reid, P. C. (2003). Plankton effect on cod recruitment in the North Sea. *Nature*, **426**, 661–4.

Beaugrand, G., Reid, P. C., Ibañez, F., Lindley, J. A., & Edwards, M. (2002). Reorganisation of North Atlantic marine copepod biodiversity and climate. *Science*, **296**, 1692–4.

Benfield, M. C., Wiebe, P. H., Stanton, T. K., Davis, C. S., Gallager, S. M., Greene, C. H. (1998). Estimating the spatial distribution of zooplankton biomass by combining Video Plankton Recorder and single-frequency acoustic data. Deep-Sea, **45**, 1175–99.

Collie, J. S., Richardson, K. and Steele, J. H. (2004). Regime shifts: can ecological theory illuminate the mechanisms? *Fisheries Oceanography, in press.*

Curry, R., Dickson, B. and Yashayaev, I. (2003). A change in the freshwater balance of the Atlantic Ocean over the past four decades. *Nature*, **426**, 826–9.

Cushing, D. H. (1982). *Climate and fisheries.* Academic Press, London.

Cushing, D. H. and R. R. Dickson. (1976). The biological response in the sea to climatic changes. *Advances in Marine Biology*, **14**, 1–122.

Dickey, T.D. (1991). The emergence of concurrent high-resolution physical and bio-optical measurements in the upper ocean and their applications. *Review of Geophys.*, **29**, 383–413.

Dickey, T., Frye, D., Jannasch, H., Boyle, E., Manov, D., Sigurdson, D., McNeil, J., Stramska, M., Michaels, A., Nelson, N., Siegel, D., Chang, G., Wu, J., Knap, A. (1998). Initial results from the Bermuda Testbed Mooring program. *Deep-Sea Research*, **45**, 771–94.

Dickson, R. R., Briffa, K. R. and Osborn, T. J. (1994). Cod and Climate: the spatial and temporal context. *ICES Marine Science Symposia*, **198**, 280–6.

Drinkwater, K. F. (2002). A review of the role of climate variability in the decline of northern cod. *American Fisheries Society Symposia*, **32**, 113–30.

Finney, B. P., Gregory-Eaves, I., Dougals, M. S. V. and Smol, J. P. (2002). Fisheries productivity in the northeastern Pacific Ocean over the past 2,200 years. *Nature* **416**, 729–33.

Finnigan, J. (2003). Earth System Science in the early Anthropocene. IGBP Newsletter, **55**, 8–11.

Fromentin, J. and Planque, B. (1996). *Calanus* and the environment in the eastern North Atlantic. II. Influence of the North Atlantic Oscillation on *Calanus finmarchicus* and *C.helgolandicus*. *Marine Ecology Progress Series*, **134**, 111–18.

GLOBEC (1997). Implementation Plan. IGBP Report 47, GLOBEC Report 13, 207pp.

Hare, S. R., and Mantua, N. J. (2000). Empirical evidence for North Pacific regime shifts in 1977 and 1989 . *Progress in Oceanography*, **47**, 103–45.

Heath, M., Boyle, P. R., Gislason, A., Gurney, W. S. C., Hay, S. J., Head, E. J. H., Holmes, S., Ingvarsdottir, A., Jonasdottir, S. H., Lindeque, P., Pollard, R. T., Rasmussen, J., Richards, K., Richardson, K., Smerdon, G. and Spiers, D. (2004). Comparative ecology of overwintering *Calanus finmarchicus* in the northern North

Atlantic, and implications for life cycle patterns. *ICES Journal of Marine Science* (in press).

Hjort, J. (1914). Fluctuations in the great fisheries of northern Europe viewed in the light of biological research. *Rapp. P.-v. Reun. cons. int. Explor. Mer.*, **20**, 1–228.

Holm, P. (2002). History of marine animal populations: a global research program of the Census of Marine Life. *Oceanologica Acta*, **25**, 207–11.

Holliday, D. V. and Pieper, R. E. (1995). Bioacoustical oceanography at high frequencies. *ICES Journal of Marine Science*, **52**, 279–96.

Hurrell, J. W., Kushnir, Y., Ottersen, G. and Visbeck, M. (2003). The North Atlantic Oscillation: Climatic Significance And Environmental Impact. *Geophysical Monograph Series*, **134**. American Geophysical Union, Washington, 279 + viii pp.

Irigoien, X., Harris, R. P., Head, R. N. and Harbour, D. (2000). North Atlantic Oscillation and spring bloom phytoplankton composition in the English Channel. *Journal of Plankton Research.* **22**, 2367–71.

Jackson, J.B.C. *et al.* (2002). Historical overfishing and the recent collapse of coastal ecosystems. *Science*, **293**, 629–38.

Manabe, S. and R.J. Stouffer. (1995). Simulation of abrupt climate change induced by freshwater input to the North Atlantic Ocean. *Nature*, **378**, 165–7.

Mantua, N. (2004). Methods for detecting regime shifts in large marine ecosystems: a review with approaches applied to North Pacific data. *Progress in Oceanography, in press.*

McGillicuddy, D. J., Lynch, D. R., Moore, A. M., Gentleman, W. C., Davis, C. S., and Meise, C. J. (1998). An adjoint data assimilation approach to diagnosis of physical and biological controls on *Pseudocalanus* sp. in the Gulf of Maine-Georges Bank region. *Fisheries Oceanography*, **7**, 205–18.

Mills, E.L. (1989). *Biological Oceanography an Early History: 1870–1960.* Ithaca, 378 pp.

Quero, J. C., Buit, M.H.Du. and Vayne, J.J. (1998). Les observations de poissons tropicaux et le rechauffement des eaux de L'Atlantic europeen. *Oceanologica Acta*, **21**, 345–51.

Rahmstorf, S. (2000). The Thermohaline Ocean Circulation: A System with Dangerous Thresholds? *Climatic Change*, **46**, 247–56.

Reid, P. C., Borges, M., and Svendsen, E. (2001). A regime shift in the North Sea circa 1988 linked to changes in the North Sea horse mackerel fishery. *Fisheries Research*, **50**, 163–71.

Reid, P. C., Colebrook, J. M., Matthews, J. B. L., Aiken, J., *et al.* (2003). The Continuous Plankton Recorder: concepts and history, from Plankton Indicator to undulating recorders. *Progress in Oceanography*, **58**, 117–73.

Rothschild, B. J. and Osborn, T. R. (1988). Small-scale turbulence and planktonic contact rates. *Journal of Plankton Research*, **10**, 465–74.

Scheffer, M., Carpenter, S. R., Foley, J. A., Folke, C. and Walker, B. (2001). Catastrophic shifts in ecosystems. *Nature*, **413**, 591–6.

Southward, A. J., Boalch, G. T., Maddock, L. (1988). Fluctuations in the herring and pilchard fisheries of Devon and Cornwall linked to change in climate since the 16th century. *Journal of the Marine Biological Association of the United Kingdom*, **68**, 423–45.

Steffen, W., Sanderson, A., Tyson, P., Jäger, J., Matson, P., Moore, B., Oldfield, F., Richardson, K., Schellnhuber, J., Turner, B. L., and Wasson, R. (2003). *The earth system: a planet under pressure.* Springer-Verlag, Berlin. 336pp.

Tande, K. S. and Miller, C. B. (2000). Population dynamics of *Calanus* in the North Atlantic: results from the Trans-Atlantic Study of *Calanus finmarchicus.* ICES Journal of Marine Science, **57**, 1527.

Taylor, A. H., Allen, J. I. and Clark, P. A. (2002). Extraction of a weak climatic signal by an ecosystem model. *Nature*, **416**, 629–32.

Townsend, D. W., Cammen, L. M., Holligan, P. M., Campbell, D. E. and Pettigrew, N. R. (1994). Causes and consequences of variability in the timing of spring phytoplankton blooms. *Deep-Sea Research*, **41**, 747–65.

Walther, G-R., Post, E., Convery, P., Menzel, A., Parmesan, C., Beebee, T. J. C., Fromentin, J-M., Hoegh-Guldberg, O. and Barlein, F. (2002). Ecological responses to recent climate change. *Nature*, **416**, 389–95.

Wiebe, P. H. and Benfield, M. C. (2003). From the Hensen net toward four-dimensional biological oceanography. *Progress In Oceanography*, 7–136

Wiebe, P. H., Beardsley, R. C., Bucklin, A., and Mountain, D. G. (2001). Coupled biological and physical studies of plankton populations in the Georges Bank region and related North Atlantic GLOBEC study sites. Deep-Sea Research II, **48**, 1–2.

Wooster, W. S. and Zhang, C. I. (2004). Regime Shifts in the N. Pacific: early indications of the 1976–1977 event. *Progress in Oceanography, in press.*

References

Chapter 1

Allan, R., Lindesay, J., and Parker, D. (1996). *El Niño southern oscillation and climate variability.* CSIRO Publishing, Collingwood, Australia.

Angstrom, A. (1935). Teleconnections of climate changes in present times. *Geographical Annals,* **17**, 242–58.

Bakun, A. (1996). *Patterns in the ocean: ocean processes and marine population dynamics.* University of California Sea Grant College System, San Diego, U.S.

Battisti, D. S., Bhatt, U. S., and Alexander, M. A. (1995). A modelling study of the interannual variability in the wintertime North Atlantic ocean. *Journal of Climate,* **8**, 3067–83.

Bjerknes, J. (1969). Atmospheric teleconnections from the equatorial pacific. *Monthly Weather Review,* **97**, 163–72.

Brander, K., Dickson, R. R., and Shepherd, J. G. (2001). Modelling the timing of plankton production and its effect on recruitment of cod (*Gadus morhua*). *ICES Journal of Marine Science,* **58**, 962–6.

Crawford, R. J. M., Shannon, L. J., and Nelson, G. (1995). Environmental change, regimes and middle-sized pelagic fish in the South-East Atlantic Ocean. *Scientia Marina,* **59**, 417–26.

Cushing, D. H. (1982). *Climate and fisheries.* Academic Press, London.

Cushing, D. H. (1990). Plankton production and year-class strength in fish populations: An update of the match/mismatch hypothesis. *Advances in Marine Biology,* **26**, 249–94.

Cushing, D. H. (1996). Towards a science of recruitment in fish populations. *Excellence in ecology,* **7**, *Inter-Research, Oldendorf, Germany,* xxi–175 pp.

Deser, C. and Blackmon, M. L. (1993). Surface climate variation over the North Atlantic ocean during winter: 1900–1989. *Journal of Climate,* **6**, 1743–53.

Dickson, R. R., Osborn, T. J., Hurrell, J. W., Meincke, J., Blindheim, J., Ådlandsvik, B., Vinje, T., Alekseev, G., and Maslowski, W. (2000). The Arctic Ocean response to the North Atlantic Oscillation. *Journal of Climate,* **13**, 2671–96.

Ellertsen, B., Fossum, P., Solemdal, P., and Sundby, S. (1989). Relation between temperature and survival of eggs and first-feeding larvae of northeast Arctic cod (*Gadus morhua* L.). *Rapports et procès-verbaux des réunions, Conseil International pour l'Exploration de la Mer,* **191**, 209–19.

Fortier, L. and Gagne, J. (1990). Larval herring (Clupea-Harengus) Dispersion, growth, and survival in the St-Lawrence estuary—match mismatch or membership vagrancy. *Canadian Journal of Fisheries and Aquatic Sciences,* **47**, 1898–912.

Furevik, T. (2001). Annual and interannual variability of Atlantic Water temperatures in the Norwegian and Barents Seas: 1980–1996. *Deep-Sea Research I,* **48**, 383–404.

Gammelsrød, T., Bartholomae, C. H., Boyer, D. C., Filipe, V. L. L., and O'Toole, M. J. (1998). Intrusion of warm surface water along the Angolan-Namibian coast in February–March 1995: the 1995 Benguela Niño. *South African Journal of Marine Science,* **19**, 41–56.

Garrison, T. (2002). *Oceanography: an invitation to marine science.* Thomson Learning, Stamford.

Harden-Jones, F. R. (1968). *Fish migration.* Edward Arnold Ltd., London.

Harris, G. P., Griffiths, F. B., and Clementson, L. A. (1992). Climate and the fisheries off Tasmania—interactions of physics, food chains and fish. *South African Journal of Marine Science,* **12**, 585–97.

Hassel, A., Skjoldal, H., Gjoseter, H., Loeng, H., and Omli, L. (1991). Impact of grazing from capelin (mallotus-villosus) on zooplankton—a case-study in the northern Barents Sea in August 1985. *Polar Research,* **10**, 371–88.

Heath, M., Backhaus, J., Richardson, K., McKenzie, E., Slagstad, D., Beare, D., Dunn, J., Fraser, J., Gallego, A., Hainbucher, D., Hay, S., Jónasdottir, S., Madden, H., Mardaljevic, J., and Schacht, A. (1999). Climate fluctuations and the spring invasion of the North Sea by *Calanus finmarchicus. Fisheries Oceanography,* **8**, 163–76.

Houghton, R. W. (1996). Subsurface quasi-decadal fluctuations in the North Atlantic. *Journal of Climate,* **9**, 1361–73.

Iselin, C. O. D. (1936). A study of the circulation of the western North Atlantic. *Papers in Physical Oceanography and Meteorology,* **4**, 101.

Izhevskii, G. K. (1964). *Forecasting of oceanological conditions and the reproduction of commercial fish.* Pishcepromizdat, Moscow.

Jones, P. D., Osborn, T. J., and Briffa, K. R. (2001). The evolution of climate over the last millennium. *Science,* **292**, 662–7.

Kawasaki, T. (1994). A decade of the regime shift of small pelagics—from the FAO Expert Consultation (1983 to the PICES III 81994). *Bulletin of the Japanese Society of Fisheries Oceanography (in Japanese)*, **59**, 321–33.

Leggett, W. C. and Deblois, E. (1994). Recruitment in marine fishes: is it regulated by starvation and predation in the egg and larval stages? *Netherland Journal of Sea Research*, **32**, 119–34.

Lluch-Belda, D., Hernandez-Vazquez, S., Lluch-Dota, D. B., Salinas-Zavala, C. A., and Schwartzlose, R. A. (1992). The recovery of the california sardine as related to global change. *Reports of California Cooperative Oceanic Fisheries Investigations*, **33**, 50–9.

Lough, R. G., Smith, W. G., Werner, F. E., Loder, J. W., Page, F. H., Hannah, C. G., Naimie, C. E., Perry, R. I., Sinclair, M. M., and Lynch, D. R. (1994). The influence of wind-driven advection on the interannual variability in cod egg and larval distributions on Georges Bank: 1982 vs. 1985. *ICES Marine Science Symposia*, **198**, 356–78.

McCartney, M. S., Curry, R. G., and Bezdek, H. F. (1996). North Atlantic's transformation pipeline chills and redistributes subtropical water—But it's not a smooth process and it mightily affects climate. *Oceanus*, **39**, 19–23.

Mysterud, A., Stenseth, N. C., Yoccoz, N. G., Langvatn, R., and Steinheim, G. (2001). Nonlinear effects of large-scale climatic variability on wild and domestic herbivores. *Nature*, **410**, 1096–9.

Mysterud, A., Stenseth, N. C., Yoccoz, N. G., Ottersen, G. and Langvatn, R. (2003). The response of terrestrial ecosystems to climate variability associated with the North Atlantic Oscillation. In *The North Atlantic Oscillation: climatic significance and environmental impact*, vol. 134 (eds J. Hurrell, Y. Kushnir, G. Ottersen, and M. Visbeck), pp. 235–262. American Geophysical Union. Washington D.C., USA.

Namias, J. and Cayan, D. R. (1981). Large-scale air-sea interactions and short-period climate fluctuations. *Science*, **214**, 869–78.

Ottersen, G., Ådlandsvik, B., Loeng, H., and Ingvaldsen, R. (2003). Temperature variability in the Northeast Atlantic. *ICES Marine Science Symposia*, **219**, 86–94

Ottersen, G., Planque, B., Belgrano, A., Post, E., Reid, P. C., and Stenseth, N. C. (2001). Ecological effects of the North Atlantic Oscillation. *Oecologia*, **128**, 1–14.

Philander, S. G. H. (1990). *El Niño, La Niña, and the Southern Oscillation*. Academic Press, New york.

Pickard, G. L. and Emery, W. J. (1990). *Descriptive physical oceanography*. Pergamon Press, New York.

Planque, B. and Fox, C. J. (1998). Interannual variability in temperature and the recruitment of Irish Sea cod. *Marine Ecology Proggress Series*, **172**, 101–5.

Pond, S. and Pickard, G. L. (1983). *Introductory dynamical oceanography*. Pergamon Press, Oxford.

Rodwell, M. J., Rowell, D. P., and Folland, C. K. (1999). Oceanic forcing of the wintertime North Atlantic oscillation and European climate. *Nature*, **398**, 320–23.

Rothschild, B. J. and Osborn, T. R. (1988). Small scale turbulence and plankton contact rates. *Journal of Plankton Research*, **10**, 465–74.

Sakshaug, E. (1997). Biomass and productivity distributions and their variability in the Barents Sea. *ICES Journal of Marine Science*, **54**, 341–50.

Sakshaug, E. and Skjoldal, H. (1989). Life at the ice edge. *Ambio*, **18**, 60–7.

Schwartzlose, R., Alheit, J., Bakun, A., Baumgartner, T., Cloete, R., Crawford, R., Fletcher, W., GreenRuiz, Y., Hagen, E., Kawasaki, T., Lluch-Belda, D., Lluch-Cota, S., MacCall, A., Matsuura, Y., Nevarez-Martinez, M., Parrish, R., Roy, C., Serra, R., Shust, K., Ward, M., and Zuzunaga, J. (1999). Worldwide large-scale fluctuations of sardine and anchovy populations. *South African Journal of Marine Science*, **21**, 289–347.

Sekine, Y. (1991). Anomalous southward intrusion of the Oyashio east of Japan. In *Long-term variability of pelagic fish populations and their environment* (eds T. Kawasaki, S. Tanaka, Y. Toba, and A. Taniguchi), pp. 61–75. Pergamon, Oxford.

Serchuk, F. M., Grosslein, M. D., Lough, R. G., Mountain, D. G., and O'Brien, L. (1994). Fishery and environmental factors affecting trends and fluctuations in the Georges Bank and Gulf of Maine Atlantic cod stocks: an overview. *ICES Marine Science Symposia*, **198**, 77–109.

Shannon, L. V., Boyd, A. J., Brundrit, G. B., and Taunton-Clark, J. (1986). On the existence of an El Niño-type phenomenon in the Benguela System. *Journal of Marine Research*, **44**, 495–520.

Sinclair, M. and Iles, T. D. (1989). Population regulation and speciation in the oceans. *Journal Conseil International pour l'Exploration de la Mer*, **45**, 165–75.

Stenseth, N. C., Chan, K. S., Tong, H., Boonstra, R., Boutin, S., Krebs, C. J., Post, E., O'Donoghue, M., Yoccoz, N. G., Forchhammer, M. C., and Hurrell, J. W. (1999). Common dynamic structure of Canada lynx populations within three climatic regions. *Science*, **285**, 1071–73.

Stenseth, N. C., Mysterud, A., Ottersen, G., Hurrell, J. W., Chan, K.-S., and Lima, M. (2002). Ecological effects of climate fluctuations. *Science*, **297**, 1292–96.

Stenseth, N. C., Ottersen, G., Hurrell, J.W., Mysterud, A., Lima, M., Chan, K.-S., Yoccoz, N. G., and Ådlandsvik, B. (2003). Studying climate effects on ecology through the use of climate indices: the North Atlantic Oscillation, El Niño Southern Oscillation and beyond. *Proceedings of the Royal Society of London (Series B.)*, **270**, 2087–96.

Sundby, S. and Fossum, P. (1990). Feeding conditions of Arcto-Norwegian cod larvae compared with the Rothschild-Osborn theory on small-scale turbulence and plankton contact rates. *Journal of Plankton Research*, **12**, 1153–62.

Sverdrup, H. U. (1953). On conditions for the vernal blooming of phytoplankton. *Journal du Conseil International pour l'Exploration de la Mer*, **11**, 287–95.

Sverdrup, H. U., Johnson, M. W., and Fleming, R. H. (1946). *The oceans, their physics, chemistry and General Biology*. Prentice-Hall, New York.

Taylor, A. H. (1995). North–South shifts of the Gulf Stream and their climatic connection with the abundance of zooplankton in the UK and its surrounding seas. *ICES Journal of Marine Science*, **52**, 711–21.

Tomczak, M. and Godfrey, J. S. (1994). *Regional oceanography: an introduction*. Pergamon Press, New York.

Walker, G. T. and Bliss, E. W. (1932). World weather V. *Mem. Royal Meteorological Society*, **4**, 53–84.

Werner, F., MacKenzie, B., Perry, R., Lough, R., Naimie, C., Blanton, B. and Quinlan, J. (2001). Larval trophodynamics, turbulence, and drift on Georges Bank: a sensitivity analysis of cod and haddock. *Scientia Marina*, **65**, 99–115.

Werner, F. E., Page, F. H., Lynch, D. R., Loder, J. W., Lough, R. G., Perry, R. I., Greenberg, D. A., and Sinclair, M. M. (1993). Influences of mean advection and simple behavior on the distributions of cod and haddock early life stages on Georges Bank. *Fisheries Oceanography*, **2**, 43–64.

Chapter 2

Aagarrd, K., Barrie, L. A., Carmack, E. C., Garrity, K., Jones, E. P., Lubin, D., Macdonald, R. W., Swift, J. H., Tucker, W. B., Wheeler, P. A., and Whritner, R. H. (1996). U.S., Canadian researchers explore the Arctic Ocean. *EOS*, **77**, 209–13.

Alexandersson, H., Schmith, T., Iden, K., and Tuomenvirta, H. (1998). Long-term variations of the storm climate over NW Europe. *The Global Ocean Atmosphere System*, **6**, 97–120.

Anon, (1996). *Report of the 3rd ICES/GLOBEC Cod and Climate Backward-facing Workshop*. Bergen, Norway, *ICES Counsil Meeting 1996/A:9*.

Bacon, S. and Carter, D. J. T. (1993). A connection between mean wave height and atmospheric pressure gradient in the North Atlantic. *International Journal of Climatology*, **13**, 423–36.

Barnston, A. G. and Livezey, R. E. (1987). Classification, seasonality, and persistence of low-frequency atmospheric circulation patterns. *Monthly Weather Review*, **115**, 1083–126.

Becker, G. A. and Pauly, M. (1996). Sea surface temperature changes in the North Sea and their causes. *ICES Journal of Marine Science*, **53**(6), 887–98.

Bersch, M., Meincke, J., and Sy, A. (1999). Interannual thermohaline changes in the northern North Atlantic. 1991–1996. *Deep-Sea Research II*, **46**, 55–75.

Bjerknes, J. (1964). Atlantic air-sea interaction. *Advances in Geophysics*, **10**, 1–82.

Blindheim, J., Borovkov, V., Hansen, B., Malmberg, S.-A., Turrell, W. R., and Osterhus, S. (2000). Upper layer cooling and freshening in the Norwegian Sea in relation to atmospheric forcing. *Deep-Sea Research I*, **47**, 655–80.

Buch, E. (1995). *A monograph on the Physical Oceanography of the Greenland Waters*. Royal Danish Administration of Navy and Hydrography. Copenhagen.

Buch, E. (2000). Oceanographic Investigations off West Greenland 1999. In *ICES WGOH Report 2000*, 9pp +13 Figs, mimeo.

Buch, E. and Hansen, H. H. (1988). Climate and cod fishery at West Greenland. In *Long term changes in marine fish populations* (eds T. Wyatt and M. G. Larraneta), pp. 345–64. Imprenta Real, Vigo.

Carmack, E. C., Aagaard, K., Swift, J. H., Macdonald, R. W., McLaughlin, F. A., Jones, E. P., Perkin, R. G., Smith, J. N., Ellis, K. M., and Kilius, L. R. (1997). Changes in temperature and tracer distributions within the Arctic Ocean: results from the 1994 Arctic Ocean Section. *Deep-Sea Research II*, **44**(8), 1487–502.

Carmack, E. C., Macdonald, R. W., Perkin, R. G., McLaughlin, F. A., and Pearson, R. (1995). Evidence for warming of Atlantic water in the southern Canadian Basin of the Arctic Ocean: results from the Larsen-93 Expedition, *Geophysical Research Letters*, **22**(9), 1061–64. (ISBN 1-880653-42-7)

Carter, D. J. T. (1999). Variability and trends in the wave climate of the North Atlantic: a review. In *Proceedings of the 9th ISOPE Conf. 30 May–4 June 1999*, Vol. III, pp. 12–18, Brest.

Cavalieri, D. J., Gloersen, P., Parkinson, C. L., Comiso, J. C., and Zwally, H. J. (1997). Observed hemispheric asymmetry in global sea ice changes. *Science*, **278**, 1104–6.

Cayan, D. R. (1992a). Latent and sensible heat flux anomalies over the Northern Oceans: the connection to monthly atmospheric circulation. *Journal of Climate*, **5**(4), 354–69.

Cayan, D. R. (1992b). Variability of latent and sensible heat fluxes estimated using Bulk Formulae. *Atmosphere-Ocean*, **30**(1), 1–42.

Cayan, D. R. (1992c). Latent and sensible heat flux anomalies over the Northern Oceans: driving the sea surface temperature. *Journal of Physical Oceanography*, **22**(8), 859–881.

Cayan, D. R. and Reverdin, G. (1994). Monthly precipitation and evaporation variability estimated over the North Atlantic and North Pacific. In *Atlantic climate change program: proceedings of PI's. meeting, Princeton, May 9–11, 1994* (ed. A.-M. Wilburn), pp. 28–32.

Chapman, W. L. and Walsh, J. E. (1993). Recent variations of sea ice and air temperature in high latitudes. *Bulletin of the American Meteorological Society*, **74**, 33–47.

Corti, S., Molteni, F., and Palmer, T. N. (1999). Signature of recent climate change in frequencies of natural atmospheric circulation regimes. *Nature*, **398**, 799–802.

Cullen, H. and deMenocal, P. B. (2000). North Atlantic influence on Tigris–Euphrates streamflow. *International Journal of Climatology*, **20**, 853–63.

Curry, R. G. and McCartney, M. S. (2001). Ocean gyre circulation changes associated with the North Atlantic Oscillation. *Journal of Physical Oceanography*, **31**, 3374–400.

Curry, R. G., McCartney, M. S., and Joyce, T. M. (1998). Oceanic transport of subpolar climate signals to mid-depth subtropical waters. *Nature*, **391**, 575–7.

Cushing, D. H. (1982). *Climate and fisheries*. Academic Press, London.

Dai, A., Fung, I. Y., and Del Genio, A. D. (1997). Surface observed global land precipitation variations during 1900–88. *Journal of Climate*, **10**, 2943–2962.

Delworth, T. and Knutson, T. R. (2000). Simulation of early 20th Century global warming. *Science*, **287**, 2246–50.

Deser, C. (2000). A note on the teleconnectivity of the 'Arctic Oscillation.' *Geophysical Research Letters*, **27**, 779–82.

Deser, C. and Timlin, M. S. (1997). Atmosphere-ocean interaction on weekly time scales in the North Atlantic and Pacific. *Journal of Climate*, **10**, 393–408.

Deser, C., Walsh, J. E., and Timlin, M. S. (2000). Arctic sea ice variability in the context of recent wintertime atmospheric circulation trends. *Journal of Climate*, **13**, 617–33.

Dickson, R. R. (1997). From the Labrador Sea to global change. *Nature*, **386**, 649–650.

Dickson, R. R. and Brander, K. M. (1993). Effects of a changing windfield on cod stocks of the North Atlantic. *Fisheries Oceanography*, **2**, 124–53.

Dickson, R. R. and Namias, J. (1976). North American influences on the circulation and climate of the North Atlantic Sector. *Monthly Weather Review*, **104**(10), 1256–65.

Dickson, R. R. and Turrell, W. R. (1999). The NAO: the dominant atmospheric process affecting oceanic variability in home, middle and distant waters of European Salmon. In *The ocean life of Atlantic salmon: environmental and biological factors influencing survival* (ed. D. Mills), pp. 92–115. Fishing News Books, Oxford.

Dickson, R. R., Meincke, J., Malmberg, Sv.-A, and Lee, A. J. (1988). The 'Great Salinity Anomaly' in the northern North Atlantic, 1968–82. *Progress in Oceanography*, **20**, 103–51.

Dickson, R. R., Briffa, K. R., and Osborn, T. J. (1994). Cod and climate: the spatial and temporal context. *ICES Marine Science Symposia*, **198**, 280–6.

Dickson, R. R., Lazier, J., Meincke, J., Rhines, P., and Swift, J. (1996). Long-term co-ordinated changes in the convective activity of the North Atlantic. *Progress in Oceanography*, **38**, 241–95.

Dickson, B., Meincke, J., Vassie, I., Jungclaus, J., and Osterhus, S. (1999). Possible predictability in overflow from the Denmark Strait. *Nature*, **397**, 243–46.

Dickson, R. R., Osborn, T. J., Hurrell, J. W., Meincke, J., Blindheim, J., Adlandsvik, B., Vigne, T., Alekseev, G., and Maslowski, W. (2000). The Arctic Ocean Response to the North Atlantic Oscillation. *Journal of Climate*, **13**, 2671–2696.

Dickson, B., Yashayaev, I., Meincke, J., Turrell, B., Dye, S., and Holfort, J. (2002). Rapid freshening of the deep North Atlantic Ocean over the past four decades. *Nature*, **416**, 832–7.

Dooley, H. D., Martin, J. H. A., and Ellett, D. J. (1984). Abnormal hydrographic conditions in the northeast Atlantic during the 1970s. *Rapports et Procès-verbaux des Réunions, Conseil international pour l'Exploration de la Mer*, **185**, 179–87.

Drinkwater, K. F., Mountain, D. B., and Herman, A. (1999). Variability in the Slope Water properties off eastern north America and their effects on the adjacent shelves. *ICES Counsil Meeting 1999/O:08*.

Dye, S. R. (1999). *A century of variability of flow through the Faroe–Shetland Channel*. Ph.D. Thesis, University of East Anglia.

Fabricius, O. (1780). *Fauna Groenlandica*. xvi+452 pp., 1PL, foannis Guttlab, Rothe Hafriae, Lipsiae.

Frank, P. (1997). Changes in the glacier area in the Austrian Alps between 1973 and 1992 derived from LANDSAT data. *Max Planck Institute Report*, **242**, 21 pp.

Fromentin, J.-M. and Planque, B. (1996). Calanus and environment in the eastern North Atlantic. II. Influence of the North Atlantic Oscillation on *C.finmarchicus* and *C.helgolandicus*. *Marine Ecology Progress Series*, **134**, 111–18.

Fyfe, J. C., Boer, G. J., and Flato, G. M. (1999). The Arctic and Antarctic oscillations and their projected changes under global warming. *Geophysical Research Letters*, **26**, 1601–1604.

Gillett, N. P., Hegerl, G. C., Allen, M. R., and Stott, P. A. (2000). Implications of changes in the Northern Hemisphere circulation for the detection of anthropogenic climate change. *Geophysical Research Letters*, **27**, 993–6.

Gillett, N. P, Allen, M. R., McDonald, R. E., Senior, C. A., Shindell, D. T., and Schmidt, G. A. (2002). How linear is the Arctic Oscillation response to greenhouse gases? *Journal of Geophysical Research*, **107**(D3): art. no. 4022.

Grotefendt, K., Logemann, K., Quadfasel, D., and Ronski, S. (1998). Is the Arctic Ocean warming? *Journal of Geophysical Research*, **103**(27), 679–87.

Hagen, J. O. (1995). Recent trends in the mass balance of glaciers in Scandinavia and Svalbard. In *Proceedings of the international symposium on environmental research in the Arctic* (ed. W. Okitsugu), pp. 343–54. National Institute of Polar Research, Tokyo, Japan.

Hansen, D. V. and Bezdek, H. F. (1996). On the nature of decadal anomalies in North Atlantic sea surface temperature. *Journal of Geophysical Research*, **101**, 8749–58.

Hansen, P. M. (1949). Studies on the biology of the cod in Greenland waters. *Rapports et Procès-verbaux des Réunions, Conseil international pour l'Exploration de la Mer*, **123**, 1–83.

Harvey, J. G. (1962). Hydrographic conditions in Greenland waters during August 1960. *Annales Biologie Copenhagen*, **17**, 14–17.

Hayden, B. P. (1981). Secular variation in the Atlantic coast extratropical cyclones. *Monthly Weather Review*, **109**, 159–67.

Helland-Hansen, B. and Nansen, F. (1909). The Norwegian Sea. *Fiskiridirektoratets Skrifter, serie Havundersøkelser*, **2**, 1–360.

Hilmer, M., Harder, M., and Lemke, P. (1998). Sea ice transport: a highly variable link between Arctic and North Atlantic. *Geophysical Research Letters*, **25**(17), 3359–62.

Hilmer, M. and Jung, T. (2000). Evidence for a recent change in the link between the North Atlantic Oscillation and Arctic sea ice export. *Geophysical Research Letters*, **27**, 989–92.

Hoerling, M. P., Hurrell, J. W., and Xu, T. (2001). Tropical origins for recent North Atlantic climate change. *Science*, **292**, 90–2.

Hurrell, J. W. (1995a). Decadal trends in the North Atlantic Oscillation regional temperatures and precipitation. *Science*, **269**, 676–79.

Hurrell, J. W. (1995b). Transient eddy forcing of the rotational flow during northern winter. *Journal of Atmospheric Science*, **52**, 2286–301.

Hurrell, J. W. (1996). Influence of variations in extratropical wintertime teleconnections on Northern Hemisphere temperatures. *Geophysical Research Letters*, **23**, 665–8.

Hurrell, J. W. and van Loon, H. (1997). Decadal variations in climate associated with the North Atlantic oscillation. *Climatic Change*, **36**, 301–26.

Hurrell, J. W., Kushnir, Y., Ottersen, G., and Visbeck, M. (2003). *The North Atlantic Oscillation: Climatic Significance and Environmental Impact. Geophysical Monograph Series*, **134**, American Geophysical Union, Washington, DC, 279 pp.

Jakobsson, J. (1992). Recent variability in fisheries of the North Atlantic. *ICES Marine Science Symposia*, **195**, 291–315.

Jenkins, W. J. (1982). On the climate of a subtropical gyre: decade timescale variations in water mass renewal in the Sargasso Sea. *Journal of Marine Research*, **40** (Suppl.), 265–90.

Joyce, T. M., Deser, C., and Spall, M. A. (2000). On the relation between decadal variability of Subtropical Mode Water and the North Atlantic Oscillation. *Journal of Climate*, **13**, 2550–69.

Kushnir, Y., Cardone, V. J., Greenwood, J. G., and Cane, M. (1997). On the recent increase in North Atlantic wave heights. *Journal of Climate*, **10**, 2107–13.

Kwok, R. and Rothrock, D. A. (1999). Variability of Fram Strait Ice Flux and the North Atlantic Oscillation. *Journal of Geophysical Research-Oceans*, **104**, 5177–89.

Lau, N.-C. and Nath, M. J. (1991). Variability of the baroclinic and barotropic transient eddy forcing associated with monthly changes in the midlatitude storm tracks. *Journal of Atmospheric Science*, **48**, 2589–613.

Lazier, J. R. N. (1980). Oceanographic conditions at Ocean Weather Ship Bravo, 1964–1974. *Atmosphere-ocean*, **18**, 227–38.

Lazier, J. R. N. (1988). Temperature and salinity changes in the deep Labrador Sea, 1962–86. *Deep-Sea Research*, **35**, 1247–53.

Lazier, J. R. N. (1995). The salinity decrease in the Labrador Sea over the past thirty years. In *Natural climate variability on decade-to-century time scales* (eds D. G. Martinson, K. Bryan, M. Ghil, M. M. Hall, T. M. Karl, E. S. Sarachik, S. Sorooshian, and L. D. Talley), pp. 296–304. National Academy Press, Washington, D.C.

Lee, A. J. (1949). Bibliography. *Rapports et Procès-verbaux des Réunions, Conseil International pour l'Exploration de la Mer*, **125**, 43–52.

Marsh, R., Petrie, B., Weidman, C. R., Dickson, R. R., Loder, J. W., Hannah, C. G., Frank, K., and Drinkwater, K. F. (1999). The 1882 tilefish kill—a cold event in shelf waters off the north-eastern United States? *Fisheries Oceanography*, **8**, 39–49.

Maslanik, J. A., Serreze, M. C., and Barry, R. G. (1996). Recent decreases in Arctic summer ice cover and linkages to atmospheric circulation anomalies. *Geophysical Research Letters*, **23**, 1677–1680.

McCartney, M. S., Curry, R. G., and Bezdek, H. F. (1997). The interdecadal warming and cooling of Labrador Sea Water. *ACCP Notes*, **IV**(1), 1–11.

McPhee, M. G., Stanton, T. P., Morison, J. H., and Martinson, D. G. (1998). *Geophysical Research Letters*, **25**, 1729–32. "Freshening of the upper ocean in the Arctic: is perennial sea ice disappearing?

Mehta, V. M., Suarez, M. J., Manganello, J. V., and Delworth, T. L. (2000). Oceanic influence on the North Atlantic Oscillation and associated Northern Hemisphere climate variations: 1959–1993. *Geophysical Research Letters*, **27**, 121–24.

Monahan, A. H., Fyfe, J. C., and Flato, G. M. (2000). A regime view of Northern Hemisphere atmospheric variability and change under global warming. *Geophysical Research Letters*, **27**, 1139–42.

Morison, J., Aagaard, K., and Steele, M. (2001). Recent environmental changes in the Arctic: a review. *Arctic*, **53**, 359–71.

Morison, J., Steele, M., and Anderson, R. (1998a). Hydrography of the upper Arctic Ocean measured from the Nuclear Submarine, USS PARGO. *Deep-Sea Research I*, **45**(1) 15–38.

Moulin, C., Lambert, C. E., Dulac, F., and Dayan, U. (1997). Control of atmospheric export of dust from North Africa by the North Atlantic Oscillation, *Nature*, **387**, 691–94.

Myers, R. A., Helbig, J., and Holland, D. (1989). Seasonal and interannual variability of the Labrador Current and West Greenland Current. *ICES Council Meeting 1989/C:16*, 18pp (mimeo).

Nakamura, H. (1996). Year-to-year and interdecadal variability in the activity of intraseasonal fluctuations in the Northern Hemisphere wintertime circulation. *Theoretical and Applied Climatology*, **55**, 19–32.

Osborn, T. J., Briffa, K. R., Tett, S. F. B., Jones, P. D., and Trigo, R. M. (1999). Evaluation of the North Atlantic oscillation as simulated by a climate model. *Climate Dynamics*, **15**, 685–702.

Quadfasel, D., Sy, A., Wells, D., and Tunik, A. (1991). Warming in the Arctic. *Nature*, **350**, 385.

Parkinson, C. L., Cavalieri, D. J., Gloersen, P., Zwally, H. J., and Comiso, J. (1998). Variability of the Arctic sea ice cover, 1978–1996. *Journal of Geophysical Research*, **104**, 20837–56.

Prinsenberg, S. J., Peterson, I. K., Narayanan, S., and Umoh, J. U. (1997). Interaction between atmosphere, ice cover, and ocean off Labrador and Newfoundland from 1962–1992. *Canadian Journal of Aquatic Science*, **54**, 30–9.

Read, J. F. and Gould, W. J. (1992). Cooling and freshening of the subpolar North Atlantic Ocean since the 1960's. *Nature*, **360**, 55–7.

Reverdin, G., Cayan, D., and Kushnir, Y. (1997). Decadal variability of hydrography in the upper northern North Atlantic in 1948–1990. *Journal of Geophysical Research*, **102**, 8505–32.

Rodwell, M. J., Rowell, D. P., and Folland, C. K. (1999). Oceanic forcing of the wintertime North Atlantic oscillation and European climate. *Nature*, **398**, 320–23.

Rogers, J. C. (1990). Patterns of low-frequency monthly sea level pressure variability (1899–1986) and associated wave cyclone frequencies. *Journal of Climate*, **3**, 1364–79.

Rogers, J. C. (1997). North Atlantic storm track variability and its association to the North Atlantic Oscillation and climate variability of northern Europe. *Journal of Climate*, **10**, 1635–47.

Rothrock, D. A., Yu, Y., and Maykut, G. A. (1999). Thinning of the Arctic sea-ice cover. *Geophysical Research Letters*, **26**(23), 3469–72.

Schopka, S. (1993). The Greenland cod (*Gadus morhua*) at Iceland 1941–90 and their impact on assessments. *NAFO Scientific Studies*, **18**, 81–5.

Schopka, S. (1994). Fluctuations in the cod stock off Iceland during the twentieth century in relation to changes in the fisheries and environment. *ICES Marine Science Symposia*, **198**, 175–93.

Serreze, M. C., Carse, F., Barry, R. G., and Rogers, J. C. (1997). Icelandic low activity: climatological features, linkages with the NAO, and relationships with recent changes in the Northern Hemisphere circulation. *Journal of Climate*, **10**, 453–64.

Shindell, D. T., Miller, R. L., Schmidt, G., and Pandolfo, L. (1999). Simulation of recent northern winter climate trends by greenhouse-gas forcing. *Nature*, **399**, 452–55.

Siggurdson, O. and Jonsson, T. (1995). Relation of glacier variations to climate changes in Iceland. *Annals of Glaciology*, **21**, 263–70.

Steele, M. and Boyd, T. (1998). Retreat of the cold halocline layer in the Arctic Ocean. *Journal of Geophysical Research*, **103**, 10419–35.

Stenseth, N. C., Ottersen, G., Hurrell, J.W., Mysterud, A., Lima, M., Chan, K.-S., Yoccoz, N. G., and Ådlandsvik, B. (2003). Studying climate effects on ecology through the use of climate indices: the North Atlantic Oscillation, El Niño Southern Oscillation and beyond. *Proceedings of the Royal Society of London (Series B.)*, **270**, 2087–96.

Stockton, C. W. and Glueck, M. F. (1999). Long-term variability of the North Atlantic oscillation (NAO). *Proceedings American Meteorological Societies Tenth Symposium on Global Change Studies, 11–15 January, 1999, Dallas, TX*, 290–93.

Sverdrup, H. U. (1953). On conditions for the vernal blooming of phytoplankton. *Journal Conseil International pour l'Exploration de la Mer*, **18**, 287–95.

Swift, J. H., Jones, E. P., Aagaard, K., Carmack, E. C., Hingston, M., Macdonald, R. W., McLaughlin, F. A., and Perkin, R. G. (1997). Waters of the Makarov and Canada Basins. *Deep-Sea Research II*, **44**(8), 1503–29.

Sy, A., Rhein, M., Lazier, J. R. N., Koltermann, K. P., Meincke, J., Putzka, A., and Bersch, M. (1997). Surprisingly rapid spreading of newly formed intermediate waters across the North Atlantic Ocean. *Nature*, **386**, 675–79.

Talley, L. D. (1996). North Atlantic circulation and variability, reviewed for the CNLS Conference, Los Alamos, May 1995. *Physica D*, 98, 625–46.

Talley, L. D. and Raymer. M. E. (1982). Eighteen Degree Water variability. *Journal of Marine Research*, **40**(Suppl.), 757–75.

Thompson, D. W. J. and Wallace, J. M. (1998). The arctic oscillation signature in the wintertime geopotential height and temperature fields. *Geophysical Research Letters*, **25**, 1297–300.

Thompson, D. W. J., Wallace, J. M., and Hegerl, G. C. (2000). Annular modes in the extratropical circulation Part II: Trends. *Journal of Climate*, **13**, 1018–36.

Ting, M. and Lau, N.-C. (1993). A diagnostic and modeling study of the monthly mean wintertime anomalies appearing in an 100-year GCM experiment. *Journal of Atmospheric Science*, **50**, 2845–67.

Toresen, R. and Ostvedt, O. J. (2000). Variation in abundance of Norwegian spring-spawned herring (*Clupea harengus, Clupeidae*) throughout the 20th Century and the influence of climatic fluctuations. *Fish and Fisheries*, **1**, 231–56.

Turrell, W. R., Slesser, G., Adams, R. D., Payne, R., and Gillibrand, P. A. (1999). Decadal variability in the composition of Faroe-Shetland Channel bottom water. *Deep-Sea Research I*, **46**, 1–25.

Ulbrich, U. and Christoph, M. (1999). A Shift of the NAO and increasing storm track activity over Europe due to anthropogenic greenhouse gas forcing. *Climate Dynamics*, **15**, 551–9.

Van Loon, H. and Rogers, J. C. (1978). The seesaw in winter temperatures between Greenland and northern Europe. Part I: General description. *Monthly Weather Review*, **106**, 296–310.

Verduin, J. and Quadfasel, D. (1999). Long-term temperature and salinity trends in the central Greenland Sea. In *European Sub-Polar Ocean Programme II, Final Scientific Report* (ed.) Jansen, E. AI, I-II, Univ. of Bergen, Norway.

Vilhjalmsson, H. (1997). Climatic changes and some examples of their effects on the marine ecology of Icelandic and Greenland waters, in particular during the present century. *Rit Fiskideildar*, **15**(1), 1–29.

Vinje, T. and Finnekasa, O. (1986). The ice transport through the Fram Strait. *Norsk Polarinstitutt. Skrifter,* **186,** 37–9.

Vinje, T., Nordlund, N., and Kvambekk, A. (1998). Monitoring ice thickness in fram strait. *Journal of Geophysical Research,* **103,** 10437–49.

Walker, G. T. and Bliss, E. W. (1932). World Weather V. *Mem. Royal Meteorological Society,* **4,** 53–84.

Wallace, J. M. and Gutzler, D. S. (1981). Teleconnections in the geopotential height field during the Northern Hemisphere winter. *Monthly Weather Review,* **109,** 784–812.

Wallace, J. M. and Lau, N.-C. (1985). On the role of barotropic energy conversions in the general circulation. *Advances in Geophysics,* **28A,** 33–74.

Wallace, J. M. (2000). North Atlantic Oscillation/annular mode: two paradigms—one phenomenon. *Quarterly Journal of the Royal Meteorological Society,* **126,** 791–805.

Walsh, J. E., Chapman, W. L., and Shy, T. L. (1996). Recent decrease of sea-level pressure in the Central Arctic. *Journal of Climate,* **9,** 480–486.

Wright, P. B., Kings, J., Parker, D. E., Folland, C. K., and Basnett, T. A. (1999). Changes to the climate of the United Kingdom. *Hadley Centre Internal Note* No. **89.**

Wunsch, C. (1992). Decade-to-Century Changes in the Ocean Circulation. *Oceanography,* **5**(2), 99–106.

Wunsch, C. (1999). The interpretation of short climate records, with comments on the North Atlantic and Southern Oscillations. *Bulletin of the American Meteorological Society,* **80,** 257–70.

Chapter 3

Alheit, J. and Hagen, E. (1997). Long-term climate forcing of European herring and sardine populations. *Fisheries Oceanography.* **6,** 130–9.

Anderson, L. A. and Robinson, A. R. (2001). Physical and biological modeling in the Gulf Stream region Part II. Physical and biological processes. *Deep-Sea Research I,* **48,** 1139–68.

Anderson, L. A. and Sarmiento, J. L. (1995). Global ocean phosphate and oxygen simulations. *Global Biogeochemical Cycles,* **9,** 621–36.

Anderson, L. A., Robinson, A. R., and Lozano, C. J. (2000). Physical and biological modeling in the Gulf Stream region: I. Data assimilation methodology. *Deep-Sea Research I,* **47,** 1787–827.

Anderson T. R. and Ducklow, H. W. (2001). Microbial loop carbon cycling in ocean environments studied using a simple steady state model. *Aquatic Microbial Ecology,* **26,** 37–49.

Armstrong, R. A., Sarmiento, J. L., and Slater, R. D. (1995). Monitoring ocean productivity by assimilating satellite chlorophyll into ecosystem models. In *Ecological time series* (eds T. M. Powell and J. H. Steele), pp. 371–90. Chapman and Hall, London.

Backhaus, J. O., Harms, I., Krause, M., and Heath, M. R. (1994). An hypothesis concerning the space-time succession of *Calanus finmarchicus* in the northern North Sea. *ICES Journal of Marine Science,* **51,** 169–80.

Baretta, J. W., Ebenhöh, W., and Ruardij, P. (1995). The European regional seas ecosystem model, a complex marine ecosystem model. *Netherlands Journal of Sea Research,* **33,** 233–46.

Bartsch, J. (1993). Application of a circulation and transport model system to the dispersal of herring larvae in the North Sea. *Continental Shelf Research,* **13,** 1335–61.

Bartsch, J. and Coombs, S. (1997). A numerical model of the dispersal of blue whiting larva, *Micromesistius poutassou* (Risso), in the eastern North Atlantic. *Fisheries Oceanography,* **6,** 141–54.

Bartsch, J. and Knust, R. (1994). Simulating the dispersion of vertically migrating sprat larvae *Sprattus sprattus* (L.) in the German Bight with a circulation and transport model system. *Fisheries Oceanography,* **3,** 92–105.

Berntsen, J., Skagen D. W., and Svendsen, E. (1994). Modeling the transport of particles in the North Sea with reference to sandeel larvae. *Fisheries Oceanography,* **3,** 81–91.

Besiktepe, S. T., Lermusiaux, P. F. J., and Robinson, A. R. (2003). Coupled physical and biogeochemical data-driven simulations of Massachusetts Bay in late summer: real-time and postcruise data assimilation. *Journal of Marine Systems,* **407,** 1–42.

Blanton, J. O., Werner, F. E., Kapolnai, A., Blanton, B. O., Knott, D., and Wenner, E. L. (1999). Wind-generated transport of fictitious passive larvae to shallow tidal estuaries. *Fisheries Oceanography,* **8,** 210–23.

Bradford, M. J. (1992). Precision of recruitment predictions from early life stages of marine fishes. *Fishery Bulletin* **90,** 439–53.

Brander, K. (1995). The effect of temperature on growth of Atlantic cod (*Gadus morhua* L.). *ICES Journal of Marine Science,* **52,** 1–10.

Brander, K. (2000). Effects of environmental variability on growth and recruitment in cod (*Gadus morhua*) using a comparative approach. *Oceanologica Acta,* **23**(4), 485–96.

Bryant, A. D., Hainbucher, D., and Heath, M. (1998). Basin-scale advection and population persistence of *Calanus finmarchicus.* *Fisheries Oceanography,* **7,** 235–44.

Carlotti, F. and Wolf, K.-U. (1998). A Lagrangian ensemble model of *Calanus finmarchicus* coupled with a 1D ecosystem model. *Fisheries Oceanography,* **7,** 191–204.

Carlotti, F., Giske, J., and Werner, F. (2000). Modeling zooplankton dynamics. In *ICES Zooplankton Methodology Manuel* (eds R. P. Harris, P. H. Wiebe, J. Lenz, H. R. Skjokdal, and M. Huntley), pp. 571–667. Academic Press, San Diego.

Cushing, D. H. (1982). *Climate and fisheries.* Academic Press, London.

Cushing, D. H. (1995). The Long-Term Relationship between Zooplankton and Fish. *ICES Journal of Marine Science*, **52**(3–4), 611–26.

Cushing, D. H. (1996). Towards a science of recruitment in fish populations. In *Excellence in Ecology* (ed. O. Kinne), 1–175. Ecology Institute, Oldendorf/Luhe.

Davis, C. S. (1984). Interaction of a copepod population with the mean circulation on Georges Bank. *Journal of Marine Research*, **42**, 573–90.

Denman, K. L. and Peña, M. A. (1999). A coupled 1-D biological/physical model of the northeast subarctic Pacific Ocean with iron limitation. *Deep-Sea Research II*, **46**, 2877–908.

de Young, B., Anderson, J., Greatbatch, R. J., and Fardy, P. (1994). Advection-diffusion modelling of larval capelin (*Mallotus villosus*) dispersion in Conception Bay, Newfoundland. *Canadian Journal of Fisheries and Aquatic Sciences*, **51**, 1297–307.

Dickey, T. D. (2003). Emerging ocean observations for interdisciplinary data assimilation systems. *Journal of Marine Systems*, **40–41**, 5–48.

Doney, S. C., Glover, D. M., and Najjar, R. G. (1996). A new coupled, one-dimensional biological-physical model for the upper ocean: applications to the JGOFS Bermuda Atlantic time-series study (BATS) site. *Deep-Sea Research II*, **43**, 591–624.

Doney, S. C., Kleypas, J. A., Sarmiento, J. L., and Falkowski, P. G. (2002). The US JGOFS synthesis and modeling project—an introduction. *Deep-Sea Research II*, **49**(1–3), 1–20.

Evans, G. T. (1999). The role of local models and data sets in the Joint Global Ocean Flux study. *Deep-Sea Research I*, **46**, 1369–89.

Fasham, M. J. R. and Evans, G. T. (1995). The use of optimization techniques to model marine ecosystem dynamics at the JGOFS station at 47°N 20°W. *Philosophical Transactions of the Royal Society of London, B*, **348**, 203–9.

Fasham, M. J. R. and Evans, G. T. (2000). Advances in ecosystem modelling within JGOFS. In *The changing ocean carbon cycle: a mid-term synthesis of the Joint Flux Study* (eds H. W. Ducklow, J. G. Field, and R. Hanson), pp. 417–46. Cambridge University Press.

Fasham, M. J. R., Ducklow, H. W., and McKelvie, S. M. (1990). A nitrogen-based model of plankton dynamics in the oceanic mixed layer. *Journal of Marine Research*, **48**, 591–639.

Fasham, M. J. R., Boyd, P. W., and Savidge, G. (1999). Modeling the relative contributions of autotrophs and heterotrophs to carbon flow at a Lagrangian JGOFS station in the Northeast Atlantic: the importance of DOC. *Limnology and Oceanography*, **44**, 80–94.

Fiksen, Ø. and Giske, J. (1995). Vertical distribution and population dynamics of copepods by dynamic optimization. *ICES Journal of Marine Science*, **52**, 483–503.

Fiksen, Ø., J. Giske, and D. Slagstad. (1995). A spatially explicit fitness-based model of capelin migrations in the Barents Sea. *Fisheries Oceanography*, **4**, 193–208.

Flierl, G. R. and Wroblewski, J. S. (1985). The possible influence of warm core Gulf Stream rings upon shelf water larval fish distributions. *Fisheries Bulletin*, **83**, 313–30.

Flynn, K. J. and Fasham, M. J. R. (2002). A modelling exploration of vertical migration by phytoplankton. *Journal of Theoretical Biology*, **218**, 471–84.

Franks, P. J. S., Wroblewski, J. S., and Flierl, G. R. (1986). Prediction of phytoplankton growth in response to the frictional decay of a warm-core ring. *Journal of Geophysical Research*, **91**, 7603–10.

Garçon, V. C., Oschlies, A., Doney, S. C., McGillicuddy, D., and Waniek, J. (2001). The role of mesoscale variability on plankton dynamics in the North Atlantic. *Deep Sea Research II*, **48**, 2199–226.

Giske, J., Aksnes, D. L., and Fiksen, Ø. (1994). Visual predators, environmental variables and zooplankton mortality risk. *Vie Milieu*, **44**, 1–9.

GLOBEC. (1999). *Report of a workshop on the assimilation of biological data in coupled physical/ecosystem models*. International GLOBEC Special Contribution No. 3, available from International GLOBEC Office, Plymouth Marine Laboratory, Prospect Place, Plymouth PL1 3DH, United Kingdom.

Gunson, J., Oschlies, A., and Garcon, V. (1999). Sensitivity of ecosystem parameters to simulated satellite ocean color data using a coupled physical-biological model of the North Atlantic. *Journal of Marine Research*, **57**(4), 613–39.

Gupta, S., Lonsdale, D. J., and Wang, D.-P. (1994). The recruitment patterns of an estuarine copepod: a biological-physical model. *Journal of Marine Research*, **52**, 687–710.

Hare, J. A., Quinlan, J. A., Werner, F. E., Blanton, B. O., Govoni, J. J., Forward, R. B., Settle, L. R., and Hoss, D. E. (1999). Larval transport during winter in the SABRE study area: results of a coupled vertical larval behaviour-three-dimensional circulation model. *Fisheries Oceanography*, **8**, 57–76.

Haury, L. R., McGowan, J. A., and Wiebe, P. H. (1978). Patterns and processes in the time-space of plankton distributions. In *Spatial pattern in plankton communities* (ed. J. H. Steele), pp. 277–327. Plenum Press, New York, USA.

Heath, M. R. (1992). Field investigations of the early-life stages of marine fish. *Advances in Marine Biology*, **28**, 1–174.

Heath, M. R. (1999). The ascent migration of the copepod *Calanus finmarchicus* from overwintering depths in the Faroe-Shetland Channel. *Fisheries Oceanography*, **8**(Suppl. 1), 84–99.

Heath, M., Backhaus, J. O., Richardson, K., McKenzie, E., Slagstad, D., Beare, D., Dunn, J., Fraser, J. G., Gallego, A., Hainbucher, D., Hay, S., Jonasdottir, S., Madden, H., Mardaljevic, J., and Schacht, A. (1999). Climate fluctuations and the spring invasion of the North Sea by *Calanus finmarchicus*. *Fisheries Oceanography*, **8**(Suppl. 1), 163–76.

Heath, M., de Young, B, Fiksen, Ø., and Werner, F. E. (2001). Secondary production in the oceans and the response to climate change. *IGBP Newsletter*, **47**, 9–12.

Helland-Hansen, B. and Nansen, F. (1909). The Norwegian Sea. *Fiskeridirektoratets Skrifter, Serie Havundersøkelser*, **2**, 1–360.

Hjort, J. (1914). Fluctuations in the great fisheries of northern Europe viewed in the light of biological research. *Rapports et procès-verbaux des Réunions, Conseil International pour l'Exploration de la Mer*, **20**, 1–228.

Hofmann, E. E. (1988). Plankton dynamics on the outer southeastern U.S. continental shelf. Part III: A coupled physical-biological model. *Journal of marine research*, **46**(4), 919–46.

Hofmann, E. E. and Friedrichs, M. A. M. (2001). Biogeochemical data assimilation. In *Encyclopedia of Ocean Sciences* (eds J. H. Steele, S. Thorpe, and K. Turekian), pp. 302–308. Academic Press, London.

Hofmann, E. E. and Friedrichs, M. A. M. (2002). Predictive Modelling for Marine ecosystems. In *The Sea, Vol. 12* (eds, A. J. Robinson, J. J. McCarthy, and B. Rothschild), pp. 537–65. John Wiley, New York.

Hofmann, E. E. and Powell, T. M. (1998). Environmental variability effects on marine fisheries: four case histories. *Ecological Applications*, **8**(1), S23–32.

Hood, R. R., Bates, N. R., Capone, D. G., and Olson, D. B. (2001). Modeling the effect of nitrogen fixation on carbon and nitrogen fluxes at BATS. *Deep-Sea Research Part II*, **48** (8–9), 1609–48.

Hurtt, G. C. and Armstrong, R. A. (1996). A pelagic ecosystem model calibrated with BATS data. *Deep-Sea Research II*, **43**, 653–83.

Hurtt, G. C. and Armstrong, R. A. (1999). A pelagic ecosystem model calibrated with BATS data and OWSI data. *Deep-Sea Research II*, **46**, 27–61.

ICES. (1993). Report of the Study Group on Spatial and temporal integration. University of Strathclyde, Glasgow, Scotland, 14–18 June 1993. ICES CM/1993/L:9.

ICES. (1993). Study group on spatial and temporal integration. *ICES Counsil Meeting 1993/L* **9**, 69.

Ishizaka, J. (1990a). Coupling of Coastal Zone Color Scanner Data to a physical-biological model of the southeastern U.S. continental shelf ecosystem 1. CZCS data description and Lagrangian particle tracing experiments. *Journal of Geophysical Research*, **95**, 20167–81.

Ishizaka, J. (1990b). Coupling of Coastal Zone Color Scanner Data to a physical-biological model of the southeastern U.S. continental shelf ecosystem 3. Nutrient and phytoplankton fluxes and CZCS data assimilation. *Journal of Geophysical Research*, **95**, 20201–12.

Ishizaka, J. and Hofmann, E. E. (1988). Plankton dynamics on the outer southeastern U.S. continental shelf. Part I: Lagrangian particle tracing experiments. *Journal of Marine Research*, **46**(4), 853–82.

Jackson, G. A. (2001). Effect of coagulation on a model planktonic food web. *Deep-Sea Research I*, **48**, 95–123.

Jackson, G. A. and Burd, A. B. (2002). A model for the distribution of particle flux in the mid-water column controlled by subsurface biotic interactions. *Deep-Sea Research II*, **49**, 193–217.

Karl, D. M. (1999). A sea of change: biogeochemical variability in the North Pacific subtropical gyre. *Ecosystems*, **2**, 181–214.

Koeve, W. and Ducklow, H. W. (2001). JGOFS synthesis and modelling: the North Atlantic Ocean. *Deep-Sea Research II*, **48**, 2121–54.

Lawson, L. M., Hofmann, E. E., and Spitz, Y. H. (1996). Time series sampling and data assimilation in a simple marine ecosystem model. *Deep-Sea Research II*, **43**, 625–51.

Leggett, W. C. and DeBlois, E. (1994). Recruitment in marine fishes: is it regulated by starvation and predation in the egg and larval stages? *Netherlands Journal of Sea Research*, **32**(2), 119–34.

Lynch, D. R., Naimie, C. E., Ip, J., Lewis, C., Werner, F. E., Luettich, Jr., R. A., Blanton, B. O., Quinlan, J. A., McGillicuddy, D., Ledwell, J., Churchill, J., Kosnyrev, V., Davis, C., Gallager, S., Ashjian, C., Lough, R. G., Manning, J., Flagg, C., Hannah, C., and Groman, R. (2001). Real-time data assimilative modelling on Georges Bank. *Oceanography*, **14**, 65–77.

Malanotte-Rizzoli, P. (ed.) (1996). Modern approaches to data assimilation in ocean modeling. *Oceanography Series, Vol. 61*. Elsevier, Amsterdam.

McGillicuddy, D. J. Jr., McCarthy, J. J., and Robinson, A. R. (1995a). Coupled physical and biological modeling of the spring bloom in the North Atlantic (I): model formulation and one dimensional bloom processes, *Deep-Sea Research I*, **42**, 1313–57.

McGillicuddy, D. J. Jr, Robinson, A. R., and McCarthy, J. J. (1995b). Coupled physical and biological modeling of the spring bloom in the North Atlantic (II): three dimensional bloom and post-bloom processes. *Deep-Sea Research I*, **42**, 1359–98.

McGillicuddy, D. J. Jr and Robinson, A. R. (1997). Eddy-induced nutrient supply and new production in the Sargasso Sea. *Deep-Sea Research I*, **44**(8), 1427–50.

McGillicuddy, D. J. Jr, Lynch, D. R., Moore, A. M., Gentleman, W. C., Davis, C. S., and Meise, C. J. (1998). An adjoint data assimilation approach to diagnosis of physical and biological controls of *Pseudocalanus* spp. in the Gulf of Maine-Georges Bank region. *Fisheries Oceanography*, **7**, 205–18.

McGillicuddy, D. J. Jr, Lynch, D. R., Wiebe, P., Runge, J., Durbin, E. G., Gentleman, W. C., and Davis, C. S. (2001). Evaluating the synopticity of the US GLOBEC Georges Bank broad-scale sampling pattern with observational system simulation experiments. *Deep-Sea Research II*, **48**, 483–99.

Miller, C. B., Lynch, D. R., Carlotti, F., Gentleman, W., and Lewis, C. V. W. (1998). Coupling of an individual-based population dynamic model of *Calanus finmarchicus* to a circulation model for the Georges Bank region. *Fisheries Oceanography*, **7**, 219–34.

Michalsen, K., Ottersen, G., and Nakken, O. (1998). Growth of North-east arctic cod (*Gadus morhua* L.) in relation to ambient temperature. *ICES Journal of Marine Science*, **55**(5), 863–77.

Moll, A. (1998). Regional distribution of primary production in the North Sea simulated by a three-dimensional model. *Journal of Marine Systems*, **16**, 151–70.

Moore, J. K., Doney, S. C., Glover, D. M., Fung, I. Y. (2002a). Iron cycling and nutrient limitation patterns in surface waters of the world ocean *Deep-Sea Research II*, **49**, 463–508.

Moore, J. K., Doney, S. C., Kleypas, J. C., Glover, D. M., and Fung, I. Y. (2002b). An intermediate complexity marine ecosystem model for the global domain. *Deep-Sea Research II*, **49**, 403–62.

Aita, M. N., Tamanaka, Y., and Kishi, M. (2003). Effects of ontogenetic vertical migration of zooplankton on annual primary production—Using NEMURO embedded in a General Circulation Model. *Fisheries Oceanography*, **12**, 284–90.

O'Connell, J. and Tunnicliffe, V. (2001). The use of sedimentary fish remains for interpretation of long-term fish population fluctuations. *Marine Geology*, **174**(1–4), 177–95.

Pinardi, N. *et al.* (1999). The Mediterranean Ocean Forecasting System: the first phase of implementation. In *Proceedings of the International Conference on the Ocean Observing System for Climate*. Saint-Raphael, France, October.

Popova, E. E., Lozano, C. J., Srokosz, M. A., Fasham, M. J. R., Haley, P. J., and Robinson, A. R. (2002). Coupled 3D physical and biological modelling of the mesoscale variability observed in North-East Atlantic in spring 1997: biological processes. *Deep-Sea Research I*, **49**, 1741–68.

Pribble, J. R., Walsh, J. J., and Dieterle, D. A. (1994). A numerical analysis of shipboard and coastal zone color scanner time series of new production within Gulf Stream cyclonic eddies in the South Atlantic Bight. *Journal of Geophysical Research*, **99**, 7513–38.

Proctor, R., Wright, P. J., and Everitt, A. (1998). Modelling the transport of larval sandeels on the north-west European shelf. *Fisheries Oceanography*, **7**, 347–54.

Radach, G. and Moll, A. (1993). Estimation of the variability of production by simulating annual cycles of phytoplankton in the central North Sea. *Progress in Oceanography*, **31**, 339–419.

Railsback, S. F., and Harvey, B. C. (2002). Analysis of habitat-selection rules using an individual-based model. *Ecology*, **83**, 1817–30.

Rice, J. A., Quirrlan, J. A., Nixon, S. W., Hettler, W. F., Warlen, S. M., and Stegmann, P. M. (1999). Spawning and transport dynamics of Atlantic menhaden: inferences from characteristics of immigrating larvae and predictions of a hydrodynamic model. *Fisheries Oceanography*, **8**, 93–110.

Robinson, A. R. and Lermusiaux, P. F. J. (2001). Data assimilation in models. In *Encyclopedia of Ocean Sciences*, (eds F. H. Steele and K. K. Turekian), pp. 623–34. Academic Press, London.

Robinson, A. R. and Lermusiaux, P. F. J. (2002). Data assimilation for modelling and predicting coupled physical-biological interactions in the sea. In *The sea, biological-physical interactions in the ocean*, Vol. 12 (eds A. R. Robinson, J. J. McCarthy, and B. J. Rothschild), pp. 475–536. John Wiley & Sons, New York, USA.

Robinson, A. R., Lermusiaux, P. F. J., and Sloan, III, N. Q. (1998). Data assimilation. In *The Sea, The Global Coastal Ocean: processes and methods*, Vol. 10 (eds A. R. Robinson and K. H. Brink), pp. 541–94. John Wiley & Sons, New York, USA.

Rose, K. A. and Summers, J. K. (1992). Relationships among long-term fish abundances, hydrographic variables, and gross pollution indicators in northeastern US estuaries. *Fisheries Oceanography*, **1**, 281–93.

Runge, J. A., Franks, P. J. S., Gentleman, W. C., Megrey, B. A., Rose, K. A., Werner, F. E., and Zakardjian, B. (2003). Diagnosis and prediction of variation in pelagic ecosystems and fish recruitment processes: developments in physical-biological modeling. *The Sea*, Vol. 13. The Global Coastal Ocean: Multi-Scale Interdisciplinary Processes, accepted.

Sambrotto, R. N., Savidge, G., Robinson, C., Boyd, P., Takahashi, T., Karl, D. M., Langdon, C., Chipman, D., Marra, J., and Codispoti, L. (1993). Elevated consumption of carbon relative to nitrogen in the surface ocean. *Nature*, **363**(6426), 248–50.

Sarmiento, J. L., Slater, R. D., Fasham, M. J. R., Ducklow, H. W., Toggweiler, J. R., and Evans, G. T. (1993). A Seasonal 3-Dimensional Ecosystem Model of Nitrogen Cycling in the North-Atlantic Euphotic Zone. *Global Biogeochemical Cycles*, **7**(2), 417–50.

Seim, H. E. (2000). Implementation of the South Atlantic Bight Synoptic Offshore Observational Network. *Oceanography*, **13**, 18–23.

Semovski, S. V., Wozniak, B., Hapter, R., and Staskiewicz, A. (1996). Gulf of Gdansk spring bloom physical, bio-optical, biological modelling and contact data assimilation. *Journal of Marine Systems*, **7**, 145–59.

Shepard, J. G., Pope, J. G., and Cousens, R. D. (1984). Variations in fish stocks and hypotheses concerning their links with climate. *Rapports et Proces-Verbaux des Reunions, Conseil International pour l'Exploration de la Mer*, **185**, 255–67.

Sinclair, M. and Page, F. (1995). Cod fishery collapses and North Atlantic GLOBEC. *U.S. GLOBEC News*, **8**(March).

Soutar, A. and Isaacs, J. D. (1974). Abundance of pelagic fish during 19th and 20th centuries as recorded in anaerobic sediment off californias. *Fishery Bulletin*, **72**(2), 257–73.

Spitz, Y. H., Moisan, J. R., Abbott, M. R., and Richman, J. G. (1998). Data assimilation and a pelagic ecosystem model: parameterization using time series observations. *Journal of Marine Systems*, **16**, 51–68.

Tremblay, M. J., Loder, J. W., Werner, F. E., Naimie, C. E., Page, F. H., and Sinclair, M. M. (1994). Drift of sea scallop larvae *Placopecten magellanicus* on Georges Bank: a model study of the roles of mean advection, larval behavior and larval origin. *Deep-Sea Research II*, **41**, 7–49.

Tyler, J. A. and Rose, K. A. (1994). Individual variability and spatial heterogeneity in fish population models. *Reviews in Fish Biology and Fisheries* **4**, 91–123.

Tyler, J. A. and Rose, K. A. (1997). Effects of individual habitat selection in a heterogenous environment on fish cohort survivorship: a modelling analysis. *Journal of Animal Ecology*, **66**, 122–36.

Vallino, J. J. (2000). Improving marine ecosystem models: use of data assimilation and mesocosm experiments. *Journal of Marine Research*, **58**, 117–64.

Walsh, J. J., Dieterle, D. A., and Meyers, M. B. (1988). A simulation analysis of the fate of phytoplankton within the Mid-Atlantic Bight. *Continental Shelf Research*, **8**, 757–87.

Walsh, J. J., Dieterle, D. A., Meyers, M. B., and Muller-Karger, F. E. (1989). Nitrogen exchange at the continental margin: a numerical study of the Gulf of Mexico. *Progress in Oceanography*, **23**, 245–301.

Werner, F. E. and Quinlan, J. A. (2002). Fluctuations in marine fish populations: physical processes and numerical modelling. *ICES Marine Science Symposia*, **215**, 264–78.

Werner, F. E., Page, F. H., Lynch, D. R., Loder, J. W., Lough, R. G., Perry, R. I., Greenberg, D. A., and Sinclair, M. M. (1993). Influences of mean advection and simple behavior on the distribution of cod and haddock early life stages on Georges Bank. *Fisheries Oceanography*, **2**, 43–64.

Werner, F. E., Perry, R. I., Lough, R. G., and Naimie, C. E. (1996). Trophodynamic and advective influences on Georges Bank larval cod and haddock. *Deep-Sea Research II*, **43**, 1793–822.

Werner, F. E., Quinlan, J. A., Blanton, B. O., and Luettich, R. A. (1997). The role of hydrodynamics in explaining variability in fish populations. *Journal of Sea Research*, **37**(3–4), 195–212.

Werner, F. E., Quinlan, J. A., Lough, R. G., and Lynch, D. R. (2001). Spatially-explicit individual based modeling of marine populations: a review of the advances in the 1990s. *Sarsia*, **86**, 411–421.

Chapter 4

Ascioti, F., Beltrami, E., Carroll, T. O., and Creighton, W. (1993). Is there chaos in plankton dynamics? *Journal of Plankton Research*, **15**, 603–17.

Beaugrand, G., Ibañez, F., Reid, P. C. (2000). Spatial, seasonal and long-term fluctuations of plankton in relation to hydroclimatic features in the English Channel, Celtic Sea and Bay of Biscay. *Marine Ecology Progress Series*, **200**, 93–102.

Becker, G. and Dooley, H. (1995). The 1989/91 high salinity anomaly. *Ocean Challenge*, **6**, 52–7.

Beffy, J. L. (1992). Application de l'analyse en composante principale à trois modes pour l'etude physico-chimique d'un écosystème lacustre d'altitude: perspective en écologie. Revue de Statistique applique, **40**, 37–56.

Belgrano, A., Lindahl, O., and Hernroth, B. (1999). North Atlantic Oscillation (NAO) primary productivity and toxic phytoplankton in the Gullmar Fjord, Sweden (1985–1996). *Proceedings of the Royal Society of London, B*, **266**, 425–30.

Borkman, D. G. (2002). Analysis and simulation of *Skeletonema costatum* (Grev.) Cleve annual abundance patterns in lower Narragansett Bay 1959 to 1996. Ph.D. Dissertation. University of Rhode Island. 395 pages.

Borkman, D. G. and Smayda, T. J. (1998). Long-term trends in water clarity revealed by Secchi-disk measurments in lower Narragansett Bay. *ICES Journal of Marine Science*, **55**, 668–79.

Borkman, D. G. and Smayda, T. J. (2003a). Sea surface temperature in lower Narragansett Bay: Analysis of a 38-year time series. (In review.)

Borkman, D. G. and Smayda, T. J. (2003b). Long-term decline in the abundance of the Boreal-Arctic diatom, *Detonula confervacea* (Cleve) Gran, in Narragansett Bay linked to climate change. (In review.)

Chang, F. H., MacKenzie, D., Till, D., Hannah, D. and Rhodes, L. (1995). The first toxic shellfish outbreaks and the associated phytoplankton blooms in early 1993 in New Zealand. In *Harmful Marine Algal Blooms* (eds P. Lassus, G. Arzul, E. Erard, P. Gentien, and C. Marcaillou), pp. 145–50. Lavoisier, Intercept Ltd., Paris.

Colebrook, J. M. (1982). Continuous plankton records: phytoplankton, zooplankton, and environment, North-East Atlantic and North Sea, 1958–1980. *Oceanologica Acta*, **5**, 473–80.

Colebrook, J. M. (1991). Continuous plankton records: from seasons to decades in the plankton of the North-East Atlantic. In *Long-term variability of pelagic fish population and their environments* (eds T. Kawasaki, S. Tanaka, Y. Toba, and A. Taniguchi). Pergamon Press, Oxford.

Conover, S. (1954). Oceanography of long island sound, 1952–1954. IV. Phytoplankton. Bulletin of the Bingham *Ocenographic Collection*, **15**, 62–112.

Cook, T., Folli, M., Klinck, J., Ford, S., and Miller, J. (1998). The relationship between increasing sea-surface temperature and the northward spread of *Perkinsus marinus* (Dermo) disease epizootics in oysters. *Estuarine, Coastal and Shelf Science*, **46**, 587–97.

Deason, E. (1980). Grazing of *Acartia hudsonica* (*A. clausi*) on *Skeletonema costatum* in Narragansett Bay (USA): Influence of food concentration and temperature. *Marine Biology*, **60**, 101–13.

Dickson, R. R. (1995). The natural history of time series. In *Ecological Time Series* (eds T. M. Powell and J. H. Steele), pp. 70–98. Chapman & Hall, NY.

Dickson, R. R., Kelly, P. M., Colebrook, J. M., Wooster, W. S., Cushing, D. H. (1988). North winds and production in the eastern North Atlantic. *Journal of Plankton Research,* **10**, 151–59.

Drinkwater. (2000). Changes in ocean climate and its general effect on fisheries: examples from the North-West Atlantic. In *The ocean life of Atlantic salmon. Environmental and biological factors influencing survival* (ed. D. Mills), pp. 116–36. Fishing News Books, Bodmin.

Durbin, A. G. and Durbin, E. G. (1992). Seasonal changes in size frequency distribution and estimated age in the marine copepod *Acartia hudsonica* during a winter-spring diatom bloom in Narragansett Bay. *Limnology and Oceanography,* **37**, 379–92.

Edwards, M., John, A. W. G., Hunt, H. G., and Lindley, J. A. (1999). Exceptional influx of oceanic species into the North Sea late 1997. *Journal of Marine Biology,* **79**, 737–39.

Edwards, M., Beaugrand, G., Reid, P. C., Rowden, A. A., and Jones, M. B. (2002). Ocean climate anomalies and the ecology of the North Sea. *Marine Ecology Progress Series,* **239**, 1–10.

Erickson, G. and Nishitani, L. (1985). The possible relationship of El Niño/Southern Oscillation events to interannual variation in *Gonyaulax* populations as shown by records of shellfish toxicity. In *El Niño North: Niño Effects in the Eastern Subarctic Pacific Ocean* (eds W. S. Wooster and D. L. Fluharty), pp. 283–90. University Washington Seagrant, Seattle, WA.

Ford, S. E. (1996). Range extension by the oyster parasite *Perkinsus marinus* into the northeastern United States: response to climate change? *Journal of Shellfish Research,* **15**, 45–56.

Fraga, S. and Bakun, A. (1993). Global climate change and harmful algal blooms: the example of *Gymnodinium catenatum* on the Galician coast. In *Toxic phytoplankton blooms in the sea* (eds T. J. Smayda and Y. Shimizu), pp. 59–65. Elsevier, Amsterdam.

Fromentin, J. M. and Planque, B. (1996). *Calanus* and environment in the eastern North Atlantic. II. Influence of the North Atlantic Oscillation on *C. finmarchicus* and *C. helgolandicus*. *Marine Ecology Progress Series,* **134**, 111–8.

Hayes, M. L., Bonaventura, J., Mitchell, T. P., Prospero, J. M., Shinn, E. A., Van Dolah, F., Barber, R. T. (2001). How are climate and marine biological outbreaks functionally linked? *Hydrobiologia,* **460**, 213–20.

Hohn, M. E. (1979). Principal component analysis of three-way tables. *Mathematical Geology,* **11**, 611–26.

Hohn, M. E. (1993). Principal component analysis of three-way data. In *Computers in geology* (eds J. C. Davis and U. C. Hertzfeld), pp. 181–194. Oxford University Press, New York.

Hurrell, J. W. (1995). Decadal trends in the North Atlantic Oscillation: regional temperatures and precipitation. *Science,* **269**, 676–79.

Karentz, D. and Smayda, T. J. (1998). Temporal patterns and variations in phytoplankton community organization and abundance in Narragansett Bay during 1959–1980. *Journal of Plankton Research,* **20**, 145–68.

Keller, A. A., Oviatt, C. A., Walker, H. A., and Hawk, J. D. (1999). Predicted impacts of elevated temperature on the magnitude of the winter-spring phytoplankton bloom in temperate coastal waters: A mesocosm study. *Limnology and Oceanography,* **44**, 344–56.

Letelier, R. M. and Karl D. M. (1996). Role of *Trichodesmium* spp. in the productivity of the subtropical North Pacific Ocean. *Marine Ecology Progress Series,* **133**, 263–73.

Li, Y. and Smayda, T. J. (1998). Temporal variability of chlorophyll in Narragansett Bay, 1973–1990. *ICES Journal of Marine Science,* **55**, 661–7.

Li, Y. and Smayda, T. J. (2001). A long-term phytoplankton biomass decline in Narragansett Bay linked to climatic forcing. (In review).

Lindahl, O., Belgrano, A., Davidsson, L. and Hernroth, B. (1998). Primary production, climatic oscillations, and physico-chemical processes: the Gullmar Fjord time-series data set (1985–1996). *ICES Journal of Marine Science,* **55**, 723–9.

Lomas, M. W., Gilbert, P. M., Shiah, F. and Smith, E. M. (2002). Microbial processes and temperature in Chesapeake Bay: current relationships and potential impacts of regional warming. *Global Change Biology* **8**, 51–70.

McGowan, J. A. (1990). Climate and change in oceanic ecosystems: the value of time-series data. *TREE,* **5**, 293–9.

Maclean, J. L. (1989). An overview of *Pyrodinium* red tides in the western Pacific. In *Biology, epidemiology and management of Pyrodinium* (eds G. M. Hallegraeff and J. L. Maclean), pp. 1–8, red tides. ICLARM Conference Proceedings, ICLARM, Manila.

Margalef, R. (1978). Life-forms of phytoplankton as survival alternatives in an unstable environment. *Oceanologica Acta,* **1**, 493–509.

Marshall, J., Johnson, H. and Goodman, J. (2001). A Study of the Interaction of North Atlantic Oscillation with Ocean Circulation. *Journal of Climate [J. Clim.]. Vol. 14,* **7**, 1399–421.

Martin, J. H. (1970). Phytoplankton-zooplankton relationships in Narragansett Bay. IV. The seasonal importance of grazing. *Limnology and Oceanography,* **15**, 413–18.

Planque, B. and Reid, P. C. (1998). Predicting *Calanus finmarchicus* abundance from a climatic signal. *Journal of Marine Biology,* **78**, 1015–18.

Pratt, D. M. (1965). The winter-spring diatom flowering in Narragansett Bay. *Limnology and Oceanography,* **10**, 173–84.

Reid, P. C. and Beaugrand, G. (2002). Interregional biological responses in the North Atlantic to hydrometeorological forcing. In *Changing states of the large marine ecosystems of the North Atlantic* (eds K. Sherman and H.-R. Skjoldal), pp. 27–48. Elsevier Science.

Reid, P. C. and Planque, B. (2000). Long-term planktonic variations and the climate of the North Atlantic. In *The ocean life of Atlantic salmon. Environmental and biological factors influencing survival* (ed. D. Mills), pp. 153–69. Fishing News Books, Bodmin.

Reid, P. C., Edwards, M., Hunt, H. G., and Warner, A. 1998a. Phytoplankton change in the North Atlantic. *Nature*, **391**, 546.

Reid, P. C., Planque, B., and Edward, M. 1998b. Is observed variability in the long-term results of the Continuous Plankton Recorder survey a response to climate change? *Fisheries Oceanography*, 7, 282–8.

Reid, P., de Fatima Borges, M., and Svendsen, E. (2001). A regime shift in the North Sea circa 1988· linked to changes in the North Sea horse mackerel fishery. *Fisheries Research*, **50**(1–2), 163–71.

Rhodes, L. L., Haywood, A. J., Ballantine, W. J., and MacKenzie, A. L. (1993). Algal blooms and climate anomalies in north-east New Zealand, August–December 1992. *New Zeeland Journal of Marine and Freshwater Research*, **27**, 419–30.

Smayda, T. J. (1969). Experimental studies on the influence of temperature, light, and salinity on cell division of the marine diatom, *Detonula confervacea* (Cleve) Gran. *Journal of Phycology*, **5**, 150–7.

Smayda, T. J. (1973). The growth of *Skeletonema costatum* during a winter-spring bloom in Narragansett Bay, Rhode Island. *Norwegian Journal of Bot*any, **20**, 219–47.

Smayda, T. J. (1980). Phytoplankton species succession. In *The physiological ecology of phytoplankton* (ed. I. Morris), pp. 493–570. Blackwell Scientific Publications, Oxford.

Smayda, T. J. (1990). Novel and nuisance phytoplankton blooms in the sea: Evidence for a global epidemic. In *Toxic marine phytoplankton* (eds E. Granéli, B. Sundström, L. Edler, and D. M. Anderson), pp. 29–40. Elsevier, NY.

Smayda, T. J. (1998). Patterns of variability characterizing marine phytoplankton, with examples from Narragansett Bay. *ICES Journal of Marine Science*, **55**, 562–73.

Stein, M., Lloret, L., and Raetz, H.-J. (1998). North Atlantic Oscillation (NAO) Index—environmental variability effects on marine fisheries? *NAFO Scientific Council Research Document*, 1998, No. **98/20**, 1–6.

Taylor, A. H. and Stephens, J. A. (1998). The North Atlantic Oscillation and the latitude of the Gulf Stream. *Tellus*, **50A**, 134–42.

Taylor, A. H., Prestidge, M. C., and Allen, J. I. (1996). Modeling seasonal and year-to-year changes in the ecosystems of the NE Atlantic Ocean and the European Shelf Seas. *Journal of Advances in Marine Science and Technology Society of Japan*, **2**, 133–50.

Thompson, D. J. (1995). The seasons, global temperature, and precession. *Science*, **268**, 59–68.

Warner, A. J. and Hays, G. C. (1994). Sampling by the Continuous Plankton Recorder survey. *Progress in Oceanography*, **34**, 237–56.

Chapter 5

Astthorsson, O. S. and Gislason, A. (1995). Long-term changes in zooplankton biomass in Icelandic waters in spring. *ICES Journal of Marine Science*, **52**, 657–68.

Backhaus, J. O., Harms, I. H., Krause, M., and Heath, M. R. (1994). An hypothesis concerning the space–time succession of *Calanus finmarchicus* in the northern North Sea. *ICES Journal of Marine Science*, **51**, 169–80.

Brown, W. S. and Beardsley, R. C. (1978). Winter circulation in the western Gulf of Maine, part 1: cooling and water mass formation. *Journal of Physical Oceanography*, **8**, 265–77.

Conversi, A., Piontkovski, S., and Hameed, S. (2001). Seasonal and interannual dynamics of *Calanus finmarchicus* in the Gulf of Maine (Northeastern US shelf) with reference to the North Atlantic Oscillation. *Deep-Sea Research II*, **48**, 519–20.

Cushing, D. H. (1982). *Climate and fisheries*. Academic Press, New York, NY.

Cushing, D. H. and Dickson, R. R. (1976). The biological response in the sea to climatic changes. *Advances in Marine Biology*, **14**, 1–22.

Dalpadado, P., Ingvaldsen, R., and A. Hassel. (2003). Zooplankton biomass variation in relation to climatic conditions in the Barents Sea. *Polar Biology*, **26**, 233–41.

Dickson, R. (1997). From the Labrador Sea to global change. *Nature*, **386**, 649–50.

Dickson, R., Lazier, J., Meincke, J., Rhines, P., and Swift, J. (1996). Long-term coordinated changes in the convective activity of the North Atlantic. *Progress in Oceanography*, **38**, 241–95.

Dickson, R. R., Osborn, T. J., Hurrell, J. W., Meincke, J., Blindheim, J., Adlandsvik, B., Vinje, T., Alekseev, G., and Maslowski, W. (2000). The Arctic Ocean response to the North Atlantic Oscillation. *Journal of Climate*, **13**, 2671–96.

Fromentin, J. and Planque, B. (1996). *Calanus* and the environment in the eastern North Atlantic. II. Influence of the North Atlantic Oscillation on *C. finmarchicus* and *C. helgolandicus. Marine Ecology Progress Series*, **134**, 111–18.

Gallego, A., Mardaljevic, J., Heath, M. R., Hainbucher, D., and Slagstad, D. (1999). A model of the spring migration into the North Sea by *Calanus finmarchicus* overwintering off the Scottish continental shelf. *Fisheries Oceanography*, **8**, 107–25.

Gamble, J. C. (1994). Long-term planktonic time series as monitors of marine environmental change. In *Long-term experiments in agriculture and ecological sciences: proceedings of a conference to celebrate the 150th anniversary of Rothamsted experimental station, held at Rothamsted, 14–17 July, 1993* (eds R. A. Leigh and A. E. Johnston), pp. 365–86. CAB International, Wallingford, U.K.

Gatien, M. (1976). A study in the slope water region south of Halifax. *Journal of the Fisheries Research Board of Canada*, **33**, 2213–17.

Glover, R. S. (1957). An ecological survey of the drift net herring survey off the northeast coast of Scotland II: the planktonic environment of the herring. *Bulletin of Marine Ecology*, **5**, 1–43.

Graybill, M. R. and Hodder, J. (1985). Effects of the 1982–83 El Niño on reproduction of six species of seabirds in Oregon. In *El Niño North: El Niño effects in the Eastern Subarctic Pacific Ocean* (eds W. S. Wooster and D. L. Fluharty), pp. 205–10. Univeristy of Washington, Seattle, WA.

Greene, C. H. and Pershing, A. J. (2001). The response of *Calanus finmarchicus* populations to climate variability in the Northwest Atlantic: basin-scale forcing associated with the North Atlantic Oscillation. *ICES Journal of Marine Science*, **57**, 1536–44.

Greene, C. H. and Pershing, A. (2003). The flip-side of the North Atlantic Oscillation and modal shifts in slope-water circulation patterns. *Limnology and Oceanography*, **48**, 319–22.

Hänninen, J., Vuorinen, I., and Hjelt, P. (2000). Climatic factors in the Atlantic control the oceanographic and ecological changes in the Baltic Sea. *Limnology and Oceanography*, **45**, 703–10.

Hansen, B., Turrell, W. R., and Østerhus, S. (2001). Decreasing overflow from the Nordic seas into the Atlantic Ocean through the Faroe Bank channel since 1950. *Nature*, **411**, 927–30.

Head, E. J. H., Harris, L. R., and Petrie, B. (1999). Distribution of *Calanus* spp. on and around the Nova Scotia Shelf in April—evidence for an offshore source of *Calanus finmarchicus* to the mid- and western regions. *Canadian Journal of Fisheries and Aquatic Science*, **56**, 2463–76.

Heath, M. R. (1999). Introduction. *Fisheries Oceanography*, **8**, vii–viii.

Heath, M. R. and Jonasdottir, S. H. (1999). Distribution and abundance of overwintering *Calanus finmarchicus* in the Faroe–Shetland Channel. *Fisheries Oceanography*, **8**, 40–60.

Heath, M. R., Backhaus, J. O., Richardson, K., McKenzie, E., Slagstad, D., Beare, D., Dunn, J., Fraser, J. G., Gellego, A., Hainbucher, D., Hay, S., Jonasdottir, S., Madden, H., Mardaljevic, J., and Schacht, A. (1999). Climate fluctuations and the spring invasion of the North Sea by *Calanus finmarchicus*. *Fisheries Oceanography*, **8**(Suppl. 1), 163–76.

Heath, M. R., Fraser, J. G., Gislason, A., Hay, S. J., Jonasdottir, S. H., and Richardson, K. (2000). Winter distribution of *Calanus finmarchicus* in the Northeast Atlantic. *ICES Journal of Marine Science*, **7**, 1628–35.

Helle, K. and Pennington, M. (1999). The relation of the spatial distribution of early juvenile cod (*Gadus morha* L.) in the Barents Sea to zooplankton density and water flux during the period 1978–84. *ICES Journal of Marine Science*, **56**, 15–27.

Holliday, N. P. and Reid, P. C. (2001). Is there a connection between high transport of water through the Rockall Trough and ecological changes in the North Sea? *ICES Journal of Marine Science*, **58**, 270–4.

Hirche, H.-J. (1996). The reproductive biology of the marine copepod *Calanus finmarchicus*—a review. *Ophelia*, **44**, 111–28.

Hurrell, J. (1995). Decadal trends in the North Atlantic Oscillation: regional temperatures and precipitation. *Science*, **269**, 676–9.

Hurrell, J. W., Kushnir, Y., Ottersen, G., and Visbeck, M. (2003). The North Atlantic Oscillation: climatic significance and environmental impact. *Geophysical Monograph Series*, **134**. American Geophysical Union, Washington, 279 + viii pp.

Jossi, J. and Goulet, J. (1993). Zooplankton trends: US north-east shelf ecosystem and adjacent regions differ from north-east Atlantic and North Sea. *ICES Journal of Marine Science*, **50**, 303–13.

Loder, J. W., Shore, J. A., Hannah, C. G., and Petrie, B. D. (2001). Decadal-scale hydrographic and circulation variability in the Scotia-Maine region. *Deep-Sea Research II*, **48**, 3–36.

Mann, K. H. and Lazier, J. R. N. (1996). *Dynamics of marine ecosystems: biological-physical interactions in the oceans*. Blackwell Science, Cambridge, MA.

Marshall, S. M. and Orr, A. P. (1955). *The biology of a marine copepod*. Oliver and Boyd, Edinburgh.

MERCINA (2001). Gulf of Maine/Western Scotian Shelf ecosystems respond to changes in ocean circulation associated with the North Atlantic Oscillation. *Oceanography*, **14**, 76–82.

Miller, C. B., Cowles, T. J., Wiebe, P. H., Copley, N. J., and Grigg, H. (1991). Phenology of *Calanus finmarchicus*—hypotheses about control mechanisms. *Marine Ecology Progress Series*, **72**, 79–91.

Miller, C. B., Lynch, D. R., Carlotti, F., Gentleman, W. C., and Lewis, C. (1998). Coupling of an individual-based population dynamical model for stocks of *Calanus finmarchicus* with a circulation model for the Georges Bank region. *Fisheries Oceanography*, **8**, 219–34.

Mills, E. L. (1989). *Biological oceanography: an early history, 1870–1960*. Cornell University Press, Ithaca, NY.

Ottersen, G. and Stenson, G. B. (2001). Atlantic climate governs oceanographic and ecological variability in the Barents Sea. *Limnology and Oceanography*, **46**, 1774–80.

Ottersen, G., Planque, B., Belgrano, A., Post, E., Reid, P. C., and Stenseth, N. C. (2001). Ecological effects of the North Atlantic Oscillation. *Oecologia*, **128**, 1–14.

Pershing, A. J. (2001). *Response of large marine ecosystems to climate variability: patterns, processes, concepts and methods*. Ph.D. Thesis, Cornell University, Ithaca, NY.

Petrie, B. and Yeats, P. (2000). Annual and interannual variability of nutrients and their estimated fuxes in the Scotian Shelf-Gulf of Maine region. *Canadian Journal of Fisheries and Aquatic Science*, **57**, 2536–46.

Petrie, B. D. and Drinkwater, K. (1993). Temperature and salinity variability on the Scotian Shelf and in the Gulf of Maine, 1945–90. *Journal of Geophysical Research*, **98**, 20,079–89.

Pickart, R., McKee, T., Torres, D., and Harrington, S. (1999). Mean structure and interannual variability of the slopewater system south of Newfoundland. *Journal of Physical Oceanography*, **29**, 2541–58.

Planque, B. and Batten, S. D. (2000). *Calanus finmarchicus* in the North Atlantic: the year of *Calanus* in the context of interdecadal change. *ICES Journal of Marine Science*, **57**, 1528–35.

Planque, B. and Reid, P. C. (1998). Predicting *Calanus finmarchicus* abundance from a climate signal. *Journal of the Marine Biological Association UK*, **78**, 1015–18.

Planque, B. and Taylor, A. H. (1998). Long-term changes in zooplankton and the climate of the North Atlantic. *ICES Journal of Marine Science*, **55**, 644–54.

Reid, P. C. and Planque, B. (1999). Long-term planktonic variations and the climate of the North Atlantic. In *The Life of the Atlantic Salmon* (ed. D. Mills), pp. 153–69. Blackwell Science, Oxford.

Reid, P. C., Borges, M. D., and Svendsen, E. (2001). A regime shift in the North Sea circa 1988 linked to changes in the North Sea fishery. *Fisheries Research*, **50**, 163–71.

Sameoto, D. D. and Herman, A. W. (1990). Life cycle and distribution of *Calanus finmarchicus* in deep basins on the Nova Scotia shelf and seasonal changes in *Calanus* spp. *Marine Ecology Progress Series*, **66**, 225–37.

Skjoldal, H. R., Gjøsæter, H., and Loeng, H. (1992). The Barents Sea ecosystem in the 1980s: ocean climate, plankton, and capelin growth. *ICES Marine Science Symposia*, **195**, 278–90.

Stephens, J. A., Jordan, M. B., Taylor, A. H., and Proctor, R. (1998). The effects of fluctuations in North Sea flows on zooplankton abundance. *Journal of Plankton Research*, **20**, 943–56.

Taylor, A. H. and Stephens, J. A. (1998). The North Atlantic Oscillation and the latitude of the Gulf Stream. *Tellus*, **50**, 134–42.

Ulbrich, U. and Christoph, M. (1999). A shift of the NAO and increasing storm track activity over Europe due to anthropogenic greenhouse gas forcing. *Climate Dynamics*, **15**, 551–9.

Viitasalo, M., Vuorinen, I., and Saesmaa, S. (1995). Mesozooplankton dynamics in the northern Baltic Sea: implications of variations in hydrography and climate. *Journal of Plankton Research*, **17**, 1857–78.

Williams, R., Conway, D. V. P., and Hunt, H. G. (1994). The role of copepods in the planktonic ecosystems of mixed and stratified waters of the European shelf seas. *Hydrobiologia*, **293**, 521–30.

Chapter 6

Alheit, J. and Hagen, E. (1997). Long-term climate forcing of European herring and sardine populations. *Fisheries Oceanography*, **6**, 130–9.

Alheit, J. and Hagen, E. (2001). The effect of climatic variation on pelagic fish and fisheries. In *History and climate memories of the future* (eds P. D. Jones, A. E. J. Ogilvie, T. D. Davies, and K. R. Briffa), pp. 247–65. Kluwer Academic/Plenum Publishers, New York.

Alheit, J. and Hagen, E. (2002). Climate variability and historical NW European fisheries. In *Climate and history in the North Atlantic realms.* (eds G. Wefer, W. Berger, K.-E. Behre, and E. Jansen). Springer-Verlag, Berlin.

Andersen, N. G., Ottersen, G., and Swain, D. (2002). *Report of the ICES/GLOBEC workshop on the dynamics of growth in cod*. ICES Cooperative Research Report 252, ICES, Copenhagen.

Anon. (2002), Report of the North-Western Working Group. *ICES CM Doc., No. ACFM: 20*, 416 p.

Antonsson, T., Gudbergsson, G., and Gudjonsson, S. (1996). Environmental continuity in fluctuation of fish stocks in the North Atlantic Ocean, with particular reference to Atlantic salmon. *North American Journal of Fisheries Management*, **16**, 540–7.

Bagge, O. and Thurow, F. (1994). The Baltic cod stock: fluctuations and possible causes. *ICES Marine Science Symposia*, **198**, 254–68.

Bakun, A., Beyer, J., Pauly, D., Pope, J. G., and Sharp, G. D. (1982). Ocean sciences in relation to living marine resources: a report. *Canadian Journal of Fisheries and Aquatic Sciences*, **39**, 1059–70.

Bardonnet, A. and Baglinière, J.-L. (2000). Freshwater habitat of Atlantic salmon (*Salmo salar*). *Canadian Journal of Fisheries and Aquatic Sciences*, **57**, 497–506.

Beverton, R. J. H. and Lee, A. J. (1965). Hydrodynamic fluctuations in the North Atlantic Ocean and some biological consequences. In *The biological significance of climatic changes in Britain* (eds C. G. Johnson and L. P. Smith), pp. 79–107. Academic Press, New York.

Binet, D. (1988). French sardine and herring fisheries: a tentative description of their fluctuations since the eighteenth century. In *Long term changes in marine fish populations* (eds T. Wyatt and M. G. Larraneta), pp. 253–72. Proceedings of an international Symposium, 18–21 November 1986, Vigo, Spain.

Bjørnsson, B. and Steinarsson, A. (2000). Growth potential of cod (*Gadus morhua* L.) fed on maximum rations: effects of temperature and size. In *Report of the ICES/GLOBEC Workshop on the Dynamics of Growth of Cod, ICES Council Meeting 2000:C:12*.

Boeck, A. (1871). *Om silden og sildefiskerierne, navnlig om det norske vaarsildfiske. Indberetning til Departementat for det Indre Christiana* (in Norwegian) (Government Report).

Bonardelli, J. C., Himmelman, J. H., and Drinkwater, K. (1996). Relation of spawning of the giant scallop, *Placopecten magellanicus*, to temperature fluctuations during downwelling events. *Marine Biology*, **124**, 637–49.

Borja, A. and Santiago, J. (2001). *Does the North Atlantic Oscillation control some processes influencing recruitment of temperate tunas?* ICCAT SCRS/01/33.

Borovkov, V. A. and Stein, M. (2001). Recruitment of West Greenland cod—modelling different cause-effect regimes. *NAFO SCR Doc.*, Serial No. N4373. 8 p.

Brander, K. (1993). Comparison of spawning characteristics of cod (*Gadus morhua*) stocks in the North Atlantic. *NAFO Scientific Council Studies*, **18**, 13–20.

Brander, K., Dickson, R. R., and Shepherd, J. G. (2001). Modelling the timing of plankton production and its effect on recruitment of cod (*Gadus morhua*). *ICES Journal of Marine Science*, **58**, 962–6.

Brander, K. M. (1994). Patterns of distribution, spawning, and growth in North Atlantic cod: the utility of interregional comparisons. *ICES Marine Science Symposia*, **198**, 406–13.

Brander, K. M. (1995). The effects of temperature on growth of Atlantic cod (*Gadus morhua* L.). *ICES Journal of Marine Science*, **52**, 1–10.

Brander, K. M. (2000). Effects of environmental variability on growth and recruitment in cod (*Gadus morhua*) using a comparative approach. *Oceanologica Acta*, **23**, 485–96.

Brett, J. R. (1979). Environmental factors and growth. *In Fish physiology*. Vol. VIII. Bioenergetics and Growth (*eds* W. S. Hoar, D. J. Randall, and J. R. Brett), pp. 599–675. Academic Press, New York.

Brodie, W. B. (1987). American plaice in divisions 3LNO—an assessment update. *NAFO Scientific Council Research Document*, **87**(40), 42 pp.

Brodie, W. B., Power, D., and Morgan, M. J. (1993). An assessment of the American plaice stock in NAFO Divisions 3LNO. *NAFO Scientific Council Research Document*, **93**(91), 60 pp.

Buch, E., Horsted, S. A., and Hovgård, H. (1994). Fluctuations in the occurrence of cod in the Greenland waters and their possible causes. *ICES Marine Science Symposia*, **198**, 158–74.

Buch, E., Nielsen, M. H., and Pedersen, S. A. (2002). Ecosystem variability and regime shift in Western Greenland waters. *NAFO Scr. Doc.* Serial No. N4617, 19 p.

Campana, S. E., Mohn, R. K., Smith, S. J., and Chouinard, G. (1995). Spatial visualization of a temperature-based growth model for Atlantic cod (*Gadus morhua*) off the eastern coast of Canada. *Canadian Journal of Fisheries and Aquatic Sciences*, **52**, 2445–56.

Chadwick, E. M. P. (1987). Causes of variable recruitment in a small Atlantic Salmon Stock. *American Fisheries Society Symposium*, **1**, 390–401.

Colton, J. B. Jr. (1959). A field observation of mortality of marine fish larvae due to warming. *Limnology and Oceanography*, **4**, 219–22.

Colton, J. B. Jr. (1972). Temperature trends and the distribution of groundfish in continental shelf waters, Nova Scotia to Long Island. *Fisheries Bulletin*, **70**, 637–57.

Corten, A. (1990). Long-term trends in pelagic fish stocks of the North Sea and adjacent waters and their possible connection to hydrographic changes. *Netherlands Journal of Sea Research*, **25**, 227–35.

Corten, A. and van de Kamp, G. (1992). Natural changes in pelagic fish stocks of the North Sea in the 1980s. *ICES Marine Science Symposia*, **195**, 402–17.

Coutant, C. C. (1977). Compilation of temperature preference data. *Journal of the Fisheries Research Board of Canada*, **34**, 739–45.

Cunjak, R. A. (1988). Behavior and microhabitat of young Atlantic salmon (*Salmo salar*) during winter. *Canadian Journal of Fisheries and Aquatic Sciences*, **45**, 2156–2160.

Cunjak, R. A., Prowse, T. D., and Parrish, D. L. (1998). Atlantic salmon (*Salmo salar*) in winter: 'the season of parr discontent'? *Canadian Journal of Fisheries and Aquatic Sciences*, **55**, 161–80, Suppl. 1.

Cushing, D. H. (1966). Biological and hydrographic changes in British Seas during the last thirty years. *Biological Review*, **41**, 221–58.

Cushing, D. H. (1969). The regularity of the spawning season of some fishes. *Journal Conseil International pour l'Exploration de la Mer*, **185**, 201–13.

Cushing, D. H. (1974). The possible density-dependence of larval mortality in fishes. In *The early life history of fish* (ed. J. H. S. Blaxter), pp. 103–11. Springer-Verlag, Berlin.

Cushing, D. H. (1982). *Climate and fisheries*. Academic Press, London.

Cushing, D. H. (1984). The gadoid outburst in the North Sea. *Journal Conseil International pour l'Exploration de la Mer*, **41**, 159–66.

Cushing, D. H. (1990). Plankton production and year-class strength in fish populations: an update of the match/mismatch hypothesis. *Advances in Marine Biology*, **26**, 249–93.

Cushing, D. H. and R. R. Dickson. (1976). The biological response in the sea to climatic changes. *Advances in Marine Biology*, **14**, 1–22.

Dalley, E. L., Anderson, J. T., and Davis, D. J. (2000). Short term fluctuations in the pelagic ecosystem of the Northwest Atlantic. *DFO Canadian Stock Assessment Sec. Research Document* 2000/101, 36 p.

de Cárdenas, E. (1996). Some considerations about annual growth rate variations in cod stocks. *NAFO Scientific Council Studies*, **24**, 97–107.

deYoung, B. and Davidson, F. (1994). Modelling retention of cod eggs and larvae (*Gadus morhua* L.) on the Newfoundland shelf. *ICES Marine Science Symposia*, **198**, 346–55.

deYoung, B. and Rose, G. A. (1993). On recruitment and distribution of Atlantic cod (*Gadus morhua*) off Newfoundland. *Canadian Journal of Fisheries and Aquatic Sciences*, **50**, 2729–41.

Devold, F. (1963). The life history of the Atlanto-Scandian herring. *Rapports et procès-verbaux des réunions, Conseil International pour l'Exploration de la Mer*, **154**, 98–108.

Dickson, R. R. and Brander, K. M. (1993). Effects of a changing windfield on cod stocks of the North Atlantic. *Fisheries Oceanography*, **2**, 124–53.

Dickson, R. R. and Turrell, W. R. (2000). The NAO: the dominant atmospheric process affecting oceanic variability in home, middle and distant waters of European salmon. In *The ocean life of Atlantic Salmon* (ed. D. Mills), pp. 92–115. Fishing News Books, London.

Dickson, R. R., Pope, J. G., and Holden, M. J. (1973). Environmental influences on the survival of North Sea cod. In *The early life history of fish* (ed. F. H. S. Blaxter), 69–80. Springer-Verlag, Heidelberg and New York.

Dickson, R. R., Briffa, K. R., and Osborn, T. J. (1994). Cod and climate: the spatial and temporal context. *ICES Marine Science Symposia*, **198**, 280–86.

Dippner, J. W. (1997). Recruitment success of different fish stocks in the North Sea in relation to climate variability. *Deutsche Hydrographische Zeitschrift*, **49** (2–3), 277–93.

Dippner, J. and Ottersen, G. (2001). Cod and climate variability in the Barents Sea. *Climate Research*, **17**, 73–82.

Dragesund, O., Hamre, J., and Ulltang, O. (1980). Biology and population dynamics of the Norwegian spring spawning herring. *Rapports et procès-verbaux des réunions, Conseil International pour l'Exploration de la Mer*, **177**, 43–71.

Drinkwater, K. F. (1999). Changes in ocean climate and its general effect on fisheries: examples from the North-west Atlantic. In *The ocean life of Atlantic Salmon-environmental and biological factors influencing survival* (ed. D. Mills), pp. 116–136. Fishing News Books, Oxford, UK.

Drinkwater, K. F. (2002). A review of the role of climate variability in the decline of northern cod. *American Fisheries Society Symposia*, **32**, 113–30.

Drinkwater, K. F. and Myers, R. A. (1987). Testing predictions of marine fish and shellfish landings from environmental variables. *Canadian Journal of Fisheries and Aquatic Sciences*, **44**, 1568–73.

Drinkwater, K. F. and Myers, R. A. (1997). Interannual variability in the atmospheric and oceanographic conditions in the Labrador Sea and their association with the North Atlantic Oscillation. *Working paper for the ICES/GLOBEC Workshop on prediction and decadal-scale ocean climate fluctuations for the North Atlantic*, Copenhagen.

Drinkwater, K. F., Harding, G. C., Mann, K. H., and Tanner, N. (1996). Temperature as a possible factor in the increased abundance of American lobster, *Homarus americanus*, during the 1980s and early 1990s. *Fisheries Oceanography*, **5**, 176–93.

Drinkwater, K. F., Lochman, S., Taggart, C., Thompson, K., and Frank, K. (2000). Entrainment of redfish (*Sebastes* sp.) larvae off the Scotian Shelf. *ICES Journal of Marine Science*, **57**, 372–82.

Drinkwater, K., Belgrano A, Borja, A., Conversi, A., Edwards, M., Greene, C., Ottersen, G., Pershing, A., and Walker, H. (2003). The response of marine ecosystems to climate variability associated with the North Atlantic Oscillation. In *The North Atlantic Oscillation: climatic significance and environmental impact* (eds J. Hurrell, Y. Kushnir, G. Ottersen, and M. Visbeck), pp. 211–34. American Geophysical Union.

Duston, J. and Saunders, R. L. (1999). Effect of winter food deprivation on growth and sexual maturity of Atlantic salmon (*Salmo salar*) in seawater. *Canadian Journal of Fisheries and Aquatic Science*, **56**, 201–7.

Dutil, J.-D. and Coutu, J.-M. (1988). Early marine life of Atlantic salmon, *Salmo salar*, postsmolts in the northern Gulf of St Lawrence. *Fisheries Bulletin*, **86**(2), 197–212.

Eggvin, J. (1932). Vannlagene på fiskefeltene (in Norwegian). *Årsberetn. vedk. Norges fiskerier*, **2**, 90–5.

Ellertsen, B., Fossum, P., Solemdal, P., and Sundby, S. (1989). Relation between temperature and survival of eggs and first-feeding larvae of northeast Arctic cod (*Gadus morhua* L.). *Rapports et procès-verbaux des réunions, Conseil International pour l'Exploration de la Mer*, **191**, 209–19.

Fiksen, O., Utne, A. C. W., Aksnes, D. L., Eiane, K., Helvik, J. V., and Sundby, S. (1998). Modelling the influence of light, turbulence and ontogeny on ingestion rates in larval cod and herring. *Fisheries Oceanography*, **7**(3–4), 355–63.

Fleming, A. M. (1960). Age, growth and sexual maturity of cod (*Gadus morhua* L.) in the Newfoundland area, 1947–1950. *Journal of Fisheries Research Board of Canada*, **17**, 775–809.

Frank, K. T., Perry, R. I., and Drinkwater, K. F. (1990). Predicted response of northwest Atlantic invertebrate and fish stocks to CO_2-induced climate change. *Transactions of the American Fisheries Society*, **119**, 353–65.

Frank, K. T., Carscadden, J. E., and Simon, J. E. (1996). Recent excursions of capelin (*Mallotus villosus*) to the Scotian Shelf and Flemish Cap during anomalous hydrographic conditions. *Canadian Journal of Fisheries and Aquatic Science*, **53**, 1473–86.

Friedland, K. D. (1998). Ocean climate influences on critical Atlantic salmon (*Salmo salar*) life history events. *Canadian Journal of Fisheries and Aquatic Science*, **55**(Suppl. 1), 119–30.

Friedland, K. D. and Haas, L. W. (1988). Emigration of juvenile Atlantic menhaden, *Brevoortia tyrannus* (Pisces: Clupeidae), from the York River estuary. *Estuaries*, **11**, 45–50.

Friedland, K. D. and Haas, R. E. (1996). Marine post-smolt growth and age at maturity of Atlantic salmon. *Journal of Fish Biology*, **48**, 1–15.

Friedland, K. D. and Reddin, D. G. (2000). Growth patterns of Labrador Sea Atlantic salmon post-smolts and the temporal scale of recruitment synchrony for North American salmon stocks. *Canadian Journal Fisheries and Aquatic Sciences*, **57**, 1181–89.

Friedland, K. D., Reddin, D. G., and Kocik, J. F. (1993). Marine survival of North American and European Atlantic salmon: effects of growth and environment. *ICES Journal of Marine Science*, **50**, 481–92.

Friedland, K. D., Haas, R. E., and Sheehan, T. S. (1996). Post-smolt growth, maturation, and survival of two stocks of Atlantic salmon. *Fisheries Bulletin*, **94**, 654–63.

Friedland, K. D., Hansen, L. P., and Dunkley, D. A. (1998a). Marine temperatures experienced by post-smolts and the survival of Atlantic salmon (*Salmo salar* L.) in the North Sea area. *Fisheries Oceanography*, **7**, 22–34.

Friedland, K. D., Reddin, D. G., Shimizu, N., Haas, R. E., and Youngson, A. F. (1998b). Strontium:Calcium ratios in Atlantic salmon otoliths and observations on growth and maturation. *Canadian Journal of Fisheries and Aquatic Sciences*, **55**, 1158–68.

Friedland, K. D., Dutil, J.-D., and Sadusky, T. (1999). Growth patterns in postmolts and the nature of the marine juvenile nursery for Atlantic salmon, *Salmo salar*. *Fisheries Bulletin*, **97**, 472–81.

Friedland, K. D., Hansen, L. P., Dunkley, D. A., and MacLean, J. C. (2000). Linkage between ocean climate, post-smolt growth, and survival of Atlantic salmon (*Salmo salar* L.) in the North Sea area. *ICES Journal of Marine Science*, **57**, 419–29.

Friedland, K. D., Reddin, D. G., and Castonguay, M. (2003a). Ocean thermal conditions in the post-smolt nursery of North American Atlantic salmon. *Journal of Marine Science*, **60**, 343–55.

Friedland, K. D., Reddin, D. G., McMenemy, J. R., and Drinkwater, K. F. (2003b). Multidecadal trends in North American Atlantic salmon (*Salmo salar*) stocks and climate trends relevant to juvenile survival. *Canadian Journal of Fisheries and Aquatic Sciences*, **60**: 563–83.

Fromentin, F.-M. and Planque, B. (1996). Calanus and environment in the eastern North Atlantic. II. Influence of the North Atlantic Oscillation on C. Finmarchicus and C. Felgolandicus. *Marine Ecology Progress Series*, **134**, 111–18.

Fry, F. E. J. (1971). The effect of environmental factors on the physiology of fish. In *Fish physiology*, Vol VI. (eds W. S. Hoar, and D. J. Randall), pp. 1–98. Academic Press, London.

Ghent, A. W. and Hanna, B. P. (1999). Statistical assessment of huntsman 3-y salmon-rainfall correlation, and other potential correlations, in the Miramichi fishery, New Brunswick. *American Midland Naturalist*, **142**, 110–28.

Gjerde, B. (1984). Response to individual selection for age at sexual maturity in Atlantic salmon. *Aquaculture*, **38**, 229–40.

Gjøsæter, H. and Loeng, H. (1987). Growth of the Barents Sea capelin, *Mallotus villosus*, in relation to climate. *Environmental Biology of Fishes*, **20**, 293–300.

Goddard, S. V., Kao, M. H., and Fletcher, G. L. (1999). Population differences in antifreeze production cycles of juvenile Atlantic cod (*Gadus morhua*) reflect adaptations to overwintering environment. *Canadian Journal of Fisheries and Aquatic Sciences*, **56**(11), 1991–99.

Gomes, M. C., Haedrich, R. L., and Villagarcia, M. G. (1995). Spatial and temporal changes in the groundfish assemblages on the north-east Newfoundland/Labrador Shelf, north-west Atlantic, 1978–91. *Fisheries Oceanography*, **4**, 85–101.

Gudjonsson, S., Einarsson, S. M., Antonsson, T., and Gudbergsson, G. (1995). Relation of grilse to salmon ratio to environmental changes in several wild stocks of Atlantic salmon (*Salmo salar*) in Iceland. *Canadian Journal of Fisheries and Aquatic Sciences*, **52**, 1385–98.

Guisande, C., Cabanas, J. M., Vergara, A. R., and Riveiro, I. (2001). Effect of climate on recruitment success of Atlantic Iberian sardine *Sardina pilchardus*. *Marine Ecology Progress Series*, **223**, 243–50.

Gulland, J. A. (1965). Survival of the youngest stages of fish and its relation to year-class strength. *ICNAF Special Publications*, **6**, 363–71.

Hänninen, J., Vuorinen, I., and Hjelt, P. (1999). Climatic factors in the Atlantic control the oceanographic and ecological changes in the Baltic Sea. *Limnology and Oceanography*, **45**(3), 703–10.

Hansen, P. M. (1949). Studies on the biology of the cod in Greenland waters. *Rapports et procès-verbaux des réunions, Conseil International pour l'Exploration de la Mer*, **123**, 1–83.

Hansen, H. and Buch, E. (1986). Prediction of year-class strength of Atlantic cod off West Greenland. *NAFO Scientific Council Studies*, **10**, 7–11.

Hansen, L. P. and Jacobsen, J. A. (2000). Distribution and migration of Atlantic salmon, *Salmo salar* L., in the sea. In *The ocean life of Atlantic salmon* (ed. D. Mills), pp. 75–87. Fishing News Books, London.

Hansen, B., Gaard, E., and Reinert, J. (1994). Physical effects on the recruitment of Faroe Plateau cod. *ICES Marine Science Symposia*, **198**, 520–8.

Helbig, J. H., Mertz, G., and Pepin, P. (1992). Environmental influences on the recruitment of Newfoundland/Lab-rador cod. *Fisheries Oceanography*, **1**, 39–56.

Helland-Hansen, B. and Nansen, F. (1909). The Norwegian Sea. *Fiskeridirektoratets Skrifter, Serie Havundersøkelser*, **2**, 1–360.

Hermann, F. (1953). Influence of temperature on strength of cod year-classes. *Annals of Biology, Copenhagen*, **9**, 1–31.

Hermann, F., Hansen, P. M., and Aa., H. S. (1965). The effect of temperature and currents on the distribution and survival of cod larvae at West Greenland. *ICNAF Special Pubications*, **6**, 389–95.

Hinrichsen, H. H., John, M. S., Aro, E., Gronkjær, P., and Voss, R. (2001). Testing the larval drift hypothesis in the Baltic Sea: retention versus dispersion caused by wind-driven circulation. *ICES Journal of Marine Science*, **58**, 973–84.

Hjort, J. (1914). Fluctuations in the great fisheries of Northern Europe viewed in the light of biological research. Rapport et Proces-verbaux des Reunions du Council international pour l'Exploration de la mer. **20**, 1–228.

Holm, M., Holst, J. C., and Hansen, L. P. (2000). Spatial and temporal distribution of post-smolts of Atlantic salmon (*Salmo salar* L.) in the Norwegian Sea and adjacent areas. *ICES Journal of Marine Science*, **57**, 955-64.

Horsted, S. A. (2000) A review of the cod fisheries at Greenland, 1910–95. *Journal of Northwest Atlantic Fisheries Science*, **28**, 1–109.

Houde, E. D. (1989). Comparative growth, mortality and energetics of marine fish larvae: temperature and implied latitudinal effects. *Fisheries Bulletin*, **87**, 471–96.

Houde, E. D. (1990). Temperature-dependent and size-dependent variability in vital rates of marine fish larvae. *ICES Council Meeting*, **90**/L:3.

Hunter, J. R., and Alheit, J. (1995). International GLOBEC small pelagic fishes and climate change program. *GLOBEC Report*, No. **8**, 72.

Huntsman, A. G. (1938). North Amerinca Atlantic salmon. *Rapports et procès-verbaux des réunions, Conseil International pour l'Exploration de la Mer*, **101**, 11–15.

Hutchings, J. A. and Myers, R. A. (1994). Timing of cod reproduction: interannual variability and the influence of temperature. *Marine Ecology Progress Series*, **108**, 21–31.

Höglund, H. (1978). Long-term variations in the Swedish herring fishery off Bohuslän and their relation to North Sea herring. *Rapports et procès-verbaux des réunions, Conseil International pour l'Exploration de la Mer*, **172**, 175–86.

ICES (1979). *ICES Counsil Meeting*, 1979/G:10.

ICES (1991). Report of the ICES study group on cod stock fluctuations. *ICES Counsil Meeting*, 1991/G:78.

ICES (1997). ICES/Assess:6.

ICES (2000). Report of the working group on the assessment of demersal stocks in the North Sea and Skagerrak, *ICES Council Meeting*, 2000/ACFM:7, ICES HQ, 11–20 October 1999.

Izhevskii, G. K. (1964). *Forecasting of oceanological conditions and the reproduction of commercial fish*. Moskva. Pishcepromizdat, Moscow.

Jacobsen, J. A. and Hansen, L. P. (2000). Feeding habits of Atlantic salmon at different life stages at sea. In *The ocean life of Atlantic salmon* (ed. D. Mills), pp. 170–92. Fishing News Books, London

Jakobsson, J. (1992). Recent variability in the fisheries of the North Atlantic. *ICES Marine Science Symposia*, **195**, 291–315.

Jahnke, C. (1997). *Heringsfang und -handel im Ostseeraum des Mittelalters*. Thesis, Kiel University, Kiel.

Jarre-Teichmann, A., Wieland, K. MacKenzie, B. R., Hinrichsen, H. H., Plikshs, M., and Aro, E. (2000). Stock-recruitment relationships for cod (*Gadus morhua* L.) in the central Baltic Sea incorporating environmental variability. *Archive of Fishery and Marine Research*, **48**(2), 97–123.

Jensen, Ad. S. and Hansen, P. M. (1931). Investigations on the Greenland cod (*Gadus callarias* L.). *Rapports et procès-verbaux des réunions*, **72**, 1–41.

Jobling, M. (1988). A review of the physiological and nutritional energetics of cod, *Gadus morhua* L., with particular reference to growth under farmed conditions. *Aquaculture*, **70**(1–2), 1–19.

Jonsson, B. and Ruud-Hansen, J. (1985). Water temperature as the primary influence on timing seaward migrations of Atlantic salmon (*Salmo salar*) smolts. *Canadian Journal of Fisheries and Aquatic Sciences*, **42**, 593–5.

Juanes, F., Letcher, B. H., and Gries, G. (2000). Ecology of stream fish: insights gained, from an individual-based approach to juvenile Atlantic salmon. *Ecology of Freshwater Fish*, **9**, 65–73.

Kendall, A. W. J. and Duker, G. J. (1998). The development of recruitment fisheries oceanography in the United States. *Fisheries Oceangraphy*, **7**(2), 69–88.

Klyashtorin, L. B. (1998). Long-term climate change and main commercial fish production ion the Atlantic and Pacific. *Fisheries Research*, **37**, 115–25.

Köster, F., Hinrichsen, H., St John, M., Schnack, D., MacKenzie, B., Tomkiewicz, J., and Plikshs, M. (2001). Developing Baltic cod recruitment models. II. Incorporation of environmental variability and species interaction. *Canadian Journal of Fisheries and Aquatic Sciences*, **58**, 1534–56.

Krohn, M. and Kerr, S. (1996). Declining weight-at-age in northern cod and the potential importance of the early-years and size-selective fishing mortality. *NAFO Scientific Council Research Document*, **96/56**, Serial No. N2732.

Kurlansky, M. (1997). *Cod. A biography of the fish that changed the world*. Jonathan Cape, London.

Lauzier, L. M. and Tibbo, S. N. (1965). Water temperature and the herring fishery of Magdalen Islands, Quebec. *ICNAF Special Publications*, **6**, 591–6.

Leggett, W. C. and Whitney, R. R. (1972). Water temperature and the migrations of American shad. *Fisheries Bulletin*, **70**, 659–70.

Lilly, G. R. (1994). Predation by Atlantic cod on capelin on the southern Labrador and Northeast Newfoundland shelves during a period of changing spatial distributions. *ICES Marine Science Symposia*, **198**, 600–11.

Lilly, G. R., Hop, H., Stansbury, D. E., and Bishop, C. A. (1994). Distribution and abundance of polar cod (*Boreogadus saida*) off southern Labrador and eastern Newfoundland. *ICES Counsil Meeting*, 1994/O:6.

Lindquist, A. (1983). Herring and sprat: fishery independent variations in abundance. *FAO Fisheries Report*, **291**, 813–21.

Loeng, H., Bjørke, H., and Ottersen, G. (1995). Larval fish growth in the Barents Sea. *Canadian Special Publication of Fisheries and Aquatic Science*, **121**, 691–8.

Lough, R. G., Smith, W. G., Werner, F. E., Loder, J. W., Page, F. H., Hannah, C. G., Naimie, C. E., Perry, R. I., Sinclair, M., and Lynch, D. R. (1994). Influence of wind-driven advection on interannual variability in cod egg and larval distributions on Georges Bank: 1982 vs 1985. *ICES Marine Science Symposia*, **198**, 356–78.

MacKenzie, B. R. and Leggett, W. C. (1991). Quantifying the contribution of small-scale turbulence to the encounter rates between larval fish and their zooplankton prey: effects of wind and tide. *Marine Ecology Progress Series*, **73**, 149–60.

MacKenzie, B. R., Miller, T. J., Cyr, S., and Leggett, W. C. (1994). Evidence for a dome-shaped relationship between turbulence and larval fish ingestion rates. *Limnology and Oceanography*, **39**, 1790–9.

MacKenzie, B. R., Hinrichsen, H., Plikshs, M., Wieland, K., and Zezera, A. S. (2000). Quantifying environmental heterogeneity: habitat size necessary for successfull development of cod *gadus morhua* eggs in the Baltic Sea. *Marine Ecology Progress Series*, **193**, 143–56.

Malmberg, S.-A. and Blindheim, J. (1994). Climate, cod, and capelin in northern waters. *ICES Marine Science Symposia*, **198**, 297–310.

Mann, K. H. and Drinkwater, K. F. (1994). Environmental influences on fish and shellfish production in the Northwest Atlantic. *Environmental Reviews*, **2**, 16–32.

Marak, R. R. and Livingstone, R. Jr. (1970). Spawning date of Georges Bank haddock. *ICNAF Research Bulletin*, **7**, 56–8.

Marsh, R., Petrie, B., Weidman, C. R., Dickson, R. R., Loder, J. W., Hannah, C. G., Frank, K., and Drinkwater, K. (1999). The Middle Atlantic Bight tilefish kill of 1882. *Fisheries Oceanography*, **8**, 39–49.

Martin, W. R. and Kohler, A. C. (1965). Variation in recruitment of cod (*Gadus morhua* L.) in southern ICNAF waters, as related to environmental changes. *ICNAF Special Publications*, **6**, 833–46.

Martin, J. H. A. and Mitchell, K. A. (1985). Influence of sea temperature upon the numbers of grilse and multi-sea winter Atlantic salmon (*Salmo salar*) caught in the vicinity of the River Dee (Aberdeenshire). *Canadian Journal of Fisheries and Aquatic Science*, **42**, 1513–21.

McKenzie, R. A. (1934). Cod and water temperature. *Biological Board of Canada, Atlantic Progress Report*, **12**, 3–6.

McKenzie, R. A. (1938). Cod take smaller bites in ice-cold water. *Fisheries Research Board of Canada, Atlantic Progress Report*, **22**, 12–14.

McLeese, D. W. and Wilder, D. G. (1958). The activity and catchability of the lobster (*Homarus americanus*) in relation to temperature. *Journal of Fisheries Reearch Board of Canada*, **15**, 1345–54.

Mejuto, J. and de la Serna, J. M. (1997). Updated standardized catch rates by age for the swordfish (*Xiphias gladius*) from the Spanish longline fleet in the Atlantic using commercial trips from the period 1983–1995. *ICCAT Colleted Volumes of Scientific Papers*, SCRS/96/141.

Messieh, S. N. (1986). The enigma of Gulf herring recruitment. *NAFO Scientific Council Research Document*, 86/103, Serial No. N1230.

Michalsen, K., Ottersen, G., and Nakken, O. (1998). Growth of North-east arctic cod (*Gadus morhua* L.) in relation to ambient temperature. *ICES Journal of Marine Science*, **55**(5), 863–77.

Miller, T. J., Herra, T., and Leggett, W. C. (1995). An individual-based analysis of the variability of eggs and their newly hatched larvae of Atlantic cod (*Gadus morhua*) on the Scotian Shelf. *Canadian Journal of Fisheries and Aquatic Science*, **52**, 1088–93.

Mills, D. (1989). *Ecology and management of Atlantic salmon*. Chapman & Hall, London.

Minns, C. K., Randall, R. G., Chadwick, E. M. P., Moore, J. E., and Green, R. (1995). Potential impact of climate change on the habitat and population dynamics of juvenile Atlantic salmon (*Salmo salar*) in Eastern Canada. In *Climate change and northern fish populations*, Vol. 121 (ed. R. J. Beamish), pp. 699–708. Canadian Special Publishing of Fisheries and Aquatic Science.

Monahan, A. H., Fyfe, J. C., and Flato, G. M. (2000). A regime view of North Hemisphere atmosphere variability and change under global warming. *Geophysical Research Letters*, **27**, 1139–42.

Montevecchi, W. A., Cairns, D. K., and Birt, V. L. (1988). Migration of postsmolt Atlantic salmon, *Salmo Salar*, off northeastern Newfoundland, as inferred by tag recoveries in a seabird colony. *Canadian Journal of Fisheries and Aquatic Sciences*, **45**(3), 568–71.

Morgan, M. J. (1992). Low-temperature tolerance of American plaice in relation to declines in abundance. *Transactions of the American Fisheries Society*, **121**, 399–402.

Morse, W. W. (1989). Catchability, growth, and mortality of larval fishes. *Fisheries Bulletin*, **87**, 417–46.

Mountain, D. B. and Murawski, S. A. (1992). Variation in the distribution of fish stocks on the northeast continental shelf in relation to their environment, 1980–89. *ICES Marine Science Symposium*, **195**, 424–32.

Muelbert, J. H., Lewis, M. R., and Kelley, D. E. (1994). The importance of small-scale turbulence in the feed of herring larvae. *Journal of Plankton Research*, **16**, 927–44.

Myers, R., (1998). When do environment-recruitment correlations work? *Reviews in Fish Biology and Fisheries*, **8**, 1–21.

Myers, R. A. and Drinkwater, K. F. (1989). The influence of Gulf Stream warm core rings on recruitment of fish in the northwest Atlantic. *Journal of Marine Research*, **47**, 635–56.

Myers, R. A. and Cadigan, N. G. (1993). Density-dependent juvenile mortality in marine demersal fish. *Canadian Journal Fisheries Aquatic Sciences*, **50**(8), 1576–90.

Myers, R. A., Drinkwater, K. F., Barrowman, N. J., and Baird, J. W. (1993). Salinity and recruitment of Atlantic cod (*Gadus morhua*) in the Newfoundland region. *Canadian Journal of Fisheries and Aquatic Sciences*, **50**, 1599–609.

Nakashima, B. S. (1996). The relationship between oceanographic conditions in the 1990s and changes in spawning behaviour, growth and early life history of capelin (*Mallutus villosus*). *NAFO Scientific Council Studies*, **24**, 55–68.

Narayanan, S., Carscadden, J., Dempson, J. B., O'Connell, M. F., Prinsenberg, S., Reddin, D. G., and Shackell, N. (1995). Marine climate off Newfoundland and its influence on Atlantic salmon (*Salmo salar*) and capelin (*Mallotus villosus*). In *Climate change and northern fish populations*, vol. 121 (ed. R. J. Beamish), pp. 461–74. Canadian Special Publication of Fisheries and Aquatic Sciences 121.

Nesterova, V. N. (1990). *Plankton biomass along the drift route of cod larve* (in Russian). Pinro, Murmansk.

O'Connell, J. and Tunnicliffe, V. (2001). The use of sedimentary fish remains for interpretation of long-term fish population fluctuations. *Marine Geology*, **174**(1–4), 177–95.

Ommundsen, A. (2002). Models of cross shelf transport introduced by the Lofoten Maelstrom. *Continental Shelf Research*, **22**, 93–113.

Orlova, E. L., Ushakov, N. G., Nesterova, V. N., and Boitsov, V. D. (2002). Food supply and feeding of capelin (*Mallotus villosus*) of different size in the central latitudinal zone of the Barents Sea during intermediate and warm years. *ICES Journal of Marine Science*, **59**(5), 968–75.

Ottersen, G. (1996). *Environmental impact on variability in recruitment, larval growth and distribution of Arcto-Norwegian cod*. University of Bergen.

Ottersen, G. and Sundby, S. (1995). Effects of temperature, wind and spawning stock biomass on recruitment of Arcto-Norwegian cod. *Fisheries Oceanography*, **4**(4), 278–92.

Ottersen, G. and Loeng, H. (2000). Covariability in early growth and year-class strength of Barents Sea cod, haddock and herring: the environmental link. *ICES Journal of Marine science*, **57**(2), 339–48.

Ottersen, G. and Stenseth, N. C. (2001). Atlantic climate governs oceanographic and ecological variability in the Barents Sea. *Limnology and Oceanography*, **46**, 1774–80.

Ottersen, G., Michalsen, K., and Nakken, O. (1998). Ambient temperature and distribution of north-east arctic cod. *ICES Journal of Marine Science*, **55**, 67–85.

Page, F. H. and Frank, K. T. (1989). Spawning time and egg stage duration in northwest Atlantic haddock (*Melanogrammus aeglefinus*) stocks with emphasis on Georges and Brown Bank. *Canadian Journal of Fisheries and Aquatic Sciences*, **46**(Suppl. 1), 68–81.

Pearcy, W. G. (1992). *Ocean ecology of North Pacific salmonids*. Washing Sea Grant Program. University of Washington Press, Seattle.

Petterson, O. (1926). Hydrography, climate and fisheries in the transition area. *Journal Conseil International pour l'Exploration de la Mer*, **1**, 305–21.

Planque, B. and Fox, C. J. (1998). International variability in temperature and the recruitment of Irish sea cod, *Marine Ecology Progress Series*, **172**, 101–5.

Planque, B. and Fredou, T. (1999). Temperature and recruitment of Atlantic cod (*Gadus morhua*). *Canadian Journal of Fisheries and Aquatic Sciences*, **56**, 2069–77.

Polocheck, T., Mountain, D., McMillan, D., Smith, W., and Berrien, P. (1992). Recruitment of the 1987 year class of Georges Bank Haddock (*Melanogrammus aeglefinus*): the influence of unusual larval transport. *Canadian Journal of Fisheries and Aquatic Sciences*, **49**, 484–96.

Power, G. (1981). Stock characteristics and catches of Atlantic salmon, *Salmo salar*, in Quebec, and Newfoundland and Labrador in relation to environmental variables. *Canadian Journal of Fisheries and Aquatic Sciences*, **38**, 1601–11.

Reddin, D. G. and Friedland, K. D. (1993). Marine environmental factors influencing the movement and survival of Atlantic salmon. In *Salmon in the Sea* (ed. D. Mills), pp. 79–103. Blackwell Scientific Publications Ltd., London, U.K.

Reddin, D. G. and Shearer, W. M. (1987). Sea-surface temperature and distribution of Atlantic salmon in the Northwest Atlantic Ocean. *American Fisheries Society Symposium on Common Strategies in Anadromous/ Catadromous Fishes*, **1**, 262–75.

Reddin, D. G. and Short, P. B. (1991). Postsmolt Atlantic salmon (*Salmo salar*) in the Labrador Sea. *Canadian Journal of Fisheries and Aquatic Sciences*, **48**, 2–6.

Ritter, J. A. (1989). Marine migration and natural mortality of North American salmon (*Salmo salar* L.). *Canadian Manuscript Report of Fisheries and Aquatic Sciences*, **2041**, 136 pp.

Rodionov, S. N. (1995). Atmospheric teleconnections and coherent fluctuations in recruitment to North Atlantic cod (*Gadus morhua*) stocks. *Canadian Special Publication of Fisheries and Aquatic Science*, **121**, 45–55.

Rose, G. A. and Leggett, W. C. (1988). Atmosphere-ocean coupling and Atlantic cod migrations: the effects of wind forced variations in sea temperatures and currents on nearshore distributions and catch rates of *Gadus morhua*. *Canadian Journal of Fisheries and Aquatic Sciences*, **45**, 1234–43.

Rose, G. A., Atkinson, B. A., Baird, J., Bishop, C. A., and Kulka, D. W. (1994). Changes in distribution of Atlantic cod and thermal variations in Newfoundland waters, 1980–1992. *ICES Marine Science Symposium*, **198**, 542–52.

Rose, G. A., deYoung, B., Kulka, D. W., Goddard, S., and Fletcher, G. L. (2000). Distribution shifts and overfishing the northern cod (*Gadus morhua*): a view from the ocean. *Canadian Journal of Fisheries and Aquatic Science*, **57**, 644–63.

Rothschild, B. J. and Osborn, T. R. (1988). Small scale turbulence and plankton contact rates. *Journal of Plankton Research*, **10**, 465–74.

Røttingen, I. (1992). Recent migration routes of Norwegian spring spawning herring. *ICES Counsil Meeting* 1992/H:18, 10p.

Sahrhage D. and Lundbeck, J. (1992). *A history of fishing*. Springer-Verlag, Berlin.

Salminen, M., Kuikka, S., and Erkamo, E. (1995). Annual variability in survival of sea ranched Baltic salmon, *Salmo salar* L.: significance of smolt size and marine conditions. *Fisheries Management and Ecology*, **2**, 171–84.

Santiago, J. (1997). The North Atlantic Oscillation and recruitment of temperate tunas. *ICCAT SCRS/97/40*.

Saunders, R. L., Duston, J., and Benfey, T. J. (1994). Environmental and biological factors affecting growth dynamics in relation to smolting of Atlantic salmon, *Salmo Salar* L. *Aquatic Fisheries Management*, **25**, 9–20.

Saunders, R. L., Henderson, E. B., Glebe, B. D., and Loudenslager, E. J. (1983). Evidence of a major environmental component in determination of grilse: large salmon ration in Atlantic salmon (*Salmo salar*). *Aquaculture*, **33**, 107–18.

Scarnecchia, D. L. (1984). Climatic and oceanic variations affecting yield of Icelandic stocks of Atlantic salmon (*Salmo salar*). *Canadian Journal of Fisheries and Aquatic Science*, **41**, 917–35.

Scarnecchia, D. L., Ísaksson, Á., and White, S. E. (1989a). Effects of oceanic variations and the West Greenland fishery on age at maturity of Icelandic west coast stocks of Atlantic salmon (*Salmo salar*). *Canadian Journal of Fisheries and Aquatic Science*, **46**, 16–27.

Scarnecchia, D. L., Ísaksson, Á, and White, S. E. (1989b). Oceanic and riverine influences on variations in yield among Icelandic stock of Atlantic salmon. *Transactions of the American Fisheries Society*, **118**, 482–94.

Scarnecchia, D. L., Ísaksson, Á., and White, S. E. (1991). Effects of the Faroese long-line fishery, other oceanic fisheries and oceanic variations on age at maturity of Icelandic north-coast stocks of Atlantic salmon (*Salmo salar*). *Fisheries Research*, **10**, 207–28.

Schopka, S. A. (1991). The Greenland cod at Iceland, 1941–1990 and its impact on assessment. *NAFO Scientific Council Research Document*, **91**,102.

Scott, J. S. (1982). Depth, temperature and salinity preferences of common fishes of the Scotian Shelf. *Journal of Northwest Atlantic Fishery Science*, **3**, 29–39.

Serchuk, F. M., Grosslein, M. D., Lough, R. G., Mountain, D. G., and O'Brien, L. (1994). Fishery and environmental factors affecting trends and fluctuations in the Georges Bank and Gulf of Maine Atlantic cod stocks: an overview. *ICES Marine Science Symposia*, **198**, 77–109.

Sette, O. E. (1950). Biology of the Atlantic Mackerel (*Scomber scombrus*) of North America, Part II-migrations and habits. *Fishery Bulletin*, **49**, 251–358.

Shackell, N. L., Frank, K. T., Stobo, W. T., and Brickman, D. (1995). Cod (*Gadus morhua*) growth between 1956 and 1966 compared to growth between 1978 to 1985, on the Scotian Shelf and adjacent areas. *ICES Counsil Meeting* 1995/P:1, 18 p.

Sharp, G. D. (1987). Climate and fisheries: cause and effect or managing the long and short of it all. *South African Journal of Marine Science*, **5**, 811–38.

Shelton, P. A., Lilly, G. R., and Colbourne, E. (1996). Patterns in the annual weight increment for 2J3KL cod and possible prediction for stock projection. *NAFO Scientific Council Research Document*, **96/47**, p. 19.

Shepherd, J. G., Pope, J. G., and Cousens, R. D. (1984). Variations in fish stocks and hypotheses concerning their links with climate. *Rapports et procès-verbaux des réunions, Conseil International pour l'Exploration de la Mer*, **185**, 255–67.

Sinclair, A. (1996). Recent declines in cod species stocks in the northwest Atlantic. *NAFO Scientific Council Studies*, **24**, 41–52.

Sinclair, A. and Currie, L. (1994). Timing of cod migration into and out of the Gulf of St.Lawrence based on commercial fisheries, 1986–1993. *DFO Atlantic Fisheries Research Document*, **94/47**, 18 pp.

Sinclair, M. (1988). *Marine populations: an essay on population regulation and speciation*. University of Washington Press, Seattle.

Sinclair, M. and Iles, T. D. (1989). Population regulation and speciation in the oceans. *Journal Conseil International pour l'Exploration de la Mer*, **45**, 165–75.

Sirabella, P., Giuliani, A., Colosimo, A., and Dippner, J. (2001). Breaking down the climate effects on cod recruitment by principal component analysis and canonical correlation. *Marine Ecology Progress Series*, **216**, 213–22.

Sissenwine, M. P. (1984). Why do fish populations vary? In *Exploitation or Marine Communities* (ed. R. M. May), 59–94. Dahlem Konferenzen 1984, Springer-Verlag, Berlin, Germany.

Skjoldal, H. R., Noji, T. T., Giske, J., Fossa, J. H., Blindheim, J., and Sundby, S. (1993). *MARE COGNITUM–Science Plan for Research on Marine Ecology of the Nordic Seas*. Institute of Marine Research, Bergen, Norway.

Smith, S. J. and Page, F. (1996). Associations between Atlantic cod (*Gadus morhua*) and hydrographic variables: implications for the management of the 4VsW cod stock. *ICES Journal of Marine Science*, **53**, 597–614.

Smith, S. J., Perry, R. I., and Fanning, L. P. (1991). Relationships between water mass characteristics and estimates of fish population abundance from trawl surveys. *Environmental Monitoring and Assessment*, **17**, 227–45.

Sogard, S. M. (1997). Size-selective mortality in the juvenile stage of teleost fishes: a review. *Bulletin of Marine Science*, **60**, 1129–57.

Solomon, D. J. (1985). Salmon stock and recruitment, and stock enhancement. *Journal of Fish Biology*, **27**(Suppl. A), 45–57.

Solomon, D. J. (1978). Some observations on salmon smolt migration in a chalkstream. *Journal of Fish Biology*, **12**, 571–74.

Southward, A. J., Boalch, G. T., and Maddock, L. (1988). Fluctuations in the herring and pilchard fisheries of Devon and Cornwall linked to change in climate since the 16th century. *Journal of the Marine Biological Association of The United Kingdom*, **68**, 423–45.

Sundby, S. (1994). The influence of bio-physical processes on fish recruitment in an arctic-boreal ecosystem. Ph.D. Thesis, University of Bergen, Bergen.

Sundby, S. (2000). Recruitment of Atlantic cod stocks in relation to temperature and advection of copepod populations. *Sarsia*, **85**, 277–98.

Sundby, S. and Fossum, P. (1990). Feeding conditions of Arcto-Norwegian cod larvae compared with the Rothschild-Osborn theory on small-scale turbulence and plankton contact rates. *Journal of Plankton Research*, **12**, 1153–62.

Sundby, S., Bjørke, H., Soldal, A. V., and Olsen, S. (1989). Mortality rates during the early life stages and year class strength of the Arcto-Norwegian cod (*Gadus morhua* L.). *Rapports et procès-verbaux des réunions Conseil International pour l'Exploration de la Mer*, **191**, 351–58.

Sundby, S., Ellertsen, B., and Fossum, P. (1994). Encounter rates between first-feeding cod larvae and their prey during moderate to strong turbulent mixing. *ICES Marine Science Symposia*, **198**, 393–405.

Svendsen, E., Aglen, A., Iversen, S. A., Skagen, D. W., and Smestad, O. M. (1995). Influence of climate on recruitment and migration of fish stocks in the North Sea. In *Climate change and northern fish populations*, vol. 121 (ed. R. J. Beamish), pp. 135–47. Canadian special publication of fisheries and aquatic sciences.

Swain, D. P. and Kramer, D. L. (1995). Annual variation in temperature selection by Atlantic cod *Gadus morhua* in the southern Gulf of St. Lawrence, Canada, and its relation to population size. Marine Ecology Progress Series, **116**,11–23.

Taggart, C. T., Anderson, J., Bishop, C., Colbourne, E., Hutchings, J., Lilly, G., Morgan, J., Murphy, E., Myers, R., Rose, G., and Shelton, P. (1994). Overview of cod stocks, biology, and environment in the Northwest Atlantic region of Newfoundland, with emphasis on northern cod. *ICES Marine Science Symposium*, **198**, 140–57.

Tanning, A. V. (1937). Some features in the migration of cod. *Journal Conseil International pour l'Exploration de la Mer*, **12**, 1–35.

Taylor, C. C., Bigelow, H. B., and Graham, H. W. (1957). Climate trends and the distribution of marine animals in New England. *Fisheries Bulletin*, **57**, 293–345.

Templeman, W. (1965). Mass mortalities of marine fishes in the Newfoundland area presumably due to low temperature. *ICNAF Special Publication*, **6**, 137–47.

Templeman, W. (1966). Marine resources of Newfoundland. *Fisheries Research Board of Canada Bulletin*, **154**, 170.

Templeman, W. (1972). Year-class success in some North Atlantic stocks of cod and haddock. *ICNAF Special Publication*, **8**, 223–41.

Thorpe, J. E. (1986). Age at fist maturity in Atlantic salmon, *Salmo salar*: freshwater period influences and conflicts with smolting. In *Salmonid Age at Maturity* (ed. D. J. Meerburg), pp. 7–14. (*Canadian special publication of fisheries and aquatic sciences*, **89**).

Thorpe, J. E. (1988). *Salmon migration*. Science Program Oxford **72**, 345–370.

Tibbo, S. N. and Humphreys, R. D. (1966). An occurrence of capelin (*Mallotus villosus*) in the Bay of Fundy. *Journal of the Fisheries Research Board of Canada*, **23**, 463–67.

Toresen, R. and Østvedt, O. J. (2000). Variation in abundance of Norwegian spring-spawning herring (*Clupea harengus*, Clupeidae) throughout the 20th century and the influence of climatic fluctuations. *Fish and Fisheries*, **1**, 231–56.

Tucker, S., Pazzia, I., Rowan, D., and Rasmussen, J. B. (1999). Detecting pan-Atlantic migration in salmon (*Salmo salar*) using ^{137}Cs. *Canadian Journal of Fisheries and Aquatic Sciences*, **56**, 2235–9.

van Loon, H. and Rogers, J. C. (1978). The seesaw in winter temperatures between Greenland and northern Europe. Part 1: General description. *Monthly Weather Review*, **106**, 296–10.

Ware, D. M. (1975). Relation between egg size, growth and natural mortality of larval fish. *Journal of the Fisheries Research Board of Canada*, **32**, 2503–12.

Ware, D. M. (1977). Spawning time and egg size of Atlantic mackerel, *Scomber scombus*, in relation to the plankton. *Journal of the Fisheries Research Board of Canada*, **34**, 2308–15.

Werner, F. E., Page, F. H., Lynch, D. R., Loder, J. W., Lough, R. G., Perry, R. I., Greenberg, D. A., and Sinclair, M. M. (1993). Influences of mean advection and simple behaviour on the distribution of cod and haddock early life stages on Georges Bank. *Fisheries Oceanography*, **2**, 43–64.

Werner, F., Perry, R. I., MacKenzie, B. R., Lough, R. G., and Naimie, C. E. (2001). Larval trophodynamics, turbulence, and drift on Georges Bank: A sensitivity analysis of cod and haddock. *Scientia Marina*, **65**, 99–115.

Whalen, K. G., Parrish, D. L., and Mather, M. E. (1999). Effect of ice formation on selection of habitats and winter distribution of post-young-of-the-year Atlantic salmon parr. *Canadian Journal of Fisheries and Aquatic Sciences*, **56**(1), 87–96.

Wieland, K. and Hovgård, H. (2002). Distribution and drift of Atlantic cod (*Gadus morhua*) eggs and larvae in Greenland offshore waters. *Journal of Northwest Atlantic Fishery Science* **30**, 61–76.

Winters, G. H., Wheeler, J. P., and Dalley, E. L. (1986). Survival of a herring stock subjected to a catastrophic event and fluctuation environmental conditions. *Journal Conseil International pour l'Exploration de la Mer*, **43**, 26–43.

Zorita, E. and Laine, A. (2000). Dependence of salinity and oxygen concentrations in the Baltic Sea on large-scale atmospheric circulation. *Climate Research*, **14**, 25–41.

Øiestad, V. (1994). Historic changes in cod stocks and cod fisheries: Northeast Arctic cod. *ICES marine Science Symposia*, **198**, 17–30.

Chapter 7

Aebischer, N. (1993). Immediate and delayed effects of a gale in late spring on the breeding of the Shag *Phalacrocorax aristotelis*. Ibis, **135**, 225–32.

Aebischer, N. and Wanless, S. (1992). Relationships between colony size, adult non-breeding and environmental conditions for Shags *Phalacrocorax aristotelis* on the Isle of May, Scotland. *Bird Study*, **32**, 43–52.

Aebischer, N., Coulson, J., and Colebrook, J. (1990). Parallel long-term trends across four marine trophic levels and weather. *Nature*, **347**, 753–5.

Ainley, D., Jacobs, S., Ribic, C., and Gaffney, I. (1998). Seabird distribution and oceanic features of the Amundsen and southern Bellingshausen seas. *Antarctic Science*, **10**(2), 111–23.

Ainley, D., Sydeman, W., Hatch, S., and Wilson, U. (1994). Seabird population trends along the west coast of North America: causes and extent of regional concordance. *Study of Avian Biology*, **15**, 119–33.

Ainley, D., Sydeman, W., and Norton, J. (1996). Apex predators indicate interannual negative and positive anomalies in the California current food web. *Marine Ecology Progress Series*, **137**, 1–10.

Allen, C., Simpson, J., and Carson, R. (1980). The structure and variability of shelf sea fronts as observed by an undulating CTD system. *Oceanologica Acta*, **3**, 59–68.

Anker-Nilssen, T. (1987). The breeding performance of the Puffins *Fratercula arctica* on Røst, northern Norway in 1979–1985. *Fauna norvegica, Series C, Cinclus*, **10**, 21–38.

Anker-Nilssen, T. (1992). Food supply as a determinant of reproduction and population development in Norwegian Puffins Fratercula arctica. Ph.D. Thesis, University of Trondheim, Department of Zoology, Trondheim, Norway.

Anker-Nilssen, T. and Aarvak, T. (2002). The population ecology of Puffins at Røst. Status after the breeding season 2001. *NINA Oppdragsmelding*, **736**, 1–40.

Ashmole, N. (1971). Seabird ecology and the marine environment. *Avian Biology*, pp. 223–86. Academic Press, London.

Baduini, C., Hyrenbach, K., Coyle, K., Pinchuk, A., Mendenhall, V., and Hunt, G. (2001). Mass mortality of short-tailed shearwaters in the south-eastern Bering Sea during summer 1997. *Fisheries Oceanography*, **10**, 117–30.

Ballance, L. and Pitman, R. (1999). Foraging ecology of tropical seabirds. *Proceedings of 22nd International Ornithology Congress Durban*, pp. 2057–71. BirdLife South Africa, Johannesburg.

Ballance, L., Pitman, R., and Reilly, S. (1997). Seabird community structure along a productivity gradient: importance of competition and energetic constraint. *Ecology*, **78**(5), 1502–18.

Barbraud, C. and Weimerskirch, H. (2001). Emperor penguins and climate change. *Nature*, **411**, 183–6.

Barbraud, C., Weimerskirch, H., Guinet, C., and Jouventin, P. (2000). Effect of sea-ice extent on adult survival of an Antarctic top predator: the snow petrel *Pagodroma nivea*. *Oecologia*, **125**(4), 483–8.

Barrett, R. and Krasnov, Y. (1996). Recent responses to changes in stocks of prey species by seabirds breeding in the southern Barents Sea. *ICES Journal of Marine Science*, **53**, 713–22.

Begg, G. and Reid, J. (1997). Spatial variation in seabird density at a shallow sea tidal mixing front in the Irish Sea. *ICES Journal of Marine Science*, **54**(4), 552–65.

Bertram, D., Jones, I., Cooch, E., Knechtel, H., and Cooke, F. (2000). Survival rate of Cassin's and Rhinoceros auklets at Triangle Island, British Columbia. *Condor*, **102**, 155–62.

Bodkin, J. and Jameson, R. (1991). Patterns of seabird and marine mammal carcass deposition along the central California coast, 1980–1986. *Canadian Journal of Zoology*, **69**, 1149–55.

Boersma, P. (1978). Breeding patterns of Galapagos penguins as an indicator of oceanographic conditions. *Science*, **200**, 1481–3.

Boersma, P. (1998). Population trends of the Galapagos penguin: impact of El Nino amd La Nina. *Condor*, **100**, 245–53.

Brown, R. G. B. (1991). Marine birds and climatic warming in northwest Atlantic. *Canadian Wildlife Service, Occasional Paper*, **68**, 49–54.

Cairns, D. (1987). Seabirds as indicators of marine food supplies. *Biological Oceanography*, **5**, 261–71.

Cairns, D. (1992). Population regulation of seabird colonies. *Current Ornithology*, **9**, 37–61.

Carey, C. (1996). Female reproductive energetics. In *Avian energetics and nutritional ecology*, (ed. C. Carey), pp. 324–74. Chapman & Hall. New York, USA.

Catard, A., Weimerskirch, H., and Cherel, Y. (2000). Exploitation of distant Antarctic waters and close shelf-break waters by white-chinned petrels rearing chicks. *Marine Ecology Progress Series*, **194**, 249–61.

Charrassin, J. and Bost, C. (2001). Utilisation of the oceanic habitat by king penguins over the annual cycle. *Marine Ecology Progress Series*, **221**, 285–97.

Chastel, O., Weimerskirch, H., and Jouventin, P. (1993). High annual variability in reproductive success and survival of an Antarctic seabird, the Snow Petrel *Pagodroma nivea*. *Oecologia*, **94**, 278–85.

Cook, M. and Hamer, K. (1997). Effects of supplementary feeding on provisioning and growth rates of nestling Puffins *Fratercula arctica*: evidence for regulation of growth. *Journal of Avian Biology*, **28**, 56–62.

Coulson, J. (1968). Differences in the quality of birds nesting in the centre and on the edges of a colony. *Nature*, **217**, 478–9.

Coyle, K. and Cooney, R. (1993). Water column sound scattering and hydrography around the Pribilof Islands, Bering Sea. *Continental Shelf Research*, **13**, 803–27.

Croll, D. and McLaren, E. (1993). Diving metabolism and thermoregulation in common and thick-billed murres. *Journal of Comparative Physiology*, **163**, 160–6.

Croxall, J. (1992). Southern ocean environmental changes: effect on seabirds, seal and whale populations. *Philosophical Transactions of the Royal Society of London, B*, **338**, 319–28.

Croxall, J., McCann, T., Prince, P., and Rothery, P. (1988). Reproductive performance of seabirds and seals on South Georgia and Sidney Island, South Orkney Island, 1897–1976: implication for southern ocean monitoring studies. In *Antarctic ocean and resources variability*, (ed. D. Sahrhage), pp. 261–85. Springer-Verlag, Berlin, Germany.

Croxall, J. P., Trathan, P. N., and Murphy, E. J. (2002). Environmental change and Antarctic seabird populations. *Science*, **297**, 1510–14.

Cushing, D. (1990). Plankton production and year-class strength in fish populations—an update of the match mismatch hypothesis. *Advances in Marine Biology*, **26**, 249–93.

Dawson, W. and O'Connor, T. (1996). Energetic features of avian thermoregulatory responses. In *Avian energetics and nutritional ecology*, (ed. C. Carey), pp. 85–124. Chapman & Hall. New York, USA.

Dearborn, D. (2001). Body condition and retaliation in the parental effort decisions of incubating great frigatebirds (*Fregata minor*). *Behavioral Ecology*, **12**(2), 200–06.

Decker, M. and Hunt, G. (1996). Foraging by murres (*Uria spp.*) at tidal fronts surrounding the Pribilof Islands, Alaska, USA. *Marine Ecology Progress Series*, **139**, 1–10.

Duffy, D. (1990). Seabirds and the 1982–84 El Nino Southern Oscillation. In *Global Ecological Consequences of the 1982–84 El Nino-Southern Oscillation*, pp. 395–415. Elsevier, Amsterdam.

Durant, J. M., Anker-Nilssen, T., and Stenseth, N. C. (2003). Trophic interactions under climate fluctuations: the Atlantic puffin as an example. *Proceedings of the Royal Society of London, B*, **270**, 1461–66.

Ellis, H. and Gabrielsen, G. (2001). Energetics of free-ranging seabirds. In *Biology of marine birds* (ed. Peter L. Lutz), pp. 359–408. CRC, Marine Biology Series, CRC Press.

Erikstad, K., Asheim, M., Fauchald, P., Dahlhaug, L., and Tveraa, T. (1997). Adjustment of parental effort in the puffin: the roles of adult body condition and chick size. *Behavioral Ecology and Sociobiology*, **40**(2), 95–100.

Erikstad, K., Fauchald, P., Tveraa, T., and Steen, H. (1998). On the cost of reproduction in long-lived birds: the influence of environmental variability. *Ecology*, **79**, 1781–88.

Finney, S., Wanless, S., and Harris, M. (1999). The effect of weather conditions on the feeding behaviour of a diving bird, the Common Guillemot *Uria aalge*. *Journal of Avian Biology*, **30**(1), 23–30.

Flint, E. (1991). Time and energy limits to the foraging radius of Sooty Terns *Sterna fuscata*. *Ibis*, **133**, 43–46.

Furness, R. and Bryant, D. (1996). Effect of wind on field metabolic rates of breeding Northern Fulmars. *Ecology*, **77**(4), 1181–8.

Gabrielsen, G., Mehlum, F., and Nagy, K. (1987). Daily energy expenditure and energy utilization of free-ranging black-legged kittiwakes. *Condor*, **89**, 126–32.

Garthe, S. (1997). Influence of hydrography, fishing activity, and colony location on summer seabird distribution in the south-eastern North Sea. *ICES Journal of Marine Science*, **54**(4), 566–77.

Guinet, C., Chastel, O., Koudil, M., Durbec, J., and Jouventin, P. (1998). Effects of warm sea-surface temperature anomalies on the blue petrel at the Kerguelen Islands. *Proceedings of the Royal Society of London, B*, **265**, 1001–6.

Guinet, C., Koudil, M., Bost, C., Durbec, J., Georges, J., Mouchot, M., and Jouventin, P. (1997). Foraging behaviour of satellite-tracked king penguins in relation to sea-surface temperatures obtained by satellite telemetry at Crozet Archipelago, a study during three austral summers. *Marine Ecology Progress Series*, **150**(1–3), 11–20.

Harrington, B. (1977). Winter distribution of juvenile and older Red-footed Boobies from Hawaiian Islands. *Condor*, **79**, 87–90.

Harris, M. (1966). Breeding biology of the Manx Shearwater *Puffinus puffinus*. *Ibis*, **108**, 17–33.

Harris, M., Murray, S., and Wanless, S. (1998). Long-term changes in breeding performance of Puffins *Fratercula arctica* on St Kilda. *Bird Study*, **45**, 371–4.

Harris, M. and Wanless, S. (1984). The effect of the wreck of seabirds in February 1983 on auk populations on the Isle of May (Fife). *Bird Study*, **31**, 103–10.

Harris, M. and Wanless, S. (1988). The breeding biology of Guillemots *Uria aalge* on the Isle of May over a six year period. *Ibis*, **130**, 172–92.

Harris, M. and Wanless, S. (1989). The breeding biology of Razorbills *Alca torda* on the Isle of May. *Bird Study*, **36**, 105–14.

Harris, M. and Wanless, S. (1996). Differential responses of Guillemots (*Uria aalge*) and Shag (*Phalacrocorax aristotelis*) to a late winter wreck. *Bird Study*, **43**, 220–30.

Hedgren, S. (1979). Seasonal variation in fledging weight of Guillemots, *Uria aalge*, *Ibis*, **121**, 356–61.

Hodum, P., Sydeman, W., Visser, G., and Weathers, W. (1998). Energy expenditure and food requirement of Cassin's Auklets provisioning nestlings. *Condor*, **100**(3), 546–50.

Hoefer, C. (2000). Marine bird attraction to thermal fronts in the California current system. *Condor*, **102**(2), 423–7.

Hohtola, E. and Visser, G. (1998). Development of locomotion and endothermy in altricial and precocial birds. In *Avian growth and development: evolution within the Altricial-Precocial spectrum*, (eds F. M. Starch & R. E. Richlefs, UK), pp. 157–173. Oxford University Press, New York, Oxford.

Hunt, G. (1990). The pelagic distribution of marine birds in a heterogeneous environment. *Polar Research*, **8**, 43–54.

Hunt, G. (1991). Marine ecology of seabirds in Polar oceans. *American Zoologist*, **31**(1), 131–42.

Hunt, G. and Schneider, D. (1987). Scale-dependant processes in the physical and biological environment of marine birds. In *Seabirds: feeding ecology and role in marine ecosystems*, (ed. F. P. Croxall, UK), pp. 7–41. Cambridge University Press, Cambridge, UK.

Hunt, G., Mehlum, F., Russell, R., Irons, D., Decker, M., and Becker, P. (1999). Physical processes, prey abundance, and the foraging ecology of seabirds. In *Proceedings of 22nd International Ornithology Congress Durban*, pp. 2040–56. BirdLife South Africa, Johannesburg.

Jouventin, P. and Mougin, J. (1981). Les stratégies adaptatives des oiseaux de mer. *Reviews in Ecology (Terre et Vie)*, **35**, 217–72.

Jouventin, P. and Weimerskirch, H. (1990). Satellite tracking of Wandering Albatrosses. *Nature*, **343**, 746–8.

Kinder, T., Hunt, G., Schneider, D., and Schumacher, J. (1983). Correlation between birds and oceanic fronts around the Pribilof Islands, Alaska. *Estuarine Coastal and Shelf Science*, **16**, 163–299.

Kitaysky, A. and Golubova, E. (2000). Climate change causes contrasting trends in reproductive performance of planktivorous and piscivorous alcids. *Journal of Animal Ecology*, **69**(2), 248–62.

Klomp, N. and Schultz, M. (1998). The remarkable foraging behaviour of short-tailed shearwaters breeding in eastern Australia. *Ostrich*, **69**, 373–4.

Koudil, M., Charrassin, J., Le Maho, Y., and Bost, C. (2000). Seabirds as monitors of upper-ocean thermal structure. King penguins at the Antarctic polar front, east of Kerguelen sector. *Comptes Rendus de l'Academie des Sciences, Serie III*, **323**, 377–84.

Konarzewski, M. and Taylor, J. R. E. (1989). The influence of weather conditions on growth of Little Auk *Alle alle* chicks. *Ornis Scandinavica*, **20**, 112–16.

Lack, D. (1968). *Ecological adaptations for breeding in birds.* Methuen. London, UK

Le Maho, Y. (1977). The Emperor Penguin: a strategy to live and breed in the cold. *American Scientist*, **65**, 680–93.

Marchant, S. and Higgins, P. J. (1990). *Handbook of Australian, New Zealand & Antarctic birds.* Oxford University Press, Melbourne, Australia.

Martin, T. (1987). Food as a limit on breeding birds: a life-history perspective. *Annual Review of Ecology and Systematics*, **18**, 453–87.

Martin, T. and Li, P. (1992). Life history traits of open- vs. cavity-nesting birds. *Ecology*, **73**, 579–92.

McGowan, J., Cayan, D., and Dorman, L. (1998). Climate-ocean variability and ecosystem response in the Northeast Pacific. *Science*, **281**, 210–17.

Mehlum, F., Nordlund, N., and Isaksen, K. (1998). The importance of the 'Polar Front' as a foraging habitat for guillemots *Uria spp.* breeding at Bjørnøya, Barents Sea. *Journal of Marine Systems*, **14**, 27–43.

Meijer, T. and Drent, R. (1999). Re-examination of the capital and income dichotomy in breeding birds. *Ibis*, **141**, 399–414.

Montevecchi, W. and Myers, R. (1995). Prey harvests of seabirds reflect pelagic fish and squid abundance on multiple spatial and temporal scales. *Marine Ecology Progress Series*, **117**(1–3), 1–9.

Montevecchi, W. and Myers, R. (1996). Dietary changes of seabirds indicate shifts in pelagic food webs. *Sarsia*, **80**, 313–22.

Montevecchi, W. and Myers, R. (1997). Centurial and decadal oceanographic influences on changes in northern gannet populations and diets in the north-west Atlantic: implications for climate change. *ICES Journal of Marine Science*, **54**(4), 608–14.

Nakashima, B. (1996). The relationship between oceanographic conditions in the 1990s and changes in spawning behaviour, growth and early life history of capelin (*Mallotus villosus*). *NAFO Scientific Council Research Document*, **94/74**, pp. 18.

Nicholls, D., Stampton, P., Klomp, N., and Schultz, M. (1998). Post-breeding flight to Antarctic waters by a short-tailed shearwater *Puffinus tenuirostris. Emu*, **98**, 79–82.

Olsen, P. and Marples, T. G. (1993). Geographic variation in egg size, clutch size and date of laying of Australian raptors (Faconiformes and Strigiformes). *Emu*, **93**, 167–79.

Ottersen, G., Planque, B., Belgrano, A., Post, E., Reid, P., and Stenseth, N. (2001). Ecological effects of the North Atlantic Oscillation. *Oecologia*, **128**, 1–14.

Øyan, H. and Anker-Nilssen, T. (1996). Allocation of growth in food-stressed Atlantic Puffin chicks. *Auk*, **113**(4), 830–41.

Perrins, C. M. (1996). Eggs, egg formation and the timing of breeding. *Ibis*, **138**, 2–15.

Piatt, J. and van Pelt, T. (1997). Mass-mortality of Guillemots (*Uria aalge*) in the Gulf of Alaska in 1993. *Marine Pollution Bulletin*, **34**, 656–62.

Prince, P., Wood, A., Barton, T., and Croxall, J. (1992). Satellite tracking of wandering albatrosses (*Diomedea exulans*) in the South Atlantic. *Science*, **4**, 31–6.

Regehr, H. and Montevecchi, W. (1997). Interactive effects of food shortage and predation on breeding failure of black-legged kittiwakes: implications for indicator species, seabird interactions and indirect effects of fisheries activities. *Marine Ecology Progress Series*, **155**, 249–60.

Regehr, H. and Rodway, M. (1999). Seabird breeding performance during two years of delayed Capelin arrival in the Northwest Atlantic: a multi-species comparison. *Waterbirds*, **22**, 60–7.

Ricklefs, R. (1968). Patterns of growth in birds. *Ibis*, **110**, 419–51.

Ricklefs, R. (1973). Patterns of growth in birds. II. Growth rate and mode of development. *Ibis*, **115**, 177–201.

Ricklefs, R. and Schew, W. (1994). Foraging stochasticity and lipid accumulation by nestling petrels. *Functional Ecology*, **8**, 159–70.

Rodway, M., Chardine, J., and Montevecchi, W. (1998). Intra-colony variation in breeding performance of Atlantic Puffins. *Colonial Waterbirds*, **21**, 171–84.

Saether, B., Andersen, R., and Pedersen, H. (1993). Regulation of parental effort in a long-lived seabird—an experimental manipulation of the cost of reproduction in the Antarctic Petrel, Thalassoica-Antarctica. *Behavioral Ecology and Sociobiology*, **33**(3), 147–50.

Sanz, J. (1999). Does daylength explain the latitudinal variation in clutch size of Pied Flycatchers *Ficedula hypoleuca*? *Ibis*, **141**, 100–08.

Schmidt-Nielsen, K. (1997). *Animal physiology*, 5th edn. Cambridge University Press, Cambridge, UK.

Schneider, D. (1990). Seabirds and fronts: a brief overview. *Polar Research*, **8**, 17–21.

Schreiber, E. (2001). Climate and weather effects on seabirds. In *Biology of Marine birds* (ed. Peter L. Lutz), pp. 179–215. CRC, Marine Biology Series, CRC Press.

Schreiber, R. and Schreiber, E. (1984). Central Pacific seabirds and the El-Niño-Southern Oscillation: 1982–1983 perspectives. *Science*, **225**, 713–16.

Sillett, T., Holmes, R., and Sherry, T. (2000). Impacts of a global climate cycle on population dynamics of a migratory songbird. *Science*, **288**, 2040–2.

Skov, H. and Durinck, J. (2000). Seabird distribution in relation to hydrography in the Skagerrak. *Continental Shelf Research*, **20**(2), 169–87.

Spear, L. and Ainley, D. (1997). Flight behaviour of seabirds in relation to wind direction and wing morphology. *Ibis*, **139**(2), 221–33.

Stahl, J., Jouventin, P., Mougin, J., Roux, J., and Weimerskirch, H. (1985). The foraging zones of seabirds in the Crozet Islands Sector of the southern ocean. In *Antarctic nutrient cycles and food webs* (eds W. R. Siegfried, P. R. Condy and R. M. Laws), 478–86. Springer-Verlag, Berlin, Germany.

Stearns, S. (1992). *The evolution of life histories*. Oxford University Press, Oxford, UK.

Stoleson, S. and Beissinger, S. (1999). Egg viability as a constraint on hatching asynchrony at high ambient temperatures. *Journal of Animal Ecology*, **68**, 951–62.

Sun, L., Xie, Z., and Zhao, J. (2000). A 3,000-year record of penguin populations. *Nature*, **407**, 858.

Thompson, P. and Ollason, J. (2001). Lagged effects of ocean climate change on fulmar population dynamics. *Nature*, **413**, 417–20.

Toresen, R. and Østvedt, O. J. (2000). Variation in abundance of Norwegian spring-spawning herring (*Clupea harengus*, Clupeidae) throughout the 20th century and the influence of climatic fluctuations. *Fish and Fisheries*, **1**, 231–56.

van Heezik, Y. (1990). Patterns and variability of growth in the yellow-eyed penguin. *Condor*, **92**, 904–12.

Veit, R., McGowan, J., Ainley, D., Wahls, T., and Pyle, P. (1997). Apex marine predators declines ninety per cent in association with changing oceanic climate. *Global Change Biology*, **3**, 23–8.

Veit, R., Pyle, P., and McGowan, J. (1996). Ocean warming and long-term change in pelagic bird abundance within the California current system. *Marine Ecology Progress Series*, **139**, 11–18.

Velando, A., Ortega-Ruano, J., and Freire, J. (1999). Chick mortality in European shag *Stictocarbo aristotelis* related to food limitations during adverse weather events. *Ardea*, **87**, 51–9.

Visser, G. (1998). Development of temperature regulation. In *Avian growth and development: evolution within the Altricial-Precocial spectrum* (eds F. M. Starch and R. L. Richlefs), 117–56. Oxford University Press, New York Oxford, UK.

Warham, J. (1990). *The petrels, their ecology and breeding systems*. Academic press, San Diego, California, USA.

Watanuki, Y., Mehlum, F., and Takahashi, A. (2001). Water temperature sampling by foraging Brünnich's Guillemots with bird-borne data loggers. *Journal of Avian Biology*, **32**, 189–93.

Weimerskirch, H. (2001). Seabird demography and its relationship with the marine environment. In *Biology of marine birds* (ed. Peter L. Lutz), pp. 115–35. CRC, Marine Biology Series, CRC Press.

Weimerskirch, H. and Cherel, Y. (1998). Feeding ecology of short-tailed shearwaters: breeding in Tasmania and foraging in the Antarctic? *Marine Ecology Progress Series*, **167**, 261–74.

Weimerskirch, H., Salamolard, M., Sarrazin, F., and Jouventin, P. (1993). Foraging strategy of wandering albatrosses through the breeding season: a study using satellite telemetry. *Auk*, **110**, 325–42.

Weimerskirch, H., Cherel, Y., Cuenot-Chaillet, F., and Ridoux, V. (1997a). Alternative foraging strategies and resource allocation by male and female Wandering Albatrosses. *Ecology*, **78**(7), 2051–63.

Weimerskirch, H., Mougey, T., and Hindermeyer, X. 1997b. Foraging and provisioning strategies of black-browed albatrosses in relation to the requirements of the chick: natural variation and experimental study. *Behavioral Ecology*, **8**(6), 635–43.

Weimerskirch, H., Fradet, G., and Cherel, Y. (1999). Natural and experimental changes in chick provisioning in a long-lived seabird, the Antarctic Prion. *Journal of Avian Biology*, **30**, 165–74.

Weimerskirch, H., Guionnet, T., Martin, J., Shaffer, S., and Costa, D. (2000a). Fat and fuel efficient? Optimal use of wind by flying albatrosses. *Proceedings of the Royal Society of London, B*, **267**, 1869–74.

Weimerskirch, H., Barbraud, C., and Lys, P. (2000b). Sex differences in parental investment and chick growth in Wandering Albatrosses: fitness consequences. *Ecology*, **81**, 309–18.

Weimerskirch, H., Chastel, O., Cherel, Y., Henden, J., and Tveraa, T. (2001). Nest attendance and foraging movements of northern fulmars rearing chicks at Bjørnøya Barents Sea. *Polar Biology*, **24**, 83–8.

Williams, G. (1966). Natural selection, the costs of reproduction, and a refinement of Lack's principle. *American Naturalist*, **100**, 687–90.

Williams, J. (1996). Energetics of avian incubation. In *Avian energetics and nutritional ecology* (ed. C. Carey), 375–416. Chapman & Hall, New York, USA.

Williams, T. (1995). *The Penguins*. Oxford University Press, Oxford, UK.

Chapter 8

Aebischer, N. J. and Coulson, J. C., and Colebrook, J. M. (1990). Parallel long-term trends across 4 marine trophic levels and weather. *Nature*, **347**, 753–5.

Barber, R. T. and Chavez, R. P. (1983). Biological consequences of El Niño. *Science*, **222**, 1203–10.

Beaugrand, G., Ibanez, F., and Reid, P. C. (2000). Spatial, seasonal and long-term fluctuations of plankton in relation to hydroclimatic features in the English Channel, Celtic Sea and Bay of Biscay. *Marine Ecology Progress Series*, **200**, 93–102.

Belgrano, A., Lindahl, O., and Hernroth, B. (1999). North Atlantic Oscillation (NAO) primary productivity and toxic phytoplankton in the Gullmar Fjord, Sweden (1985–1996). *Proceedings of the Royal Society of London, B*, **266**, 425–30.

Belgrano A., Malmgren, B., and Lindahl, O. (2001). The use of artificial neural network for predicting primary productivity. *Journal of Plankton Research*, **23**, 651–8.

Berryman, A. A. (1999). *Principles of population dynamics and their applications*. Stanley Thornes Publishers Ltd., Cheltenham, UK.

Bjørnstad, O. N., Begon, M., Stenseth, N. C., Falck, W., Sait, S. M., and Thompson, D. J. (1998). Population dynamics of the Indian meal moth: demographic stochasticity and delayed regulatory mechanisms. *Journal of Animal Ecology*, **67**, 110–26.

Bjørnstad, O. N., Fromentin, J.-M., Stenseth, N. C., and Gjøsæter, J. (1999). Cycles and trends in cod populations. *Proceedings of National Academy of Science, Washington*, **96**, 5066–5071.

Colebrook, J. M. (1986). Environmental influences on long-term variability in marine plankton. *Hydrobiologia*, **142**, 309–25.

Conell, J. H. (1978). Diversity in tropical rain forest and coral reefs. *Science*, **199**, 1302–10.

Dickson, R. R. and Reid, P. C. (1983). Local effects of wind speed and direction on the phytoplankton of the Southern Bight. *Journal of Plankton Research*, **5**, 441–54.

Edwards, M., Reid, P. C., and Planque, B. (2001). Long-term and regional variability of phytoplankton biomass in the Northeast Atlantic (1960–1995). *ICES Journal of Marine Science*, **58**, 39–49.

Edwards, M., Beaugrand, G., Reid, P. C., Rowden, A. A., and Jones, M. B. (2002). Ocean climate anomalies and the ecology of the North Sea. *Marine Ecology Progress Series*, **239**, 1–10.

Green, P. J. and Silvermann, B. W. (1994). *Nonparametric regression and generalized linear models—a roughness penalty approach*. Chapman & Hall, London.

Harris, G. P. (1983). Mixed layer physics and phytoplankton populations. In *Studies in equilibrium and non-equilibrium ecology* (eds F. E. Round and D. Chapman), pp. 1–52. Progress in Phycological Research, 2. Elsevier, Amsterdam.

Harris, G. P. (1986). *Phytoplankton ecology*. Chapman & Hall, London.

Hastie, T. and Tibshirani, R. (1990). *Generalized additive models*. Chapman & Hall, London.

Huisman, J. and Weissing, F. J. (1999). Biodiversity of plankton by species oscillations and chaos. *Nature*, **402**, 407–10.

Hutchinson, G. E. (1967). *A treatise on limnology*, Vol. II. Wiley.

Jassby, A. D., Koseff, J. R., and Monismith, S. G. (1996). Processes underlying phytoplankton variability in San Francisco Bay. In *San Francisco Bay: the ecosystem* (ed. J.T. Hollibaugh), pp. 325–350. Pacific Division American Association for Advance in Science.

Lange, C. B., Burke, S. K., and Berger, W. H. (1990). Biological production off southern California is linked to climate change. *Climatic Change*, **41**, 1052–62.

Lehman, P. W. (1992). Environmental factors associated with long term changes in chlorophyll concentration in the Sacramento San Joaquin Delta and Suisun Bay, California. *Estuaries*, **15**, 335–348.

Lehman, P. W. (1996). Changes in Chl a concentration and phytoplankton community composition with water-year type in the upper San Francisco Bay Estuary. In *San Francisco Bay: the ecosystem* (ed. J.T. Hollibaugh), pp. 351–74. Pacific Division American Association for Advance in Science.

Lehman, P. W. (2000). The influence of climate on phytoplankton community biomass in San Francisco Bay Estuary. *Limnology and Oceanography*, **45**(3), 580–90.

Lindahl, O. (1987). Plankton community dynamics in relation to water exchange in the Gullmar Fjord, Sweden. Ph.D. Thesis, Department of Zoology University of Stockholm, Sweden.

Lindahl, O., Belgrano, A., Davidsson, L., and Hernroth, B. (1998). Primary production, climatic oscillations, and physico-chemical processes: the Gullmar Fjord time-series data set (1985–1996). *ICES Journal of Marine Science*, **55**, 723–9.

May, R. M. (1975). Patterns of species abundance and diversity. In *Ecology and evolution of communities* (eds D. R. Strong, D. Simberloff, L. G. Abele, and A. B. Thistle), pp. 3–18. Princeton University Press, NJ.

Reid, P. C., Edwards, M., Hunt, H. G., and Warner, J. (1998). Phytoplankton change in the North Atlantic. *Nature*, **391**, 546.

Ricker, W. (1954). Stock and recruitment. *Journal of the Fisheries Research Board of Canada*, **211**, 559–663.

Rodwell, M. J., Rowell, D. P., and Folland, C. K. (1999). Oceanic forcing of the wintertime North Atlantic Oscillation and European climate. *Nature*, **398**, 320–3.

Stenseth, N. C., Falck, W., Bjørnstad, O. N., and Krebs, C. J. (1997). Population regulation in snowshoe hare and Canadian lynx: asymmetric food web configurations between hare and lynx. *Proceedings of National Academy of Science, Washington*, **94**, 5147–52.

Simpson, E. H. (1949). Measurement of diversity. *Nature*, **163**, 688.

Tiselius, P. and Kuylenstierna, M. (1996). Growth and decline of a diatom spring bloom: phytoplankton species composition, formation of marine snow and the role of heterotrophic dinoflagellates. *Journal of Plankton Research*, **2**, 133–55.

Tont, S. A. (1989). Climatic change: response of diatoms and dinoflegellates. In *Aspects of climate variability in the Pacific and Western Americas* (ed. D. H. Peterson), pp. 161–163. Geophysical Monographs 55. American Geophysical Union.

Chapter 9

Alve, E. (1996). Benthic foraminiferal evidence of environmental change in the Skagerrak over the past six decades. *Norges Geogiske Undersøkelsers Bulletin*, **430**, 85–93.

Aebisher, N. J., Coulson, J. C., and Colebrook, J. M. (1990). Parallel long-term trends across four marine trophic levels and weather. *Nature*, **347**, 753–5.

Andersin, A.-B., Lassig, J., Parkkonen, L., and Sandler, H. (1978). Long-term fluctuations of the soft bottom macrofauna in the deep areas of the Gulf of Bothnia 1954–1974; with special reference to *Pontoporeia affinis* Lindstöm (Amphipoda). *Finnish Marine Research*, **244**, 137–44.

Austen, M. C., Buchanan J. B., Hunt, H. G., Josefson, A. B., and Kendall, M. A. (1991). Comparison of long-term trends in benthic and pelagic communities of the North Sea. *Journal of the Marine Biological Association, United Kingdom*, **71**, 179–90.

Baumgartner, T. M., Soutar, A., and Ferreira-Bartrina, V. (1992). Reconstruction of the history of Pacific sardine and northern anchovy populations over the past two millennia from sediments of the Santa Barbara basin, California. *Reports of California Cooperative Oceanic Fisheries Investigations*, **33**, 24–40.

Beamish, R. J., Riddell, B. E., Neville, C.-E. M., Thomson, B. L., and Zhang, Z. (1995). Declines in chinook salmon catches in the Strait of Georgia in relation to shifts in the marine environment. *Fisheries Oceanography*, **4**, 243–56.

Botsford, L. W., Meethot, R. D., and Wilen, J. E. (1982). Cyclic covariation in the California king salmon, *Oncorhynchus tshawytscha*, siver salmon, *O. kisutch*, and dungeness crab, *Cancer magister*, fisheries. *Fish Biology*, **80**, 791–801.

Brey, T. (1986). Increase in macrozoobenthos above the halocline in Kiel Bay compairing the 1960s with the 1980s. *Marine Ecology Progress Series*, **28**, 299–302.

Beukema, J. J. (1990). Expected effects of changes in winter temperatures on benthic animals living in soft sediments in coastal North Sea areas. In *Expected effects of climatic change on marine coastal ecosystems* (eds J. J. Beukema, W. J. Wolff, and J. W. N. Brouns), pp. 83–92. Kluwer Academic Press, Dordrecht.

Beukema, J. J. (1992). Expected changes in the Wadden Sea benthos in a warmer world: lessons from periods with milder winters. *Netherlands Journal of Sea Research*, **30**, 73–9.

Beukema, J. J., Essink, K., and Michaelis, H. (1996). The geographic scale of synchronized fluctuation patterns in zoobenthos as a key to underlying factors: climatic or man-induced. *ICES Journal of Marine Science*, **53**, 964–71.

Buchanan, J. B. (1993). Evidence of benthic pelagic coupling at a station off the Northumberland coast. *Journal of Experimental Marine Biology and Ecology*, **172**, 1–10.

Buchanan, J. B. and Moore, J. J. (1986). Long-term studies at a benthic station off the coast of Northumberland. *Hydrobiologia*, **142**, 121–7.

Cedervall, H. and Elmgren, R. (1980). Biomass increase of benthic macrofauna demonstrates eutrophication of the Baltic Sea. *Ophelia*, **1** (Suppl.), 287–304.

Cushing, D. H. (1981). Temporal variability in production systems. In *Analysis of marine ecosystems* (ed. A.R. Longhurst), pp. 443–471. Academic Press, New York.

Francis, R. C. and Hare, S. R. (1994). Decadal-scale regime shifts in the large marine ecosystems of the North-east Pacific: a case for historical science. *Fisheries Oceanography*, **3**, 279–91.

Fromentin, J.-M. and Ibanez, F. (1994). Year-to-year changes in meteorological features of the French coast area during the last half-century. Examples of two biological responses. *Oceanologica Acta*, **17**, 285–96.

Glémarec, M. (1993). Variabilité des système littoraux. *Journal de Recherche Océanographique*, **1/2**, 19–23.

Gray, J. S. and Christie, H. (1983). Predicting long-term changes in marine benthic communities. *Marine Ecology Progress Series*, **13**, 87–94.

Hagberg, J., Jonzén, N., Lundberg, P., and Ripa, J. (2003). Uncertain biotic and abiotic interactions in benthic communities. *Oikos*, **100**, 353–61.

Hagberg, J. and Tunberg, B. G. (2000). Studies on the covariation between physical factors and the long-term variation of the marine soft bottom macrofauna in western Sweden. *Estuarine Coastal and Shelf Science*, **50**, 373–85.

Hänninen, J., Vuorinen, I., and Hjelt, P. (2000). Climatic factors in the Atlantic control the oceanography and ecological changes in the Baltic Sea. *Limnology and Oceanography*, **45**, 703–10.

Hurrell, J. W. and Van Loon, H. (1997). Decadal variations in climate associated with the North Atlantic Oscillation. *Climate Change*, **36**, 301–3.

Josefson, A. B. (1987). Large-scale patterns of dynamics in subtidal macrozoobenthic assemblages in the Skagerrak: effects of a production-related factor? *Marine Ecology Progress Series*, **38**, 13–23.

Josefson, A. B., Jensen, J. N., and Ærtebjerg, G. (1993). The benthos community structure anomaly in the late 1970s and early 1980s—a result of a major food pulse? *Journal of Experimental Marine Biology and Ecology*, **172**, 31–45.

Kröncke, I., Dippner, J. W., Heyen, H., and Zeiss, B. (1998). Long-term changes in the makrofauna communities off Norderney (East Frisia, Germany) in relation to climate variability. *Marine Ecology Progess Series*, **167**, 25–36.

Kröncke, I., Zeiss, B., and Rensing, C. (2001). Long-term variability in macrofauna species composition off the island of Norderney (East Frisia, Germany), in relation to changes in climatic and environmental conditions. *Senckenbergiana maritima*, **31**, 65–82.

Laine, A. O., Sandler, H., Andersin, A. B., and Stigzelius, J. (1997). Long-term changes of macrozoobenthos in the eastern Gotland Basin and the Gulf of Finland (Baltic Sea) in relation to the hydrohraphic regime. *Journal of Sea Research*, **38**, 135–59.

Ljøen, R. and Svansson, A. (1972). Long-term variations of subsurface temperatures in the Skagerrak. *Deep Sea Research*, **19**, 277–88.

Magurran, A. E. (1988). *Ecological diversity and its measurements*. Princeton University Press, Princeton, NJ.

Nordberg, K., Gustafsson, M., and Krantz, A. L. (2000). Decreasing oxygen concentrations in the Gullmar Fjord, Sweden, as confirmed by benthic foraminifera, and the possible association with NAO. *Journal of Marine Systems*, **23**, 303–16.

Paeth, H. and Hense, A. (1999). Climate change signals in the North Atlantic Oscillation. *CLIVAR Exchanges*, **4**, 25–9.

Pearson, T. H. and Mannvik, H. P. (1998). Long-term changes in the diversity and faunal structure of benthic communities in the northern North Sea: natural variability or induced instability? *Hydrobiologia*, **375/376**, 317–29.

Planque, B. and Taylor, A. H. (1998). Long-term changes in zooplankton and the climate of the North Atlantic. *ICES Journal of Marine Science*, **55**, 644–54.

Reid, C. P., Edwards, M., Hunt, H. G., and Warner, J. (1998). Phytoplankton change in the North Atlantic. *Nature*, **391**, 546.

Reise, K. (1993). Die verschwommene Zukunft der Nordseewatten. In *Klimaänderung und Küste* (eds H. J. Schellnhuber and H. Sterr), pp. 223–32. Springer-Verlag, Berlin.

Rogers, J. C. (1984). The association between the North Atlantic Oscillation and the Southern Oscillation in the northern hemisphere. *Monthly Weather Review*, **112**, 1999–2015.

Rosenberg, R. (1995). Benthic marine fauna structured by hydrodynamic processes and food availability. *Netherlands Journal of Sea Research*, **34**(4), 303–17.

Segerstråle, S. G. (1969). Biological fluctuations in the Baltic Sea. *Progress in Oceanography*, **5**, 169–84.

Siegismund, F. (2001). Long-term changes in the flushing time of the ICES-boxes. In *Burning issues of North Sea ecology: proceedings of the 14th international Senckenberg conference 'North Sea 2000'* (eds I. Kröncke, M. Türkay, and J. Sündermann) *Senckenbergiana maritima*, **31**.

Siegismund, F. and Schrum, C. (2001). Decadal changes in the wind forcing over the North Sea. *Climate Research*, **18**, 39–45.

Simpson, E. H. (1949). Measurements of diversity. *Nature*, **163**, 688.

Stein, M. and Lloret, J. (1995). Stability of water masses—impact on cod recruitment off West Greenland. *Fisheries Oceanography*, **4**, 230–7.

Svansson, A. (1975). Physical and chemical oceanography of the Skagerrak and the Kattegat. 1. Open sea conditions. *Report of Fishery Board of Sweden, Institute of Marine Research*, No. **1**.

Tunberg, B. G. and Nelson, W. G. (1998). Do climatic oscillations influence cyclical patterns of soft bottom macrobenthic communities on the Swedish west coast? *Marine Ecology Progress Series*, **170**, 85–94.

Ware, D. M. (1995). A century and a half of change in the climate of the NE Pacific. *Fisheries Oceanography*, **4**, 267–77.

Wieking, G. and Kröncke, I. (2001). Decadal changes in macrofaunal communities on the Dogger Bank caused by large-scale climate variability. In: *Burning issues of North Sea ecology: proceedings of the 14th international Senckenberg conference 'North Sea 2000'*. (eds I. Kröncke, M. Türkay, and J. Sündermann) *Senckenbergiana maritima*, **31**, 125–41.

Witbaard, R. (1996). Growth variations in Arctica islandica L. (Mollusca): a reflection of hydrography-related food supply. *ICES Journal of Marine Science*, **53**, 981–7.

Zeiss, B. and Kröncke, I. (1997). Long-term macrofaunal studies in a subtidal habitat off Norderney (East Frisia, Germany) from 1978 to 1994. I. The late winter samples. *Oceanologica Acta*, **20**(1), 311–18.

Ziegelmeyer, E. (1964). Einwirkungen des kalten Winters 1962/63 auf das Makrobenthos im Ostteil der Deutschen Bucht. *Helgoländer Meeresunters*, **10**, 276–82.

Chapter 10

Alheit, J. and Hagen, E. (1997). Long-term climate forcing of European herring and sardine populations. *Fisheries Oceanography*, **6**, 130–9.

Botsford, L. W., Castilla, J. C., and Peterson C. H. (1997). The management of fisheries and marine ecosystems. *Science*, **277**, 509–15.

Brockwell, P. J. and Davies, R. A. (1991). *Time series, theory and methods*, 2nd edn. Springer-Verlag Inc., New York.

Brown, J. H., Ernest, S. K. M., Parody, J. M., and Haskell, J. P. (2001). Regulation of diversity, maintenance of species richness in changing environments. *Oecologia*, **126**, 321–32.

Burnham, K. P. and Overton, W. S. (1979). Robust estimation of population size when capture probabilities vary among animals. *Ecology*, **60**, 927–36.

Colton, J. B. Jr. (1972). Temperature trends and the distribution of ground fish in continental shelf waters, Nova Scotia to Long Island. *Fisheries Bulletin*, **70**, 637–57.

Cushing, D. H. (1982). *Climate and fisheries*. Academic Press, London.

Duarte, C. M. (2000). Marine biodiversity and ecosystem services, an elusive link. *Journal of Experimental Marine Biology and Ecology*, **250**, 117–31.

Emmerson, M., Solan, C. M., Emes, C., Paterson, D. M., and Raffaelli, D. (2001). Consistent patterns and the idiosyncratic effects of biodiversity in marine ecosystems. *Nature*, **411**, 73–77.

Emmerson, M. C. and Raffaelli, D. G. (2000). Detecting the effects of diversity on measures of ecosystem function, experimental design, null models and empirical observations. *Oikos*, **91**, 195–203.

Fogarty, M. J. and Murawski, S. A. (1998). Large-scale disturbance and the structure of marine system, Fishery impacts on Georges Bank. *Ecological Applications*, **8**, S6–22.

Fromentin, J. M., Stenseth, N. C., Gjøsæter, J., Bjørnstad, O. N., Falck, W., and Johannessen, T. (1997). Spatial patterns of the temporal dynamics of three gadoid species along the Norwegian Skagerrak coast. *Marine Ecology Progress Series*, **155**, 209–22.

Fromentin, J. M., Stenseth, N., Gjøsæter, J., Johannessen, T., and Planque, B. (1998). Long-term fluctuations in cod and pollack along the Norwegian Skagerrak coast. *Marine Ecology Progress Series*, **162**, 265–78.

Gjøsæter, J., Stenseth, N. C., Ottersen, G., Lekve, K., Dahl, E., Danielssen, D. S., and Christie, H. (2004a). The key fish species along the Norwegian Skagerrak coast with an emphasis on immature cod, their environment and their interactions. Pages xx–xx in *Dynamics of the cod along the Norwegian Skagerrak coast* (eds N. C. Stenseth, J. Gjøsæter, and K. Lekve), Blackwell, Cambridge.

Gjøsæter, J., Stenseth N. C., Sollie, A., and Lekve, K. (2004b). The Flødevigen beach seine surveys, a unique long-term monitoring program. In *Dynamics of the cod along the Norwegian Skagerrak coast* (eds N. C. Stenseth, J. Gjøsæter, and K. Lekve), Blackwell, Cambridge.

Gulland, J. A. and Garcia, S. (1984). Observed patterns in multispecies fisheries. In (ed. R. M. May), *Exploitation of marine communities, report of the Dahlem Workshop on Exploitation of Marine Communities*, pp. 155–90. Springer-Verlag, Berlin.

Hines, J. E., Boulinier, T., Nichols, J. D., Sauer, J. R., Pollock, K. H., (1999). COMDYN: software to study the dynamics of animal communities using a capture-recapture approach. *Bird study*, **46**, 209–17.

Hjort, J. (1914). Fluctuations in the great fisheries of northern Europe viewed in the light of biological research. Journal *Conseil International pour l'Exploration de la mer*, **20**, 1–228.

Hughes, J. B. and Roughgarden, J. (1998). Aggregate community properties and the strength of species' interactions. *Proceedings of National Academy of Science, Washington*, **95**, 6837–42.

Hughes, J. B. and Roughgarden, J. (2000). Species diversity and biomass stability. *American Naturalist*, **155**, 618–27.

Hughes, R. G. (1986). A model of the structure and dynamics of benthic marine invertebrate communities. *Marine Ecology Progress Series*, **15**, 1–11.

Hurrell, J. W. (1995). Decadal trends in the North Atlantic Oscillation, regional temperatures and precipitation. *Science*, **269**, 676–9.

Hurrell, J. W. (1996). Influence of variations in extratropical wintertime teleconnections on Northern Hemisphere temperature. *Geophysical Research Letters*, **23**, 665–8.

Hurrell, J. W. and van Loon, H. (1997). Decadal variations in climate associated with the north Atlantic Oscillation. *Climatic Change*, **36**, 301–26.

Kirchner, J. W. (2002). Evolutionary speed limits inferred from the fossil record. *Nature*, **415**, 65–8.

Lawton, J. H. (2000). Biodiversity and ecosystem processes, theory, achievments and future directions. In *The Biology of biodiversity* (ed. M. Kato), pp. 119–131. Springer, Tokyo.

Lekve, K. (2001). Spatio-temporal community dynamics of fish in the coastal zone of the Norwegian Skagerrak coast—Patterns and processes. Doctoral thesis. University of Oslo, Oslo.

Lekve, K., Boulinier, T., Stenseth, N. C., Gjøsåter, J., Fromentin, J.-M., Hines, J. E., and Nichols, J. D. (2002a). Spatio-temporal dynamics of species richness in coastal fish communities. *Proceedings of the Royal Society of London, B*, **1502**, 1781–9.

Lekve, K., Ottersen, G., Stenseth, N. C., and Gjøsåter, J. (2002b). Length dynamics in juvenile coastal Skagerrak cods, effects of biotic and abiotic processes. *Ecology*, **86**, 1676–88.

Lekve, K., Stenseth, N. C., Johansen, R., Lingjårde, O. C., and Gjøsåter, J. (2003). Richness-dependent and climatic factors regulating the dynamics of coastal fish diversity. *Ecological Letters*, **6**, 428–39.

Lewin, R. (1986). Supply-side ecology. *Science*, **234**, 25–7.

Loreau, M. (2000). Are communities saturated? On the relationship between alpha, beta and gamma diversity. *Ecology Letters*, **3**, 73–6.

MacArthur, R. H. and Wilson, E. O. (1967). *The theory of island biogeography*. Princeton University Press, Princeton, NJ.

Mann, K. H. and Lazier, J. R. N. (1991). *Dynamics of marine ecosystems*. Blackwell, Cambridge.

May, R. A. (ed.) (1984). Exploitation of marine communities, report of the Dahlem Workshop on Exploitation of Marine Communities, Springer-Verlag, Berlin.

McCann, K., Hastings, A., and Huxel, G. R. (1998). Weak trophic interactions and the balance of nature. *Nature*, **395**, 794–8.

McCullagh, P. and Nelder, J. A. (1989). *Generalized linear models*. Chapman & Hall, London.

Menge, B. A. (1992). Community regulation, under what conditions are bottom-up factors important on rocky shores. *Ecology*, **73**, 755–65.

Micheli, F., Cottingham, K. L, Bascompte, J., Bjørnstad, O. N., Eckert, G. L., Fisher, J. M., Keitt, T. H., Kendall, B. E., Klug, J. L., and Rusak, J. A. (1999). The dual nature of community variability. *Oikos*, **85**, 161–9.

Murawski, S. A. and Idoine, J. S. (1992). Multispecies size composition, a conservative property of exploited fishery systems? *Journal of Northwest Atlantic Fisheries Science*, **14**, 79–85.

Myers, R. A., Hutchings, J. A., and Barrowman, N. J. (1997). Why do fish stocks collapse? The example of cod in Atlantic Canada. *Ecological Applications*, **7**, 91–106.

Naeem, S. and Li, S. B. (1997). Biodiversity enhances ecosystem reliability. *Nature*, **390**, 507–9.

Ottersen, G., Planque, B., Belgrano, A., Post, E., Reid, P. C., and Stenseth, N. C. (2001). Ecological effects of the North Atlantic Oscillation. *Oecologia*, **128**, 1–14.

Ottersen, G. and Stenseth, N. C. (2001). Atlantic climate governs oceanographic and ecological variability in the Barents Sea. *Limnology and Oceanography*, **46**, 1774–80.

Ottersen, G. and Sundby, S. (1995). Effects of temperature, wind and spawning stock biomass on recruitment of Arcto-Norwegian cod. *Fisheries Oceanography*, **4**, 278–92.

Paine, R. T. (1966). Food web complexity and species diversity. *American Naturalist*, **100**, 65–75.

Parody, J. M., Cuthbert, F. J., and Decker, E. H. (2001). The effect of 50 years of landscape change on species richness and community composition. *Global Ecology and Biogeography*, **10**, 305–13.

Pfisterer, A. B. and Schmid, B. (2002). Diversity-dependent production can decrease the stability of ecosystem functioning. *Nature*, **416**, 84–6.

Raffaelli, D. and Hawkins, S. (1996). *Intertidal ecology*. Chapman & Hall, London.

Rose, G. A., deYoung, B., Kulka, D. W., Goddard, S. V., and Fletcher, G. L. (2000). Distribution shifts and overfishing the northern cod (*Gadus morhua*), a view from the ocean. *Canadian Journal of Fisheries and Aquatic Sciences*, **57**, 644–63.

Roughgarden, J., Gaines, S., and Possingham, H. (1988). Recruitment dynamics in complex life cycles. *Science*, **241**, 1460–6.

Royama, T. (1992). *Analytic population dynamics*. Chapman & Hall, London.

Russel, F. S. (1973). A summary of the observations on the occurrence of planktonic stages of fish off Plymouth 1924–1972. *Journal of the Marine Biological Association of The United Kingdom*, **53**, 347–55.

Safina, C. (1995). The world's imperiled fish. *Scientific American*, **273**, 46–53.

Schwartzlose, R., Alheit, J., Bakun, A., Baumgartner, T., Cloete, R., Crawford, R., Fletcher, W., GreenRuiz, Y., Hagen, E., Kawasaki, T., Lluch-Belda, D., Lluch-Cota, S., MacCall, A., Matsuura, Y., Nevarez-Martinez, M., Parrish, R., Roy, C., Serra, R., Shust, K., Ward, M., and Zuzunaga, J. (1999). Worldwide large-scale fluctuations of Sardine and anchovy populations. *South African Journal of Marine Science*, **21**, 289–347.

Silvert, W. and Crawford, R. J. M. (1988). The periodic replacement of one fish stock by another. In *Long term changes in marine fish populations* (ed. M. G. Larrañeta), pp. 161–80. Instituto de Investigaciones Marinas, Vigo.

Southward, A. J., Hawkins, S. J., and Burrows, M. T. (1995). Seventy years' observations of changes in distribution and abundance of zooplankton and intertidal organisms in the Western English channel in relation to rising sea temperature. *Journal of Thermal Biology*, **20**, 127–55.

Stenseth, N. C., Lekve, K., and Gjøsæter, J. (2002). Modelling species richness controlled by community-intrinsic and community-extrinsic processes, coastal fish communities as an example. *Population Ecology*, **44**, 165–78.

Stevens, J. D., Bonfil, R., Dulvy, N. K., and Walker, P. A. (2000). The effects of fishing on sharks, rays, and chimaeras (chondrichthyans), and the implications for marine ecosystems. *ICES Journal of Marine Science*, **57**, 476–94.

Taylor, C. C., Bigelow, H. B., and Graham, H. W. (1957). Climate trends and the distribution of marine animals in New England. *Fisheries Bulletin*, **57**, 293–345.

Taylor, A. H. and Stepens, J. A. (1998). The North Atlantic Oscillation and the latitude of the Gulf Stream. *Tellus Series a-Dynamic Meteorology and Oceanography*, **50**, 134–42.

Tilman, D., Naeem, S., Knops, J., Reich, P., Siemann, E., Wedin, D., Ritchie, M., and Lawton, J. (1997). Biodiversity and ecosystem properties. *Science*, **278**, 1866–7.

Turchin, P. (1995). Population regulation, old arguments and a new synthesis. In *Population dynamics* (ed. P. Price), pp. 19–40. Academic Press, New York.

Underwood, A. J. and Fairweather, P. G. (1989). Supply-side ecology and benthic marine assemblages. *Trends in Ecology and Evolution*, **4**, 16–20.

Venables, W. N. and Ripley, B. D. (1997). *Modern applied statistics with S-PLUS*. Springer, New York.

Wootton, R. J. (1990). *Ecology of teleost fishes*. Chapman & Hall, London.

Chapter 11

Alexander, M. A., Timlin, M. S., and Scott J. D. (2001). Winter-to-winter recurrence of sea surface temperature, salinity and mixed-layer depth anomalies. *Progress in Oceanography*, **49**, 41–61.

Au, D. W. and Cayan, D. R. (1999). North Pacific albacore catches and decadal-scale climatic shifts. SCTB 12, 16–23 June, Tahiti. *Working Paper*, Alb-2.

Bertignac, M., Lehodey, P., and Hampton, J. (1998). A spatial population dynamics simulation model of tropical tunas using a habitat index based on environmental parameters. *Fisheries Oceanography*, **7**(3/4), 326–35.

Brodeur, R. D. and Ware, D. M. (1992). Long-term variability in zooplankton biomass in the subarctic Pacific Ocean. *Fisheries Oceanography*, **1**, 32–8.

Chai, F., Dugdale, R. C., Peng, T.-H., Wilkerson, F. P., and Barber, R. T. (2002). One dimensional ecosystem model of the equatorial pacific upwelling system, Part I: model development and silicon and nitrogen cycle. *Deep-Sea Research* II, **49**(13–14), 2713–45.

Chai, F., Jiang, M., Barber, R. T., Dugdale, R. C., and Chao, Y. (2003). Interdecadal variation of the transition zone chlorophyll front, a physical-biological model simulation between 1960 and 1990. *Journal of Oceanography*, **59**(4), 461–75.

Delcroix, T., Eldin, G., and Radenac, M.-H. (1992). Variation of the western equatorial Pacific Ocean, 1986–1988. *Journal of Geophysical Research*, **97**(C4), 5423–45.

Enfield, D. B. and Mayer, D. A. (1997). Tropical Atlantic sea surface temperature variability and its relation to El Niño-Southern Oscillation. *Journal of Geophysical Research*, **102**(C1), 929–45.

Fournier, D. A., Hampton, J., and Sibert, J. R. (1998). MULTIFAN-CL: a length-based, age-structured model for fisheries stock assessment, with application to South Pacific albacore, *Thunnus alalunga*. *Canadian Journal of Fisheries and Aquatical Science*, **55**(9), 2105–16.

Francis, R. C. and Hare, S. R. (1994). Decadal-scale regime shifts in the large marine ecosystems of the North-east Pacific: a case for historical science. *Fisheries Oceanography*, **3**, 279–91.

Gammelsrød, T., Bartholomae, C. H., Boyer, D. C., Filipe, V. L. L., and O'Toole, M. J. (1998). Intrusion of warm surface water along the Angolan–Namibian coast in February–March 1995: the 1995 Benguela Niño. *South African Journal of Marine Science*, **19**, 41–56.

Gershunov, A. and Barnett, T. P. (1998). Interdecadal modulation of ENSO teleconnections. *Bulletin of the American Meteorological Society*, **79**, 2715–25.

Gershunov, A., Barnett, T., and Cayan, D. (1999). North Pacific interdecadal oscillation seen as factor in ENSO-related north American climate anomalies. *EOS*, **80**, 25–30.

Hampton, J. and Fournier, D. A. (2001). Stock assessment of skipjack tuna in the western and central Pacific Ocean. Fourteenth Standing Committee on Tuna and Billfish, Secretariat of the Pacific Community, Noumea, New Caledonia. *Working Paper*, SKJ-1.

Hare, S. R. and Mantua, N. J. (2000). Empirical evidence for North Pacific regime shifts in 1977 and 1989. *Progress in Oceanography*, **47**(2–4), 103–46.

Jin, F.-F. (1996). Tropical ocean-atmosphere interaction, the Pacific Cold Tongue, and the El Niño-southern oscillation. *Science*, **274**, 76–8.

Jin F.-F., (2001). Low frequency modes of tropical ocean dynamics. *Journal of Climate*, **14**, 3874–81.

Latif, M. and Barnett, T. P. (1994). Causes of decadal climate variability over the north Pacific and North America. *Science*, **266**, 634–7.

Kirtman, B. P. and Schopf, P. S., (1998). Decadal variability in ENSO predictability and prediction. *Journal of Climate*, **11**, 2804–22.

Lau, N. C. and Nath, M. J. (1996). The role of the 'atmospheric bridge' in linking tropical Pacific ENSO events to extratropical SST anomalies. *Journal of Climate*, **9**, 2036–57.

Lehodey, P. (2001). The pelagic ecosystem of the tropical Pacific Ocean: dynamic spatial modelling and biological consequences of ENSO. *Progress in Oceanography*, **49**, 439–68.

Lehodey, P., Chai, F., and Hampton J., (2003). Modelling climate-related variability of tuna populations from a coupled ocean-biogeochemical-populations dynamics model. *Fisheries Oceanography*, **12**(4/5), 483–94.

Lehodey P., Bertignac, M., Hampton, J., Lewis, A., and Picaut, J. (1997). El Nino Southern Oscillation and tuna in the western Pacific. *Nature*, **389**, 715–18.

Lehodey P., Andre, J.-M., Bertignac, M., Hampton, J., Stoens, A., Menkes, C., Memery, L., Grima, N. (1998). Predicting skipjack tuna forage distributions in the equatorial Pacific using a coupled dynamical bio-geochemical model. *Fisheries Oceanography*, **7**(3–4), 317–25.

Lukas, R. and Lindström, E. (1991). The mixed layer of the western equatorial Pacific Ocean. *Journal of Geophysical Research*, **96**, 3343–457.

Mantua, N. J., Hare, S. R., Zhang, Y., Wallace, J. M., and Francis, R. C. (1997). A Pacific decadal climate oscillation with impacts on salmon. *Bulletin of the American Meteorological Society*, **78**, 1069–79.

Mantua, N. J. and Hare S. R. (2002). The pacific decadal oscillation. *Journal of Oceanography*, **58**(1), 35–44.

McCabe, G. J. and Dettinger, M. D. (1999). Decadal variations in the strength of ENSO teleconnections with precipitation in the western United States. *International Journal of Climatology*, **19**, 1399–410.

McPhaden, M. J. and Picaut, J. (1990). El Niño-Southern Oscillation Index displacements of the western equatorial Pacific warm pool. *Science*, **50**, 1385–88.

McPhaden, M. J. and Zhang, D. (2002). Slowdown of the meridional overturning circulation in the upper Pacific Ocean. *Nature*, **415**, 603–8.

Miller A. J. and Schneider, N. (2000). Interdecadal climate regime dynamics in the North Pacific Ocean: theories, observations and ecosystem impacts. *Progress in Oceanography*, **47**, 355–79.

Minobe, S. (1997). A 50–70 year climatic oscillation over the North Pacific and North America. *Geophysical Research Letters*, **24**, 683–6.

Minobe, S. (1999). Resonance in bidecadal and pentadecadal climate oscillations over the North Pacific: role in climatic regime shifts. *Geophysical Research Letters*, **26**, 855–8.

Murtugudde, R. G., Signorini, S. R., Christian, J. R., Busalacchi, A. J., McClain, C. R., and Picaut, J. (1999). Ocean color variability of the tropical Indo-Pacific Basin observed by Sea WIFS during 1977–1998. *Journal of Geophysical Research*, **104**, 18351–66.

Philander, S. G. H. (1990). *El Niño, La Niña, and the Southern Oscillation*. Academic, San Diego, CA, USA.

Picaut, J., Ioualalen, M., Delcroix, T., Masia, F., Murtuggude, R., and Vialard, J. (2001). The oceanic zone of convergence on the eastern edge of the Pacific warm pool: a synthesis of results and implications for El Niño-Southern Oscillation and biogeochemical phenomena. *Journal of Physical Research*, **106**(C2), 2363–86.

Polovina, J. J., Mitchum, G. T., and Evans, G. T. (1995). Decadal and basin-scale variation in mixed layer depth and the impact on biological production in the Central and North Pacific, 1969–88. *Deep Sea Research I*, **42**(10), 1701–16.

Polovina, J. J., Mitchum, G. T., Graham, N., Craig, M. P., Demartini, E. E., and Flint, E. N. (1994). Physical and biological consequences of a climate event in the central North Pacific. *Fisheries Oceanography*, **3**(1), 15–21.

Rudnick, D. L. and Davis, R. E. (2003). Red noise and regime shifts. *Deep Sea Research, Part I*, **50**(6), 691–9.

Schaefer, K. M. (1998). Reproductive biology of yellowfin tuna (*Thunnus albacares*) in the eastern Pacific Ocean. *Bulletin. Inter-American Tropical Tuna Commission/Bulletin*, **21**(5), 205–72. Bayliff, W. H. La Jolla, California, USA.

Schwartzlose, R. A., Alheit, J., Bakun, A., Baumgartner, T. R., Cloete, R., Crawford, R. J. M., Fletcher, W. J., Green-Ruiz, Y., Hagen, E., Kawasaki, T., Lluch-Belda, D., Lluch-Cota, S. E., MacCall, A. D., Matsuura, Y., Nevarez-Martinez, M. O., Parrish, R. H., Roy, C., Serra, R., Shust, K. V., Ward, M. N., and Zuzunaga, J. Z. (1999). Worldwide large-scale fluctuations of sardine and anchovy populations. *South African Journal of Marine Science*, **21**, 289–347.

Signorini, S. R., Murtugudde, R. G., McClain, C. R., Christian, J. R., Picaut, J., and Busalacchi, A. J. (1999). Biological and physical signatures in the tropical and subtropical Atlantic. *Journal of Geophysical Research*, **104**(C8), 18367–82.

Trenberth, K. E. (1990). Recent observed interdecadal climate changes in the northern hemisphere. *Bulletin of the American Meteorological Society*, **71**, 988–93.

Trenberth, K. E. and Hurrell, J. W. (1994). Decadal atmosphere-ocean variations in the Pacific. *Climate Dynamics*, **9**, 303.

Venrick, E. L., McGowan J. A., Cayan D. R., and Hayward, T. L. (1987). Climate and chlorophyll a: long-term trends in the central north Pacific ocean. *Science*, **238**, 70–2.

Vialard, J. and Delecluse, P. (1998). An OGCM study for the TOGA decade. Part I: role of salinity in the physics of the western Pacific fresh pool. *Journal of Physical Oceanography*, **28**, 1071–88.

Zhang, Y., Wallace, J. M., and Battisti, D. S. (1997). ENSO-like interdecadal variability: 1900–93. *Journal of Climate*, **10**, 1004–20.

Chapter 12

Anderson, P. J. and Piatt, J. F. (1999). Community reorganization in the Gulf of Alaska following ocean climate regime shift. *Marine Ecology Progress Series*, **189**, 117–23.

Bailey, K. M., Macklin, S. A., Reed, R. K., Brodeur, R. D., Ingraham, W. J., Piatt, J. F., Shima, M., Francis, R. C., Anderson, P. J., Royer, T. C., Hollowed, A. B., Somerton, D. A., and Wooster, W. S. (1995). The 1991–92 ENSO in the Gulf of Alaska and effects on selected marine fisheries. *Reports of California Cooperative Oceanic Fisheries Investigations*, **36**, 78–96.

Bailey, K. M. (2000). Shifting control of recruitment of walleye pollock (*Theragra chalcogramma*) after a major climate and ecosystem change. *Marine Ecology Progress Series*, **198**, 215–24.

Bailey, K. M. (2002). Complex processes in the survival of walleye pollock larvae and forecasting implications. *Journal of Fisheries Science*, **68**, 200–5.

Bailey, K. M. and Incze, L. (1985). El Niño and the early life history and recruitment of fishes in temperate marine waters. In *El Niño North* (eds W. S. Wooster and D. L. Fluharty), pp. 143–65. Washington Sea Grant, Seattle.

Bailey, K. M. and Macklin, S. A. (1994). Analysis of patterns in larval walleye pollock *Theragra chalcogramma* survival and wind mixing events in Shelikof Strait, Gulf of Alaska. *Marine Ecology Progress Series*, **113**, 1–13.

Bailey, K. M. and Picquelle, S. (2002). Larval distribution patterns of offshore spawning flatfish in the Gulf of Alaska: sea valleys as transport pathways and enhanced inshore transport during ENSO events. *Marine Ecology Progress Series*, **236**, 205–17.

Bailey, K. M., Brodeur, R. D., and Hollowed, A. B. (1996). Cohort survival patterns of walleye pollock, *Theragra chalcogramma*, in Shelikof Strait, Alaska: a critical factor approach. *Fisheries Oceanography*, **5**(Suppl. 1), 179–88.

Baumgartner, T. R., Soutar, A., and Ferreira-Bartrina, V. (1992). Reconstruction of the history of Pacific sardine and northern anchovy populations over the past two millennia from sediments of the Santa Barbara Basin, California. *Reports of California Cooperative Oceanic Fisheries Investigations*, **33**, 24–40.

Beamish, R. (1994). Climate change and exceptional fish production off the west coast of North America. *Canadian Journal of Fisheries and Aquatic Sciences*, **50**, 2270–91.

Bond, N. A. and Harrison, D. E. (2000). The Pacific decadal oscillation, air-sea interaction and central north Pacific winter atmospheric regimes. *Geophysical Research Letters*, **27**(5), 731–4

Brodeur, R. D. and Bailey, K. (1995). Predation on the early life history of marine fishes: a case study of walleye pollock in the Gulf of Alaska. In *Survival strategies in early life stages of marine resources* (eds Y. Watanabe, Y. Yamashita, and Y. Oozeki), pp. 245–59. Balkema Press, Rotterdam, Netherlands.

Canino, M. F., Bailey, K. M., and Incze, L. S. (1991). Temporal and geographic differences in feeding and nutritional condition of walleye pollock larvae *Theragra chalcogramma* in Shelikof Strait, Gulf of Alaska. *Marine Ecology Progress Series*, **79**, 27–35.

Davis, M. W. (2001). Behavioral responses of walleye pollock, *Theragra chalcogramma*, larvae to experimental gradients of sea water flow: implications for vertical distribution. *Environmental Biology of Fishes*, **61**, 253–60.

Drinkwater, K. F. and Myers, R. M. A. (1987). Testing predictions of marine fish and shellfish landings from environmental variables. *Canadian Journal of Fisheries and Aquatic Sciences*, **44**, 1568–73.

Finney, B. P., Gregory-Evans, I., Sweetman, J., Douglas, M. S. V., and Smol, J. P. (2000). Impacts of climatic change and fishing on Pacific salmon abundance over the past 300 years. *Science*, **290**, 795–9.

Francis, R. C., Hare, S. R., Hollowed, A. B., and Wooster, W. S. (1998). Effects of interdecadal climate variability on the oceanic ecosystems of the northeast Pacific. *Fisheries Oceanography*, **7**, 1–20.

Hare, S. R. and Mantua, N. J. (2000). Empirical evidence for North Pacific regime shifts in 1977 and 1989. *Progress in Oceanography*, **47**, 103–45.

Helle, K., Bogstad, B., Marshall, C. T., Michalsen, K., Ottersen, G., and Pennington, M. (2000). An evaluation of recruitment indices for Arcto-Norwegian cod (*Gadus morhua* L.). *Fisheries Research*, **48**, 55–67.

Herbert, T. D., Schuffert, J. D., Andreasen, D., Heusser, L., Lyle, M., Mix, A., Ravelo, A. C., Stott, L. D., and Herguera, J. C. (2001). Collapse of the California Current during glacial maxima linked to climate change on land. *Science*, **293**, 71–6.

Hinckley, S., Bailey, K. M., Picquelle, S., Schumacher, J. D., and Stabeno, P. J. (1991). Transport, distribution and abundance of larval and juvenile walleye pollock (*Theragra chalcogramma*) in the western Gulf of Alaska. *Canadian Journal of Fisheries and Aquatic Sciences*, **48**, 91–8.

Hollowed, A. B., Bailey, K. M., and Wooster, W. S. (1987). Patterns in recruitment of marine fishes in the northeast Pacific ocean. *Biological Oceanography*, **5**, 99–131.

Hollowed, A. B., Hare, S. R., and Wooster, W. S. (2001). Pacific basin climate variability and patterns of Northeast Pacific marine fish production. *Progress in Oceanography*, **49**, 257–82.

Jackson, J. B. C., Kirly, M. X., Berger, W. H., Bjorndal, K. A., Botsford, L.W., Bourque, B.J., Bradbury, R. H., Cooke, R., Erlandson, J., Estes, J. A., Hughes, T. P., Kidwell, S., Lange, C. B., Lenihan, H. S., Pandolfi, J. M., Peterson, C. H., Steneck, R. S., Tegner, M. J., and Warner, R. R. (2001). Historical overfishing and the recent collapse of coastal ecosystems. *Science*, **293**, 629–38.

Kendall, A. W., Incze, L. S., Ortner, P. B., Cummings, S., and Brown, P. K. (1994). Vertical distribution of eggs and larvae of walleye pollock, *Theragra chalcogramma*, in Shelikof Strait, Gulf of Alaska. *Fisheries Bulletin, U.S.*, **92**, 540–54.

Lea, R. N. and Rosenblatt, R. H. (2000). Observations on fishes associated with the 1997–98 El Niño off California. *Reports of California Cooperative Oceanic Fisheries Investigations*, **41**, 117–29.

Mantua, N. J., Hare, S. R., Zhang, Y., Wallace, J. M., and Francis, R. C. (1997). A Pacific interdecadal climate oscillation with impacts on salmon production. *Bulletin of the American Meteorological Society*, **78**, 1069–79.

Olla, B. L., Davis, M. W., Ryer, C. H., and Sogard, S. M. (1996). Behavioral determinants of distribution and survival in early stages of walleye pollock, *Theragra chalcogramma*: a synthesis of experimental studies. *Fisheries Oceanography*, **5**(Suppl. 1), 167–78.

Omori, M. and Kawasaki, T. (1995). Scrutinizing the cycles of worldwide fluctuations in the sardine and herring populations by means of singular spectrum analysis. *Bulletin of the Japanese Society of Fisheries Oceanography*, **59**, 361–70.

Overland, J. E., Adams, J. M., and Bond, N. A. (1999). Decadal variability of the Aleutian low and its relation to high-latitude circulation. *American Meteorological Society*, **12**, 1542–8.

Pearcy, W. G. and Schoener, A. (1987). Changes in the marine biota coincident with the 1982–1983 El Niño in the northeastern subarctic Pacific Ocean. *Journal of Geophysical Research*, **92**, 14417–28.

Sogard, S. M. and Olla, B. L. (1993). Effects of light, thermoclines and predator presence on vertical distribution and behavioral interactions of juvenile walleye pollock, *Theragra chalcogramma* Pallas. *Journal of Experimental Marine Biology and Ecology*, **167**, 179–95.

Soutar, A. and Isaacs, J. D. (1974). Abundance of pelagic fish during the 19th and 20th centuries as recorded in anaerobic sediment of the Californias. *Fisheries Bulletin, U.S.*, **72**, 257–73.

Southward, A. J., Boalch, G. T., and Maddock, L. (1988). Fluctuations in the herring and pilchard fisheries of Devon and Cornwall linked to change in climate since the 16th Century. *Journal of the Marine Biological Association of the United Kingdom*, **68**, 423–5.

Theilacker, G. H., Bailey, K. M., Canino, M. F., and Porter, S. M. (1996). Variations in larval walleye pollock feeding and conditions: a synthesis. *Fisheries Oceanography*, **5**(Suppl. 1), 112–23.

Thompson, D. W. J. and Wallace, J. M. (1998). The arctic oscillation signature in the wintertime geopotential height and temperature fields. *Geophysical Research Letters*, **25**, 1297–300.

Tunnicliffe, V., O'Connell, J. M., and McQuoid, M. R. (2001). A holocene record of marine fish remains from the northeastern Pacific. *Marine Geology*, **174**, 177–95.

Wyllie-Echeverria, T. and Wooster, W. S. (1998). Year-to-year variations in Bering Sea ice cover and some consequences for fish distributions. *Fisheries Oceanography*, **7**, 159–70.

Chapter 13

Adrian, R. (1997). Calanoid-cyclopoid interactions: evidence from an 11-year field study in a eutrophic lake. *Freshwater Biology*, **38**, 315–25.

Adrian, R., Walz, N., Hintze, T., Hoeg, S., and Rusche, R. (1999). Effects of ice duration on the plankton succession during spring in a shallow polymictic lake. *Freshwater Biology*, **41**, 621–3.

Andersen, V. (1998). Salp and pyrosomid blooms and their importance in biogeochemical cycles. In *The biology of pelagic tunicates* (ed. Q. Bone), pp. 125–37. Oxford University Press, Oxford.

Bathmann, U. V. (1988). Mass occurrence of *Salpa fusiformis* in the spring of 1984 off Ireland: implication for sedimentation processes. *Marine Biology*, **97**, 127–35.

Caron, D. A. and Goldman, J. C. (1990). Protozoan nutrient regeneration. In *Ecology of marine protozoa* (ed. G. M. Capriulo), pp. 283–306. Oxford University Press, Oxford.

Carrick, H. J., Fahnenstiel, G. L., Stoermer, E. F., and Wetzel, R. G. (1991). The importance of zooplankton-protozoan trophic couplings in Lake Michigan. *Limnology and Oceanography*, **36**, 1335–45.

Chmielewski, F.-M. and Rötzer, T. (2002). Annual and spatial variability of the beginning of growing season in Europe in relation to air temperature changes. *Climate Research*, **19**, 257–64.

Cushing, D. H. (1990). Plankton production and year-class strength in fish populations: an update of the match/mismatch hypotheses. *Advances in Marine Biology*, **26**, 249–93.

Drinkwater, K. F., Belgrano, A., Borja, A., Conversi, A., Edwards, M., Greene, C. H., Ottersen, G., Pershing, A. J., and Walker, H. (2003). The response of marine ecosystems to climate variability associated with the North Atlantic Oscillation. In *The North Atlantic Oscillation: climatic significance and environmental impact* (eds J. Hurrell, Y. Kushnir, G. Ottersen, and M. Visbeck), pp. 211–34. American Geophysical Union.

Edwards, M., Reid, P. C., and Planque, B. (2001). Long-term and regional variability of phytoplankton biomass in the Northeast Atlantic (1960–1995). *ICES Journal of Marine Science*, **58**, 39–49.

Ferrier-Pagès, C. and Rassoulzadegan, F. (1994). Seasonal impact of the microzooplankton on pico- and nanoplankton growth rates in the northwest Mediterranean Sea. *Marine Ecology Progress Series*, **108**, 283–94.

Gaedke, U. and Straile, D. (1994). Seasonal changes of the quantitative importance of protozoans in a large lake. An ecosystem approach using mass-balanced carbon flow diagrams. *Marine Microbial Food Webs*, **8**, 163–88.

Gaedke, U., Ollinger, D., Bäuerle, E., and Straile, D. (1998). The impact of the interannual variability in hydrodynamic conditions on the plankton development in Lake Constance in spring and summer. *Archiv für Hydrobiologie. Special Issues: Advances in Limnology*, **53**, 565–85.

Gaedke, U., Hochstädter, S., and Straile, D. (2002). Interplay between energy limitation and nutritional deficiency: empirical data and food-web models. *Ecological Monographs*, **72**, 251–70.

George, D. G. and Hewitt, D. P. (1999). The influence of year-to-year variations in winter weather on the dynamics of *Daphnia* and *Eudiaptomus* in Estwaite Water, Cumbria. *Functional Ecology*, **13**(Suppl. 1), 45–54.

Gerten, D. and Adrian, R. (2000). Climate-driven changes in spring plankton dynamics and the sensitivity of shallow polymictic lakes to the North Atlantic Oscillation. *Limnology and Oceanography*, **45**, 1058–66.

Gerten, D. and Adrian, R. (2001). Differences in the persistency of the North Atlantic Oscillation signal among lakes. *Limnology and Oceanography*, **46**, 448–55.

Lampert, W., Fleckner, W., Rai, H., and Taylor, B. E. (1986). Phytoplankton control by grazing zooplankton: a study on the spring clear-water phase. *Limnology and Oceanography*, **31**, 478–90.

Livingstone, D. M. (1999). Ice break-up on southern Lake Baikal and its relationship to local and regional air temperatures in Siberia and to the North Atlantic Oscillation. *Limnology and Oceanography*, **44**, 1486–97.

Madin, L. P. and Deibel, D. (1998). Feeding and energetics of *Thaliacea*. In (ed. Q. Bone), pp. 81–103. Oxford University Press, Oxford.

Müller, H., Schöne, B., Pinto-Coelho, R. M., Schweizer, A., and Weisse, T. (1991). Seasonal succession of ciliates in Lake Constance. *Microbial Ecology*, **21**, 119–38.

Mysterud, A., Stenseth, N. C., Yoccoz, N. G., Ottersen, G., and Langvatn, R. (2003). The response of terrestrial ecosystems to climate variability associated with the North Atlantic Oscillation. In *The North Atlantic Oscillation: climatic significance and environmental impact* (eds J. Hurrell, Y. Kushnir, G. Ottersen, and M. Visbeck), pp. 235–62. American Geophysical Union. Washington, D.C.

Neuer, S. and Cowles, T. J. (1994). Protist herbivory in the Oregon upwelling system. *Marine Ecology Progress Series*, **113**, 147–62.

Ottersen, G. and Stenseth, N. C. (2001). Atlantic climate governs oceanographic and ecological variability in the Barents Sea. *Limnology and Oceanography*, **46**, 1774–80.

Ottersen, G., Planque, B., Belgrano, A., Post, E., and Stenseth, N. C. (2001). Ecological effects of the North Atlantic Oscillation. *Oecologia*, **128**, 1–18.

Scheffer, M., Straile, D., van Nes, E. H., and Hosper, H. (2001). Climatic warming causes regime shifts in lake food webs. *Limnology and Oceanography*, **46**, 1780–83.

Stenseth, N. C., Ottersen, G., Hurrell, J.W., Mysterud, A., Lima, M., Chan, K.-S., Yoccoz, N. G., and Ådlandsvik, B. (2003). Studying climate effects on ecology through the use of climate indices: the North Atlantic Oscillation, El Niño Southern Oscillation and beyond. *Proceedings of the Royal Society of London (Series B.)*, **270**, 2087–96.

Straile, D. (2000). Meteorological forcing of plankton dynamics in a large and deep continental European lake. *Oecologia*, **122**, 44–50.

Straile, D. (2002). Phenology of a food-web response—North Atlantic Oscillation synchronises and speeds up plankton succession in lakes across Central Europe. *Proceedings of the Royal Society of London, B*, **269**, 391–5.

Straile, D. and Adrian, R. (2000). The North Atlantic Oscillation and plankton dynamics in two European lakes—two variations on a general theme. *Global Change Biology*, **6**, 663–70.

Straile, D. and Geller, W. (1998). The response of *Daphnia* to changes in trophic status and weather patterns: a case study from a Lake Constance. *ICES Journal of Marine Science*, **55**, 775–82.

Straile, D., Joehnk, K., and Rossknecht, H. (2003a). Complex effects of winter warming on the physico-chemical characteristics of a deep lake. *Limnology and Oceanography*, **48**, 1432–8.

Straile, D., Livingstone, D. M., Weyhenmeyer, G. A., and George, D. G. (2003b). The response of freshwater ecosystems to climate variability associated with the North Atlantic oscillation. In *The North Atlantic Oscillation: climatic significance and environmental impact* (eds J. Hurrell, Y. Kushnir, G. Ottersen, and M. Visbeck), pp. 263–79. American Geophysical Union.

Weyhenmeyer, G. A. (2001). Warmer winters—are planktonic algal populations in Swedens largest lakes affected? *Ambio*, **30**, 565–71.

Weyhenmeyer, G. A., Blenckner, T., and Pettersson, K. (1999). Changes of the plankton spring outburst related to the North Atlantic Oscillation. *Limnology and Oceanography*, **44**, 1788–92.

Chapter 14

Aas, E. Høkedal, J., Højerslev, N. K., Sandvik, R., and Sakshaug, E. (2002). Spectral properties and UV-attenuation in Arctic marine waters. In *UV-Radiation and Arctic ecosystems* (ed. D. O. Hessen), pp. 23–56. Ecological Studies 153, Springer-Verlag, Berlin.

Borgstrøm, R. (2001). Relationship between spring snow depth and growth of Brown trout Salmo trutta in an alpine lake: predicting consequences of climate change. *Arctic, Antarctic and Alpine Research*, **33**, 476–80.

Botrell, H. H., Duncan, A., Gliwicz, Z. M., Grygierek, E., Herzig, A., Illbricht-Ilkowska, A., Kurasawa, H., Larsson, P., and Weglenska, T. (1976). A review of some problems in zooplankton production studies. *Norwegian Journal of Zoology*, **24**, 419–56.

DeStasio, B. T. Jr., Hill, D. K., Kleinhans, J. M., Nibbelink, N. P., and Magnuson, J. J. (1996). Potential effects of global climate change on small north-temperate lakes: physics, fish, and plankton. *Limnology and Oceanography*, **41**, 1136–49.

Falkowski, P. G. (1997). Evolution of the nitrogen cycle and its influence on the biological sequestration of CO_2 in the ocean. *Nature*, **387**, 272–5.

Falkowski, P. G., Scholes, R. J., Boyle, E., Canadell, J., Canfield, D., Elser, J., Gruber, N., Hibbard, K., Hogberg, P., Linder, S., MacKenzie, F. T., Moore, B., Pedersen, T., Rosenthal, Y., Seitzinger, S., Smetacek, V., and Steffen, W. (2000). The global carbon cycle: a test of our knowledge of Earth as a system. *Science* **290**, 291–6.

Findlay, D. L., Kasian, S. E. M., Stainton, M. P., Beaty, K., and Lyng, M. (2001). Climatic influences on algal populations of boreal forest lakes in the Experimental Lakes Area. *Limnology and Oceanography*, **46**, 1784–93.

Forsberg, C. (1992). Will an increased greenhouse impact in Fennoscandia give rise to more humic and colored lakes? *Hydrobiologia*, **229**, 51–8.

Freeman, C., Evans, C. D., Monteith, D. T., Reynolds, B., and Fenner, N. (2001). Export of organic carbon from peat soils. *Nature*, **412**, 785.

Galloway, J. N., Howarth, R. W., Michaels, A. F., Nixon, S. W., Prospero, J. M., and Dentener, F. J. (1996). Nitrogen and phosphorus budget of the North Atlantic Ocean and its watersheds *Biogeochemistry*, **35**, 3–25.

Ganeshram, R. S., Pedersen, T. F., Calvert, S. E., McNeill, G. W., and Fontugne, M. R. (2000). Glacial-interglacial variability in denitrification in the world's oceans: causes and consequences. *Paleoceanography*, **15**, 361–76.

Granéli, E., Wallström, K., Larsson, U., Granéli, W., and Elmgren, R. (1990). Nutrient limitation of primary production in the Baltic Sea area. *Ambio* **19**, 142–151.

Hanssen-Bauer, I. and Førland, E. J. (1998). Long term trends in precipitation and temperature in the Norwegian Arctic: can they be explained by changes in atmospheric circulation patterns? *Climate Research*, **10**, 143–53.

Hessen, D. O. (1999). Catchment properties and the transport of major elements to estuaries. *Advances in Ecological Research*, **29**, 1–41.

Hessen, D. O. and Wright, R. (1993). Climatic effects on freshwaters: model predictions on acidification, nutrient loading, and eutrophication. In *Impact of climatic change on natural ecosystems with emphasis on boreal and arctic/alpine areas* (eds W. Oechel, J. Holten, and G. Paulsen), pp. 154–67. NINA/DN-report. Trondheim/Norway.

Hessen, D. O., Hindar, A., and Holtan, G. (1997). The significance of nitrogen runoff for eutrophication of freshwater and marine recipients. *Ambio*, **26**, 321–5.

Howarth, R. W., Billen, G., Swaney, D., Townsend, A., Jaworski, N., Lajtha, K., Downing, J. A., Elmgren, R., Caraco, N., Jordan, T., Berendse, F., Freney, J., Kudeyarov, V., Murdoch, P., and Zhao-Liang, Z. (1996). Regional nitrogen budgets and riverine N & P fluxes for the drainages to the North Atlantic Ocean: natural and human influences. *Biogeochemistry*, **35**, 75–139.

IPCC. (2001). Climate change 2001: Working group I: the Scientific Basis www.grida.no/climate/ipcc_tar/wg1/index.htm.

Lampert, W. (1977). Studies on the carbon balance of daphnia pulex de Geer as related to environmental conditions. II. The dependence of carbon assimilation on animal size, temperature, food conditions and diet species. *Archiv fur Hydrobiolgie Supplementband*, **48**, 310–35.

Lenton, T. (2001). The role of land plants, phosphorus weathering and fire in the rise and regulation of atmospheric oxygen. *Global Change Biology*, **7**, 613–29.

Lenton, T. and Watson, A. J. (2000). Redfield revisited: 1. Regulation of nitrate, phosphate and oxygen in the ocean. *Global Biochemical Cycles*, **14**, 225–48.

Maestrini, S. Y. and Graneli, E. (1991). Environmental conditions and ecophysiological mechanisms which led to the 1988 *Chrysochromulina polylepis* bloom—an hypothesis. *Oceanologica Acta*, **14**, 397–413.

Majorowicz, J. A. (1996). Accelerating ground warming in the Canadian Prairie Provinces: is it a result of global warming? *Pure and Applied Geophysics*, **147**, 1–24.

Meybeck, M. (1982). Carbon, nitrogen and phosphorus transport by world rivers. *American Journal of Science* **282**, 401–50.

Meybeck, M. (1993). C, N, P and S in rivers: from sources to global inputs. In *Interactions of C, N, P and S biogeochemical cycles and global change* (eds R. Wollast, F. T. Mackenzie, and L. Chou), pp. 163–93. Springer-Verlag, Berlin.

Nixon, S. W., Ammerman, J., Atkinson, L., Berounsky, V., Billen, G., Boicourt, W., Boynton, W., Church, T., Di Toro, Elmgren, R., Garber, J., Giblin, A., Jahnke, R., Owens, N., Pilson, M. E. Q., and Seitzinger, S. (1996). The fate of nitrogen and phosphorus at the land-sea margin of the North Atlantic Ocean. *Biogeochemistry*, **35**, 141–80.

Opsahl, S., Benner, R., and Amon, R. M. W. (1999). Major flux of terrigenous dissolved organic matter through the Arctic Ocean. *Limnology and Oceanography*, **44**, 2017–23.

Pettersson, C., Allard, B., and Boren, H. (1997). River discharge of humic substances and humic-bound metals to the Gulf of Bothnia. *Estuarine Coastal and Shelf Science*, **44**, 533–41.

Scheffer, M., Straile, D., van Nes, E. H., and Hosper, H. (2001). Climatic warming causes regime shifts in lake food webs. *Limnology and Oceanography*, **46**, 1780–3.

Schindler, D. W., Bayley, S. E., Parker, B. R., Beaty, K. B., Cruikshank, D. R., Fee, E. J., Schindler, E. U., and Stainton, M. P. (1996). The effects of climatic warming on the properties of boreal lakes and streams at the Experimental Lakes Area, northwestern Ontario. *Limnology and Oceanography*, **41**, 1004–17.

Smayda, T. J. (1990). Novel and nuisance phytoplankton blooms in the sea: evidence for a global epidemic. In *Toxic marine phytoplankton* (eds E. Granéli, B. Sundström, L. Edler, and D. M. Anderson), pp. 29–40. Elsevier, NY.

Stefan, H. G., Hondzo, M., Eaton, J. G., and McCormick, J. H. (1995). Validation of a fish habitat model for lakes. *Ecological Modelling*, **82**, 211–24.

Straile, D. and Adrian, R. (2000). The North Atlantic oscillation and plankton dynamics in two European lakes— two variations on a general theme. *Global Change Biology*, **6**, 663–70.

Straile, D., Livingstone, D. M., Weyhenmeyer, G. A., and George, D. G. (2003). The response of freshwater ecosystems to climate variability associated with the North Atlantic oscillation. In *The North Atlantic Oscillation: climatic significance and environmental impact* (eds J. Hurrell, Y. Kushnir, G. Ottersen, and M. Visbeck), pp. 263–79. American Geophysical Union.

Vitousek, P. M., Aber, J., Howarth, R. W., Likens, G. E., Matson, P. A., Schindler, D. W., Schlesinger, W. H., and Tilman, G. D. (1997). Human alterations of the global nitrogen cycle: sources and consequences. *Ecological Applications*, **7**, 737–50.

Watson, A. J., Bakker, D. C. E., Ridgwell, A. J., Boyd, P. W., Law, C. S. (2000). Effect of iron supply on Southern Ocean CO_2 uptake and implications for glacial atmospheric CO_2. *Nature*, **407**, 730–3.

Chapter 15

Forchhammer, M. C. (2001). Terrestrial ecological responses to climate change in the Northern Hemisphere. In *Climate change research: Danish contributions*. Gads Forlag, Copenhagen.

Forchhammer, M. C., Post, E., and Stenseth, N. C. (1998). Breeding phenology and climate. *Nature*, **391**, 29–30.

Forchhammer, M. C., Clutton-Brock, T. H., Lindström, J., and Albon, S. D. (2001). Climate and density induce long-term cohort variation in a northern ungulate. *Journal of Animal Ecology*, **70**, 721–9.

Fromentin, J.-M. and Planque, B. (1996). Calanus and environment in the eastern North Atlantic. II. Influence of the North Atlantic Oscillation on *C. finmarchicus* and *C. helgolandicus*. *Marine Ecology Progress Series*, **134**, 111–18.

Oba, G., Post, E., and Stenseth, N. C. (2001). Sub-saharan desertification and productivity are linked to hemispheric climate variability. *Global Change Biology*, **7**, 1–6.

Ottersen, G., Planque, B., Belgrano, A., Post, E., Reid, P. C., and Stenseth, N. C. (2001). Ecological effects of the North Atlantic Oscillation. *Oecologia*, **128**, 1–14.

Post, E. and Forchhammer, M. C. (2001). Pervasive influence of large-scale climate in the dynamics of a terrestrial vertebrate community. *BMC Ecology*, **1**, 5.

Post, E. and Stenseth, N. C. (1999). Climatic variability, plant phenology, and northern ungulates. *Ecology*, **80**, 1322–39.

Post, E., Stenseth, N. C., Langvatn, R., and Fromentin, J.-M. (1997). Global climate change and phenotypic variation among red deer cohorts. *Proceedings of the Royal Society of London, B*, **264**, 1317–24.

Post, E., Peterson, R. O., Stenseth, N. C., and McLaren, B. E. (1999a). Ecosystem consequences of wolf behavioural response to climate. *Nature*, **401**, 905–7.

Post, E., Forchhammer, M. C., and Stenseth, N. C. (1999b). Population ecology and the North Atlantic Oscillation (NAO). *Ecological Bulletins*, **47**, 117–25.

Post, E., Forchhammer, M. C., Stenseth, N. C., and Callaghan, T. V. (2001). The timing of life history events in a changing climate. *Proceedings of the Royal Society of London, B*, **268**, 15–23.

Thompson, P. M. and Ollason, J. C. (2001). Lagged effects of ocean climate change on fulmar population dynamics. *Nature*, **413**, 417–20.

Chapter 16

Ainley, D. G., Carter, H. R., Anderson, D. W., Briggs, K. T., Coulter, M. C., Cruz, F., Cruz, J. B., Valle, C. A., Fefer, S. I., Hatch, S. A., Schreiber, E. A., Schreiber, R. W., and Smith, N. G. (1987). Effects of the 1982–83 El Niño-Southern Oscillation on Pacific Ocean bird populations. *Proceeding International Ornithology Congress*, **19**, 1747–58.

Ainley, D. G., Sydeman, W. J., and Norton, J. (1995). Upper trophic level predators indicate interannual negative and positive anomalies in the California Current food web. *Marine Ecology Progress Series*, **118**, 69–79.

Andrewartha, H. G. and Birch, L. C. (1954). *The distribution and abundance of animals*. University of Chicago Press, Chicago, IL.

Armesto, J. A., Vidiella, P. E., and Gutiérrez, J. R. (1993). Plant communities of the fog-free coastal desert of Chile: plant strategies in a fluctuating environment. *Revista Chilena de Historia Natural*, **66**, 271–82.

Arntz, W. E. and Fahrbach, E. (1991). *El Niño: Klimaexperiment der Natur Physikalische Ursachen und biologische Fulgen*. Birkhauser Verlag.

Arntz, W. E. and Fahrbach, E. (1996). *El Niño: experimento climático de la naturaleza*. Fondo de Cultura Económica, Ciudad de México.

Berryman, A. A. (1999). *Principles of population dynamics and their application*. Stanley Thornes Publishers Ltd., Cheltenham, UK.

Coulson, T., Catchpole, E. A., Albon, S. D., Morgan, B. J. T., Pemberton, J. M., Clutton-Brock, T. H., Crawley, M. J., and Grenfell, B. T. (2001). Age, sex, density, winter weather, and population crashes in soay sheep. *Science*, **292**, 1528–31.

Dillon, M. O. and Rundel, P. W. (1990). The botanical response of the Atacama and Peruvian Desert floras to the 1982–83 El Niño event. In *Global ecological consequences of the 1982–83 El Niño Southern Oscillation* (ed. P. W. Glynn), pp. 487–504. Elsevier Oceanographic Series 52, Amsterdam.

Duffy, D. C. (1990). Seabirds and the 1982–1984 El Niño-Southern Oscillation. In *Global ecological consequences of the 1982–83 El Niño Southern Oscillation* (ed. P. W. Glynn), pp. 395–415. Elsevier Oceanographic Series 52, Elsevier, Amsterdam.

Duffy, D. C. and Merlen, G. (1986). Seabird densities and aggregations during the 1983 El Niño in the Galapagos Islands. *Wilson Bulletin*, **98**, 588–91.

Forchhammer, M. C., Stenseth, N. C., Post, E., and Langvatn, R. (1998). Population dynamics of Norwegian red deer: density-dependence and climatic variation. *Proceedings of the Royal Society of London, B*, **265**, 341–50.

Fuentes, E. R. and Campusano, C. (1985). Pest outbreaks and rainfall in the semi-arid region of Chile. *Journal of Arid Environment*, **8**, 67–72.

Fulk, G. W. (1975). Population ecology of rodents in the semiarid shrublands of Chile. *Texas Tech University, The Museum, Occasional Papers*, **33**, 1–40.

Gibbs, H. L. and Grant, P. R. (1987). Ecological consequences of an exceptionally strong El Niño event on Darwin's finches. *Ecology*, **68**, 1735–46.

Grant, P. R. and Grant, B. R. (1987). The extraordinary El Niño event of 1982–83: effects on Darwin's finches on Isla Genovesa, Galápagos. *Oikos*, **49**, 55–66.

Grant, P. R. and Grant, B. R. (1993). Evolution of Darwin's finches caused by a rare climatic event. *Proceedings of the Royal Society of London, B*, **251**, 111–17.

Guerra, C. G., Fitzpatrick, L. C., Aguilar, R. E., and Venables, B. J. (1988). Reproductive consequences of El Niño Southern Oscillation in Gray Gulls (*Larus modestus*). *Colonial Waterbirds*, **11**, 170–5.

Gutiérrez, J. R. and Meserve, P. L. (2003). El Niño effects on soil seed bank dynamics in north-central Chile. *Oecologia*, **134**, 511–17.

Gutiérrez, J. R., Meserve, P. L., Jaksic, F. M., Contreras, L. C., Herrera, S., and Vásquez, H. (1993). Structure and dynamics of vegetation in a Chilean arid thornscrub community. *Acta Oecologica*, **14**, 271–85.

Gutiérrez, J. R., Arancio, G., and Jaksic, F. M. (2000). Variation in vegetation and seed bank in a Chilean semi-arid community affected by ENSO 1997. *Journal of Vegetation Science*, **11**, 641–8.

Hall, G. A., Gibbs, H. L., Grant, P. R., Botsford, L. W., and Butcher, G. S. (1988). Effects of El Niño-Southern Oscillation (ENSO) on terrestrial birds. *Proceedings of International Ornithology Congress*, **19**, 1759–75.

Hamman, O. (1985). The El Niño influence on the Galapagos vegetation. In *El Niño en las Islas Galapagos: el evento de 1982–1983* (eds G. Robinson and E. M. del Pino), pp. 299–330. Fundación Charles Darwin para las Islas Galápagos, Quito, Ecuador.

Harvell, C. D., Kim, K., Burkholder, J. M., Colwell, R. R., Epstein, P. R., Grimes, D. J., Hofmann, E. E., Lipp, E. K., Osterhaus, A., Overstreet, R. M., Porter, J. W., Smith, G. W., and Vasta, G. R. (1999). Review: marine ecology—emerging marine diseases—climate links and anthropogenic factors. *Science*, **285**, 1505–10.

Holmgren, M., Scheffer, M., Ezcurra, E., Gutierrez, J. R., and Mohren, G. M. J. (2001). El Niño effects on the dynamics of terrestrial ecosystems. *Trends in Ecology & Evolution*, **16**, 89–94.

Jaksic, F. M. (1998). The multiple facets of El Niño/Southern Oscillation in Chile. *Revista Chilena de Historia Natural*, **71**, 121–31.

Jaksic, F. M. (2001a). Ecological effects of El Niño in terrestrial ecosystems of western South America. *Ecography*, **24**, 241–50.

Jaksic, F. M. (2001b). Spatiotemporal variation patterns of plants and animals in San Carlos de Apoquindo, central Chile. *Revista Chilena de Historia Natural*, **74**, 477–502.

Jaksic, F. M. and Lazo, I. (1999). Response of a bird assemblage in semiarid Chile to the 1997–1998 El Niño. *Wilson Bulletin*, **111**, 527–35.

Jaksic, F. M., Feinsinger, P., and Jiménez, J. E. (1993). A long-term study on the dynamics of guild structure among predatory vertebrates at a semi-arid neotropical site. *Oikos*, **67**, 87–96.

Jaksic, F. M., Feinsinger, P., and Jiménez, J. E. (1996). Ecological redundancy and long-term dynamics of vertebrate predators in semiarid Chile. *Conservation Biology*, **10**, 252–62.

Jaksic, F. M., Silva, S. I., Meserve, P. L., and Gutiérrez, J. R. (1997). A long-term study of vertebrate predator responses to an El Niño (ENSO) disturbance in western South America. *Oikos*, **78**, 341–54.

Jiménez, J. E., Feinsinger, P., and Jaksic, F. M. (1992). Spatiotemporal patterns of an irruption and decline of small mammals in north-central Chile. *Journal of Mammalogy*, **73**, 356–64.

Lima, M. and Jaksic, F. M. (1998). Delayed density-dependent and rainfall effects on reproductive parameters of an irruptive rodent in semiarid Chile. *Acta Theriologica*, **43**, 225–36.

Lima, M. and Jaksic, F. M. (1999a). Population dynamics of three Neotropical small mammals: time series models and the role of delayed density-dependence in population irruptions. *Australian Journal of Ecology*, **24**, 25–34.

Lima, M. and Jaksic, F. M. (1999b). Population rate of change in the leaf-eared mouse: the role of density-dependence, seasonality and rainfall. *Australian Journal of Ecology*, **24**, 110–16.

Lima, M., Marquet, P. A., and Jaksic, F. M. (1999a). El Niño events, precipitation patterns, and rodent outbreaks are statistically associated in semiarid Chile. *Ecography*, **22**, 213–18.

Lima, M., Keymer, J. E., and Jaksic, F. M. (1999b). ENSO-driven rainfall variability and delayed density-dependence cause rodent outbreaks in western South America: linking demography and population dynamics. *American Naturalist*, **153**, 476–91.

Lima, M., Stenseth, N. C., and Jaksic, F. M. (2002a). Food web structure and climate effects on the dynamics of small mammals and owls in semiarid Chile. *Ecology Letters*, **5**, 273–84.

Lima, M., Stenseth, N. Chr., and Jaksic, F. M. (2002b). Population dynamics of a South American small rodent: seasonal structure interacting with climate, density-dependence and predator effects. *Proceedings of the Royal Society of London, B*, **269**, 2579–86.

Limberger, D. (1990). El Niño's effect on South American pinnipeds In *Global ecological consequences of the 1982–83 El Niño Southern Oscillation* (ed. P. W. Glynn), pp. 417–32. Elsevier Oceanographic Series 52, Elsevier, Amsterdam.

Lindsey, G. D., Pratt, T. K., Reynolds, M. H., and Jacobi, J. D. (1997). Response of six species of Hawaiian forest birds to a 1991–1992 El Niño drought. *Wilson Bulletin*, **109**, 339–43.

Majluf, P. (1991). El Niño effects on pinnipeds in Peru. In *Pinnipeds and El Niño: responses to environmental stress* (eds F. Trillmich and K. A. Ono), pp. 55–65. Springer-Verlag, Berlin.

Massey, B. W., Bradley, D. W., and Atwood, J. L. (1992). Demography of a California Least Tern colony including effects of the 1982–83 El Niño. *Condor*, **94**, 976–83.

Meserve, P. L. and Le Boulengé, E. (1987). Population dynamics and ecology of small mammals in the northern Chilean semiarid region. *Fieldiana Zoology, New Series*, **39**, 413–31.

Meserve, P. L., Yunger, J. A., Gutiérrez, J. R., Contreras, L. C., Milstead, W. B., Lang, B. K., Cramer, K. L., Herrera, S., Lagos, V. O., Silva, S. I., Tabilo, E. L., Torrealba, M. A., and Jaksic, F. M. (1995). Heterogeneous responses of small mammals to an El Niño Southern Oscillation event in northcentral semiarid Chile and the importance of ecological scale. *Journal of Mammalogy*, **76**, 580–95.

Meserve, P. L., Gutiérrez, J. R., Yunger, J. A., Contreras, L. C., and Jaksic, F. M. (1996). Role of biotic interactions in a small mammal assemblage in semiarid Chile. *Ecology*, **77**, 133–48.

Meserve, P. L., Milstead, W. B., Gutiérrez, J. R., and Jaksic, F. M. (1999). The interplay of biotic and abiotic factors in a semiarid Chilean mammal assemblage: results of a long-term experiment. *Oikos*, **85**, 364–72.

Miskelly, C. M. (1990). Effects of the 1982–83 El Niño event on two endemic landbirds on the Snares Islands, New Zealand. *Emu*, **90**, 24–27.

Murúa, R., González, L., and Lima, M. (2003). Population dynamics of rice rats (a Hantavirus reservoir) in southern Chile: feedback structure and non-linear effects of climatic Oscillations. *Oikos*, **102**, 137–45.

Mysterud, A., Stenseth, N. C., Yoccoz, N. G., Langvatn, R., and Steinheim, G. (2001). Nonlinear effects of large-sclae climatic variability on wild and domestic herbivores. *Nature*, **410**, 1096–9.

Mysterud, A., Stenseth, N. C., Yoccoz, N. G., Ottersen, G., and Langvatn, R. (2003). The response of terrestrial ecosystems to climate variability associated with the North Atlantic Oscillation. In *The North Atlantic Oscillation: climatic significance and environmental impact* (eds J. Hurrell, Y. Kushnir, G. Ottersen, and M. Visbeck), pp. 235–62. American Geophysical Union.

Nicholls, N. (1991). The El Niño/Southern Oscillation and Australian vegetation. *Vegetatio*, **91**, 23–36.

Pearson, O. P. (1975). An outbreak of mice in the coastal desert of Peru. *Mammalia*, **39**, 375–86.

Péfaur, J. E., Yáñez, J. L., and Jaksic, F. M. (1979). Biological and environmental aspects of a mouse outbreak in the semi-arid region of Chile. *Mammalia*, **43**, 313–22.

Polis, G. A., Hurd, S. D., Jackson, C. T., and Sánchez-Piñero, F. (1997). El Niño effects on the dynamics and control of an island ecosystem in the Gulf of California. *Ecology*, **78**, 1884–97.

Post, E., Peterson, R. O., Stenseth, N. C., and McLaren, B. E. (1999). Ecosystem consequences of wolf behavioral response to climate. *Nature*, **401**, 905–7.

Royama, T. (1992). *Analytical population dynamics*. Chapman & Hall, London, UK.

Sæther, B.-E., Tufto, J., Engen, S., Jerstad, K., Røstad, O. W., and Skåtan, J. E. (2000). Population dynamical consequences of climate change for a small temperate songbird. *Science*, **287**, 854–6.

Sætre, G.-P., Post, E., and Král, M. (1999). Can environmental fluctuation prevent competitive exclusion in sympatric flycatchers? *Proceedings of the Royal Society of London, B*, **266**, 1247–51.

Schreiber, E. A. and Schreiber, R. W. (1989). Insights into seabird ecology from a global 'natural experiment'. *National Geographic Research*, **5**, 64–81.

Schreiber, R. W. and Schreiber, E. A. (1984). Central Pacific seabirds and the El Niño Southern Oscillation: 1982 to 1983 perspectives. *Science*, **225**, 713–16.

Silva, S. I., Lazo, I., Silva-Aránguiz, E., Jaksic, F. M., Meserve, P. L., and Gutiérrez, J. R. (1995). *Numerical and functional response of Burrowing Owls to long-term mammal fluctuations in Chile. The Journal of Raptor Research*, **29**, 250–5.

Soto, R. (1985). Efectos del fenómeno El Niño 1982–83 en ecosistemas de la I Región. *Investigacion Pesquera (Chile)*, **32**, 199–206.

Spear, L. B. (1993). Dynamics and effect of western gulls feeding in a colony of guillemots and Brandt's cormorants. *Journal of Animal Ecology*, **62**, 399–414.

Stenseth, N. C. (1999). Population cycles in voles and lemmings: density-dependence and phase dependence in a stochastic word. *Oikos*, **87**, 427–61.

Stenseth, N. C., Chan, K.-S., Tong, H., Boonstra, R., Boutin, S., Krebs, C. J., Post, E., O'Donoghue, M., Yocozz, N. G., Forchhammer, M. C., and Hurrell, J. W. (1999). Common dynamic structure of Canada lynx populations within three climate regions. *Science*, **285**, 1071–3.

Stenseth, N. C., Ottersen, G., Hurrell, J.W., Mysterud, A., Lima, M., Chan, K.-S., Yoccoz, N. G., and Ådlandsvik, B. (2003). Studying climate effects on ecology through the use of climate indices: the North Atlantic Oscillation, El Niño Southern Oscillation and beyond. *Proceedings of the Royal Society of London (Series B.)*, **270**, 2087–96.

Torres, D. (1985). Presencia del lobo fino sudamericano (*Arctocephalus australis*) en el norte de Chile, como consecuencia de El Niño 1982–83. *Investigacion Pesquera (Chile)*, **32**, 225–33.

Tovar, H., Guillén, V., and Cabrera, D. (1987). Reproduction and population levels of Peruvian guano birds, 1980 to 1986. *Journal of Geophysical Research*, **92**, 14445–8.

Trillmich, F. and Dellinger, T. (1991). The effects of El Niño on Galapagos pinnipeds In *Pinnipeds and El Niño: responses to environmental stress* (eds F. Trillmich and K. A. Ono), pp. 66–74. Springer-Verlag, Berlin.

Trillmich, F. and Limberger, D. (1985). Drastic effects of El Niño on Galapagos pinnipeds *Oecologia*, **67**, 19–22.

Turchin, P. (1995). Population regulation: old arguments and a new synthesis. In *Population dynamics: new approaches and synthesis* (eds N. Capuccino and P. W. Price), pp. 19–40. Academic Press, San Diego, CA.

Villagrán, C. (1993). Una interpretación climática del registro palinológico del último ciclo glacial-postglacial en Sudamérica. Bulletin de l'Institut français d'Etudes andine **22**, 243–58.

Wilson, U. (1991). Responses of three seabird species to El Niño events and other warm episodes on the Washington coast, 1979–1990. *Condor*, **93**, 853–8.

Glossary of Species names

Abra alba (Bivalvia)
Acartia hudsonica (Copepoda)
Aethia cristatella (Auklet)
Albacore tuna (*Thunnus alalunga*)
American lobster (*Homarus americanus*)
American plaice (*Hippoglossoides platessoides*)
American Shad (*Alosa sapidissima*)
Amphiura brachiata (Ophiuroidea, Brittle star)
Anchovies (*Engraulis* spp.)
Arctic Cod (*Boreogadus saida*)
Arctic tern (*Sterna paradisaea*)
Arctica islandica (Bivalvia)
Arrowtooth flounder (*Atheresthes stomias*)
Asterionellopsis glacialis (Diatom)
Atlantic Cod (*Gadus morhua*)
Atlantic mackerel (*Scomber scombrus*)
Atlantic menhaden (*Brevoortia tyrannus*)
Atlantic puffin (*Fratercula arctica*)
Atlantic salmon (*Salmo salar*)

Balsam fir (*Abies balsamea*)
Barn owl (*Tyto alba*)
Black-baked gull (*Larus marinus*)
Black-headed gull (*Larus ridibundus*)
Black-legged kittiwake (*Rissa tridactyla*)
Black-throated blue warbler (*Dendrocia caerulescens*)
Bluefin tuna (*Thunnus thynnus*)
Blue-petrel (*Halobaena caerulea*)
Butterfish (*Peprilus triancanthus*)
Brown trout (*Salmo trutta*)

Calanus finmarchicus (Copepoda)
Calanus furca (Copepoda)
Calanus fusu (Copepoda)
Calanus helgolandicus (Copepoda)
Calanus socialis (Copepoda)
Calanus tripos (Copepoda)
Capelin (*Mallotus villosus*)
Cassin's auklet (*Ptychramphus aleuticus*)
Centropages hamatus (Copepoda)

Centropages typicus (Copepoda)
Ceratium furca (Dinoflagellate)
Ceratium fusus (Dinoflagellate)
Ceratium macroceros (Dinoflagellate)
Ceratium tripos (Dinoflagellate)
Cerianthus lloydii (Anthozoa, Sea anemones)
Chaetoceros socialis (Diatom)
Clausocalanus spp. (Copepoda)
Clione limacina (Gastropda)
Common guillemot (*Uria aalge*)
Common gull (*Larus canus*)
Common tern (*Sterna hirundo*)
Corophium crassicorne (Amphipoda)
Corymorpha nutans (Hydrozoa)
Cyclops vicinus (Copepoda)
Cyclorhinchus psittacula (Auklet)

Daphnia (Copepoda)
Detonula confervacea (Diatom)
Dinophysis (Dinoflagellate)

Echinocyamus pusillus (Echinoidea, Sea urchins)
Elephant seal (*Mirounga Angustirostris*)
Emperor penguin (*Aptenodytes forsteri*)
Eudiaptomus gracilis (Copepoda)
European shag (*Phalacrocorax aristotelis*)

Fratercual corniculata (Charadriiformes, Puffin)

Galapagos penguin (*Spheniscus mendiculus*)
Giant Scallop (*Placopecten magellanicus*)
Green Crab (*Carcinus maenas*)
Greenland halibut (*Reinhardtius hippoglossoides*)
Gymnodinium catenatum (Dinoflagellate)

Haddock (*Melanogrammus aeglefinus*)
Halibut (*Hippoglossus hippoglossus*)
Herring (*Clupea harengus*)
Herring gull (*Larus argentatus*)

Jack (horse) mackerel (*Thrachurus* spp.)

Leach's storm petrel (*Oceanodroma leucorhoa*)
Leaf-eared mouse (*Phyllotis darwini*)
Limacina spp. (Gastropoda)
Little auk (*Alle alle*)

Manx shearwater (*Puffinus puffinus*)
Megaluropus agilis (Amphipoda)
Metridia lucens (Copepoda)
Moose (*Alces alces*)
Mouse-opossum (*Thylamys elegans*)

Nitzschia delicatissima (Diatom)
Northern fulmar (*Fulmarus glacialis*)
Northern gannet (*Sula bassana*)
Nuculoma tenuis (Bivalvia)

Oceanic anchovies (*Encrasicholinus punctifer*)
Oithona spp. (Copepoda)
Olivaceous field mouse (*Akodon olivaceus*)
Ophelia borealis (Polychaeta)

Pacific cod (*Gadus macrocephalus*)
Pacific hake (*Merluccius productus*)
Pacific halibut (*Hippoglossus stenolepis*)
Paracalanus spp. (Copepoda)
Para-Pseudocalanus spp. (Copepoda)
Penilia avirostris (Cladocera)
Phalacrocorax spp. (Pelecaniformes)
Pollock (*Pollachius pollachius*)
Pontoporeia affinis (Amphipoda)
Pontoporeia affinis (Amphipoda)
Pseudocalanus spp. (Copepoda)
Pyrodinium bahamense (Dinoflagellate)

Razorbill (*Alca torda*)
Redfish (*Sebastes* spp.)

Sagitta elegans (Chaetognatha, Arrow Worms)
Sagitta setosa (Chaetognatha, Arrow Worms)
Sardine (*Sardina pilchardus*)
Sardine (*Sardinop* spp.)
Sardine (*Sardinops sagax*)
Sea lion (*Zalophus californianus*)
Short-tailed shearwater (*Puffinus tenuirostris*)
Silver hake/Whiting (*Merluccius bilinearis*)
Siphonocoetes kroyeranus (Amphipoda)
Skeletonema costatum (Diatom)
Skipjack tuna (*Katsuwonus pelamis*)
Soay sheep (*Ovis aries*)
Sole (*Solea vulgaris*)
Sooty shearwater (*Puffinus griseus*)
Spheniscus spp. (Sphenisciformes, Penguins)
Spiophanes bombyx (Polychaeta)
Sprat (*Sprattus sprattus*)
Squid (*Illex illecebrosus*)
Swordfish (*Xiphias gladius*)

Thalassionema nitzschoides (Diatom)
Thalassiosira nordenskioeldii (Diatom)
Thalassiothrix longissima (Diatom)
Tilefish (*Lopholatilus chamaeleonticeps*)
Lunda cirrhata (Puffin)
Trichodesmium spp. (Cyanobacterium)

Walleye Pollock (*Theragra chalcogramma*)
Winter Flounder (*Pseudopleuronectes americanus*)

Yellow-eyed penguin (*Megadytes antipodes*)
Yellowfin tuna (*Thunnus albacares*)
Yellowtail flounder (*Limanda ferruginea*)

Author Index

Subject Index